AF361704

State of the Art Geo-Energy Technology in China

State of the Art Geo-Energy Technology in China

Editors

Sheng-Qi Yang
Min Wang
Qi Wang
Wen Zhang
Kun Du
Chun Zhu

MDPI • Basel • Beijing • Wuhan • Barcelona • Belgrade • Manchester • Tokyo • Cluj • Tianjin

Editors

Sheng-Qi Yang
China University of Mining and Technology
China

Min Wang
China University of Petroleum (EastChina)
China

Qi Wang
Shandong University
China

Wen Zhang
Jilin University
China

Kun Du
Central South University
China

Chun Zhu
Hohai University
China

Editorial Office
MDPI
St. Alban-Anlage 66
4052 Basel, Switzerland

This is a reprint of articles from the Topical Collection published online in the open access journal *Energies* (ISSN 1996-1073) (available at: https://www.mdpi.com/journal/energies/topical_collections/geo_energy_technology_China).

For citation purposes, cite each article independently as indicated on the article page online and as indicated below:

LastName, A.A.; LastName, B.B.; LastName, C.C. Article Title. *Journal Name* **Year**, *Volume Number*, Page Range.

ISBN 978-3-0365-4379-6 (Hbk)
ISBN 978-3-0365-4380-2 (PDF)

Contents

Article

Characteristics of Airflow Reversal of Excavation Roadway after a Coal and Gas Outburst Accident

Junhong Si [1], Lin Li [1,*], Jianwei Cheng [2], Yiqiao Wang [1], Wei Hu [1], Tan Li [1] and Zequan Li [1]

[1]   School of Emergency Technology and Management, North China Institute of Science and Technology, Beijing 101601, China; sijunhong@ncist.edu.cn (J.S.); 201908522442wyq@ncist.edu.cn (Y.W.); 202008570051hw@ncist.edu.cn (W.H.); 202008570059lt@ncist.edu.cn (T.L.); 2017012571zq@ncist.edu.cn (Z.L.)

[2]   School of Safety Engineering, China University of Mining and Technology, Xuzhou 221008, China; ChengJianwei@cumt.edu.cn

*   Correspondence: 2019085224661l@ncist.edu.cn

Abstract: Determining the influence scope of the airflow disorder is an important problem after coal and gas outburst accidents in ventilation systems. This paper puts forward the indexes of airflow disorder, including the length of the excavation roadway, the outburst pressure, the pressure difference of the air door, and the air quantity of the auxiliary fan. Using the orthogonal table of L_9 (3^4) and numerical simulation method, the characteristics of airflow reversal are studied, and the outburst airflow reversal degree is calculated should the ventilation facility fail. Furthermore, on the basis of fuzzy comprehensive optimization theory, the comprehensive evaluation model of the airflow disorder is established. The results show that the length of the excavation roadway is the most important factor affecting the stability of the ventilation system, followed by the outburst pressure, pressure difference of the air door, and air quantity of the auxiliary fan. The influence of a gas outburst accident on the return air system is greater than that on the inlet air system, and a larger air velocity has a greater impact on the ventilation system, especially the air inlet part. Moreover, the airflow reversal degree of the ventilation system increases with the increase of the outburst pressure or decreases with the length of the excavation roadway. This paper provides a basis for the prevention of gas outburst accidents.

Keywords: airflow reversal; gas outburst; mine ventilation system; orthogonal experiment; numerical simulation

Citation: Si, J.; Li, L.; Cheng, J.; Wang, Y.; Hu, W.; Li, T.; Li, Z. Characteristics of Airflow Reversal of Excavation Roadway after a Coal and Gas Outburst Accident. *Energies* **2021**, *14*, 3645. https://doi.org/10.3390/en14123645

Academic Editors: Sheng-Qi Yang, Min Wang, Qi Wang, Wen Zhang, Kun Du and Chun Zhu

Received: 17 May 2021
Accepted: 15 June 2021
Published: 18 June 2021

Publisher's Note: MDPI stays neutral with regard to jurisdictional claims in published maps and institutional affiliations.

1. Introduction

A coal and gas outburst is an accident that is caused by a large amount of coal and gas being ejected into the underground roadway in a single moment [1]. In an outburst accident, the high-energy shock wave destroys the roadway facilities instantly and changes the roadway resistance and the structure of the ventilation network [2]. After the dynamic effect of the shock wave disappears, the gas continues to flow and diffuse unsteadily [3], generating ventilation pressure and additional force [4], causing airflow disorder [5], casualties, or secondary gas explosion accidents.

A coal and gas outburst is studied from three aspects: shock wave [6–8], outburst gas [9–12], and outburst coal rock [13,14]. The theoretical analysis [15] and numerical simulation are popular methods [16,17] in a single roadway. After the outburst power disappears in the upwind roadway, Feng et al. [18] concluded, the influencing factors of the airflow reversal in the parallel branch include the pressure of the main fan, the height difference of roadway, the initial air velocity, length, and the sectional area of roadway. Yu et al. [19] pointed out that the airflow is affected by the seam inclination in the inlet roadway. However, the underground mine ventilation network is formed by connected roadways, and the study is often not limited to a single roadway. Zhou et al. [20] established 45° and 135° crossover tunnels to simulate the flow and attenuation rules of shock waves and gas in a gas outburst accident. The influence of air pressure increases with the increase of the angle between the excavation roadway and the adjacent excavation roadway, but the intersection of the roadway has no effect on the airflow reversal. They also found that the outburst pressure could not only lead to the airflow reversal in the downward ventilation roadway, but also may lead to the airflow reversal in parallel branches of the upward ventilation roadway [21,22]. The phenomenon of gas flow restriction and retention appears in the roadways with the air door and windshield, so draining the gas immediately is the key to prevent secondary disasters in the underground roadways.

In recent years, scholars have conducted a series of studies on the disaster process of the whole mine ventilation system during the outburst period. Due to the complexity of the ventilation system after an gas outburst accident, the tunneling face is usually seen as a unidirectional burst source, and a three-dimensional simulation method is adopted to simulate the flow process of air and gas in the ventilation system during the disaster period [23]. The result shows that the disturbance of the ventilation system is affected by the gas force, natural wind pressure, and ventilation power. Then, the characteristics of bidirectional gas outburst in the coal mining face are studied [24]. Different from the unidirectional outburst mode, the impact strength of this type is reduced. Ventilation facilities are the weak point of the ventilation system. The research mainly focuses on the characteristics of air door failure [25,26], the reasonable position, the strength of the air door [27], as well as the automatic air door [28,29]. There is still a lack of research on the influencing factors of airflow disorders in ventilation systems when ventilation facilities fail. There are many factors that affect the ventilation system of a coal and gas outburst [30–32], it is particularly important to establish the index system of the airflow disorder and determine relatively important influencing factors when ventilation facilities fail, which could provide theoretical support for disaster prevention.

The whole paper is structured as follows: Section 2 introduces the ventilation system. In Section 3, the influencing factors of the airflow disorder induced by gas outburst accident are put forward. Using the orthogonal experiment analysis and numerical simulation method, the sensitivity factors of the airflow disorder when the ventilation facility fails are analyzed in Section 4. Section 5 conducts a comprehensive evaluation of the airflow disorder on the basis of fuzzy comprehensive optimization theory.

2. Ventilation System and Its Simplification

Most of the coal and gas outburst accidents occur in the heading face [24]. The ventilation system is formed by the excavation roadway, heading face, auxiliary fan, and air ducts. Figure 1 illustrates a forced ventilation system, which is the most commonly used ventilation form. It has the advantages of a small air leakage and large air supply. The fresh air is transported from the intake airway to the heading face through the auxiliary fan and air duct, and the polluted air enters the return airway through the excavation roadway. The air door obstructs fresh air and polluted air, and the air duct needs to pass through the air door. The blue arrow indicates the airflow direction of the fresh air, and the red one indicates the polluted air.

In general operation conditions, the order of the air gauge pressure is $P_{20} > P_{21} > P_{22}$, $P_{10} > P_{11} > P_{12} > P_{13}$, $P_{11} > P_{31} > P_{21}$.

According to graph theory, the air duct can be considered as a separate roadway for the air supply on the heading face, so it is an independent branch. There are nine branches and nine nodes in a general ventilation system. The ventilation system is simplified, as shown in Figure 2.

In a gas outburst accident, the air pressure of the roadway is redistributed as the gas concentration increases rapidly in the ventilation system, and the shock wave usually destroys the ventilation facilities. As a result, if the air door is invalid in a tunneling ventilation system, the outburst gas will flow into the intake airway. Then, the airflow will be reversed if the air pressure at the upwind side is lower than the outburst pressure.

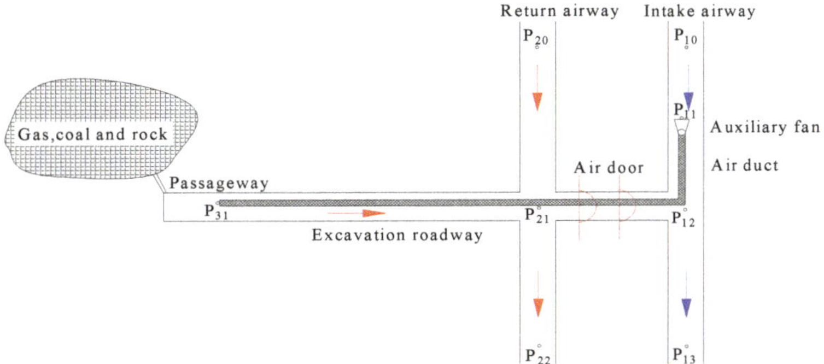

Figure 1. Sketch map of the ventilation system.

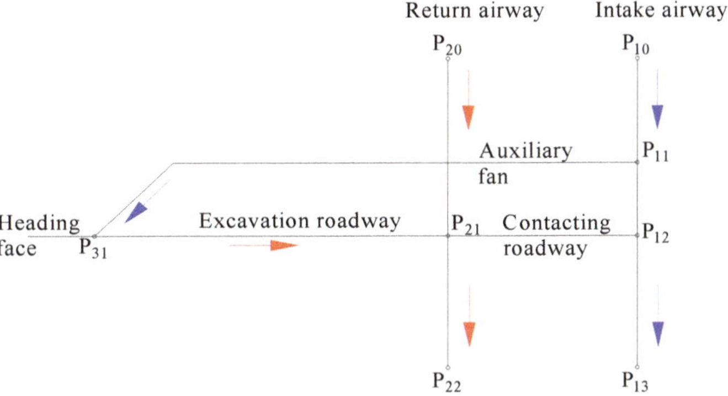

Figure 2. Diagram of the simplified ventilation network in a general ventilation system.

The outburst airflow reversal degree is defined as the influence of outburst gas on the airflow disorder in an airway, which is the ratio of the air quantity variation after the gas outburst accident and the original value. A positive reversal degree means that the airflow direction of the outburst gas is the same as the original airflow, and a negative reversal degree means that the airflow is reversed.

3. Indexes of Airflow Disorder

There are 3 factors that affect the safety and stability of the mine ventilation system, as shown in Figure 3.

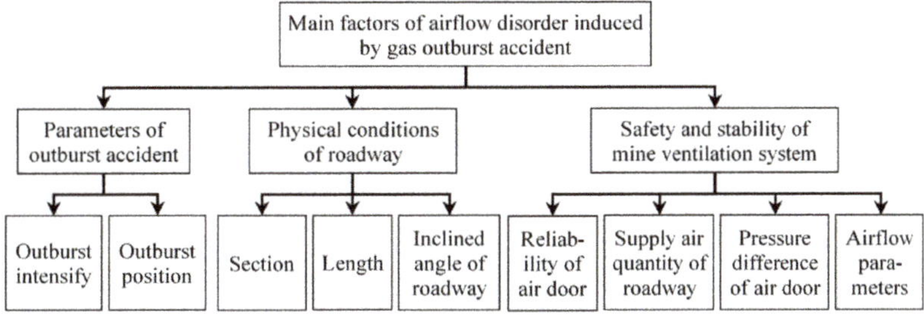

Figure 3. Hierarchical structure of factor indicators.

3.1. Parameters of Outburst Accident

The outburst intensity indicates the amount of coal and rock thrown into the roadway and the amount of gas emissions in a coal and gas outburst accident. Because it is difficult to calculate the amount of gas emission, the outburst intensity is represented by the amount of outburst coal and rock. Usually, an outburst accident is divided into a small outburst accident (<100 t), medium outburst accident (<500 t), large outburst accident (<1000 t), and extra-large outburst accident ($\geq$1000 t).

The outburst position refers to the place where the outburst accident occurs. If the air pressure of the outburst position is low, the outburst gas can be faster exhausted to the ground, as the outburst position is closer to the air shaft. On the contrary, it has a great impact on the ventilation system, especially the air intake roadway. Therefore, we can reduce the impact of the outburst accident on the ventilation system by reducing the air resistance of the return airway or increasing the intake airway.

3.2. Physical Conditions of Roadway

The characteristics of the roadway are determined by the physical conditions, including the section and length of roadway. Specifically, the outburst gas is easy to exhaust from the large section or the shorter length of roadway.

3.3. Safety and Stability of Mine Ventilation System

The high-pressure outburst gas suddenly flows into the roadway space in a gas outburst accident, which has a great impact on the mine ventilation system. It is easy to cause a large range of airflow disorders in the mine ventilation system after the shock wave destroys the air doors.

Among the above factors, the outburst pressure, length of roadway, pressure difference of the air door, and supply air quantity of the roadway are chosen to analyze the laws of the airflow disorder and the degree of influence in a mine ventilation system.

4. Orthogonal Experiment

4.1. Orthogonal Design

The orthogonal experiment is used for the analysis of the mutual influence of multiple factors [33]. Using the orthogonal table, an appropriate amount of representative points is selected from a large number of test points. Through the overall design, comprehensive comparison, and statistical analysis, the balanced sampling within the range of factor change is carried out. Then, the analysis of the representative results is achieved by a small number of tests.

In the orthogonal design, the table of L_9 (3^4) is adopted to study the effect of 4 factors, i.e., A is the length of roadway, B is the outburst pressure, C is the pressure difference of air door, and D is the air quantity of the auxiliary fan on the airflow disorder in the mine ventilation system. The specific parameters of each test are illustrated in Table 1.

Table 1. Initial parameters of the orthogonal experiment.

Test Number	A/m	B/MPa	C/Pa	D/m$^3\cdot$min^{-1}
L_1	200	0.1	500	240
L_2	200	0.2	1000	1200
L_3	200	0.3	1500	600
L_4	1000	0.2	1500	240
L_5	1000	0.1	1000	600
L_6	1000	0.3	500	1200
L_7	2000	0.1	1500	1200
L_8	2000	0.2	500	600
L_9	2000	0.3	1000	240

From Table 1, the level of each factor is in the common value range, such that the length of roadway is 200 m, 1000 m, and 2000 m, the outburst pressure is 0.1 MPa, 0.2 MPa, and 0.3 MPa respectively, the pressure difference of the air door is 500 Pa, 1000 Pa, and 1500 Pa, the air quantity of the auxiliary fan is 240 m^3/min, 600 m^3/min, and 1200 m^3/min. A total of nine combined schemes are designed.

4.2. Mathematical Experiment Modeling

4.2.1. Physical Model and Meshing

According to the orthogonal tests, it is necessary to establish three physical models with different roadway lengths. The spatial relationship model and grid meshing of the ventilation system are shown in Figure 4. The parameters of each test model are shown in Table 2.

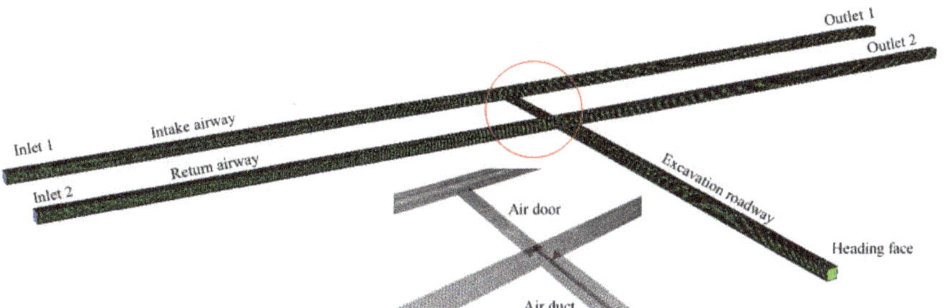

Figure 4. Physical model and its meshing.

Table 2. Physical model of the roadway.

Test	Length of Excavation Roadway/m	Model Size				Number of Grids
		Excavation Roadway Length × Width × Height/m	Air Duct Length × Radium/m	Connecting Roadway Length × Width × Height/m	Intake and Return Airway Length × Width × Height/m	
L1, L2, L3	200	200 × 5 × 5	200 × 0.5			237,342
L4, L5, L6	1000	1000 × 5 × 5	1000 × 0.5	45 × 5 × 2	500 × 5 × 5	463,984
L7, L8, L9	2000	2000 × 5 × 5	2000 × 0.5			764,081

4.2.2. Mathematical Model

There are three basic equations of the fluid numerical simulation, which includes the mass conservation equation, the momentum conservation equation, and the energy conservation equation. After the gas outburst accident in the excavation roadway, the airflow should also be satisfied by gas migration. Furthermore, a chemical component exchange exists in this numerical simulation, i.e., CH_4, O_2, and N_2, so the component transport equation should be introduced.

A turbulent flow is a common flow phenomenon in nature, and the fluid is always in the state of turbulence in most engineering problems. The realizable k-ε model is adopted because its good performance for solving the adverse pressure gradient and vortex problems.

4.2.3. Boundary Conditions

The boundary conditions of the model are shown in Table 3.

From Table 3, the physical model area is the Fluid Type, and the pressure of operation condition is 101.325 kPa. The inlet boundary condition includes Inlet 1 of the intake airway, Inlet 2 of the return airway, outburst position, and auxiliary fan. The inlet boundary type of auxiliary fan inlet is the Velocity-inlet, and the other is the Pressure-inlet. The species of the outburst position is 100% CH_4, and the other is 20% oxygen and 80% nitrogen.

The outlet boundary includes Outlet 1 of the intake airway and Outlet 2 of the return airway, the outlet boundary type is also the Pressure-outlet. Furthermore, the species are 20% oxygen and 80% nitrogen either.

The gauge pressure of each boundary condition is calculated as follows: Assuming that both the sectional area of the return airway and excavation roadway is 25 m^2, the friction resistance coefficient is 100×10^{-4}, the inlet air quantity of the intake airway and return airway $Q_{10} = Q_{20}$ is 75 m^3/s, and the air quantity of the auxiliary fan Q_{fan} is 4 m^3/s, 10 m^3/s, and 20 m^3/s, respectively. The outlet of the return airway is $Q_{21} = Q_{20} + Q_{fan}$, $Q_{12} = Q_{10} - Q_{fan}$.

Table 3. Boundary conditions of each test.

Test	Inlet Boundary				Outlet Boundary	
	Inlet 2/Pa	Inlet 1/Pa	Auxiliary Fan Inlet/m·s^{-1}	Outburst Position/MPa	Outlet 2/Pa	Outlet 1/Pa
L_1	37.97	537.97	5.10	0.1	0	503.84
L_2	46.88	1046.88	25.48	0.2	0	1019.20
L_3	41.12	1541.12	12.74	0.3	0	1509.60
L_4	37.97	1537.97	5.10	0.2	0	1503.84
L_5	41.12	1041.12	12.74	0.1	0	1009.60
L_6	46.88	546.88	25.48	0.3	0	519.20
L_7	46.88	1546.88	25.48	0.1	0	1519.20
L_8	41.12	541.12	12.74	0.2	0	509.60
L_9	37.97	1037.97	5.10	0.3	0	1003.84

The relationship between the air resistance and the friction resistance coefficient of the roadway is

$$R = \frac{LU\alpha}{S^3} \tag{1}$$

where R represents the air resistance, kg/m^7; L represents the length of the roadway, m; U represents the perimeter of the roadway section, m; α represents coefficient of frictional resistance, kg/m^3; S represents the sectional area of the roadway, m^2.

From Equation (1), the air resistance value of the return airway is $R_{10} = R_{12} = R_{20} = R_{21} = 0.0032$ kg/m^7, the excavation roadway from the air duct outlet to the return airway is $R_{31} = 0.0256$ kg/m^7. The air pressure $H_{10} = 16.13$ Pa, $H_{12} = H_{20} = 18$ Pa, $H_{21} = 19.97$ Pa and the air pressure of the excavation roadway are calculated by the air quantity of the auxiliary fan and the length of the excavation roadway.

Furthermore, the initial gauge total pressure at Inlet 1 is the sum of the pressure difference of the air door and the air pressure of the upper wind side of the intake airway H_{10}, and the initial pressure of Outlet 1 is the sum of the pressure difference of air door and H_{12}. Suppose the operation pressure is the standard atmospheric pressure (101.325 kPa), the gauge total pressure at Outlet 2 of the return airway is 0 Pa. Then, the air pressure of each inlet and outlet are calculated as shown in Table 3.

4.2.4. Controlling Parameters of Fluid Dynamics

(1) Solver

In the Pressure-based type, the absolute velocity formulation and Pressure-velocity coupling algorithm are usually adopted to solve the problem. Furthermore, using the SIMPLE scheme to calculate the mathematical model, it computes the mass conservation and obtains the pressure field by the mutual correction of pressure and velocity.

(2) Convergence accuracy

A reasonable accuracy is an important parameter to ensure the convergence of the model. The convergence residuals are 10^{-6} in this model, and the number of iterations is 500, so the calculation will be finished when the residuals of each variable are less than

10^{-6} or the iterations reach to 500 steps. Furthermore, the convergence of the calculation results can be dynamically monitored by checking the iterative residual of each variable.

4.3. Results of Numerical Simulation

Using Fluent numerical simulation software, the flow field distributions of the outburst gas of the above nine tests are simulated. Figure 5 illustrates the path-line near the air door of each test on the horizontal plane 1 m away from the floor, which is colored by the velocity magnitude.

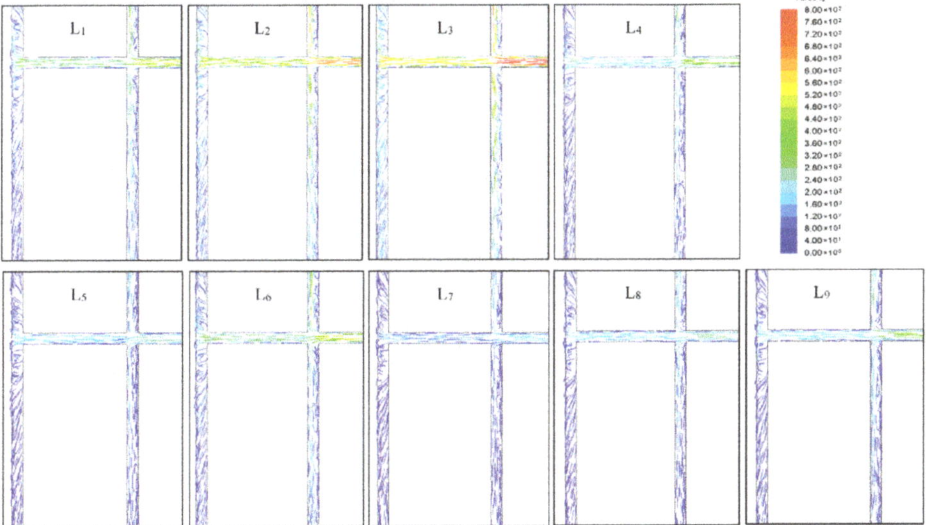

Figure 5. Color map of velocity path-line near the air door.

From Figure 5, the air velocity of L_3 is the highest near the air door after the gas burst accident, followed by L_2 and L_1, and L_7 is the smallest. The excavation roadway length of L_1, L_2, and L_3 is 200 m, which has the greatest air velocity magnitude. The larger air velocity has a greater impact on the ventilation system, especially in the air inlet part. Figures 6 and 7 show the air velocity distributions of the two inlet boundaries.

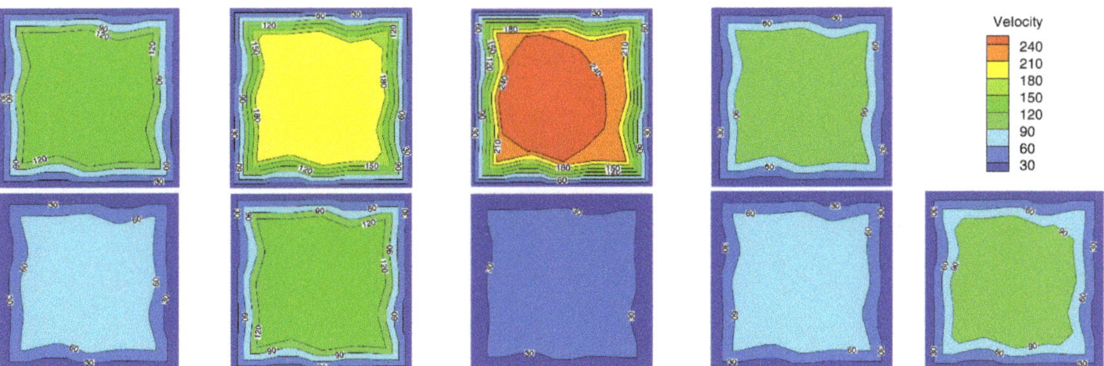

Figure 6. Air velocity of Inlet 2 of the return airway.

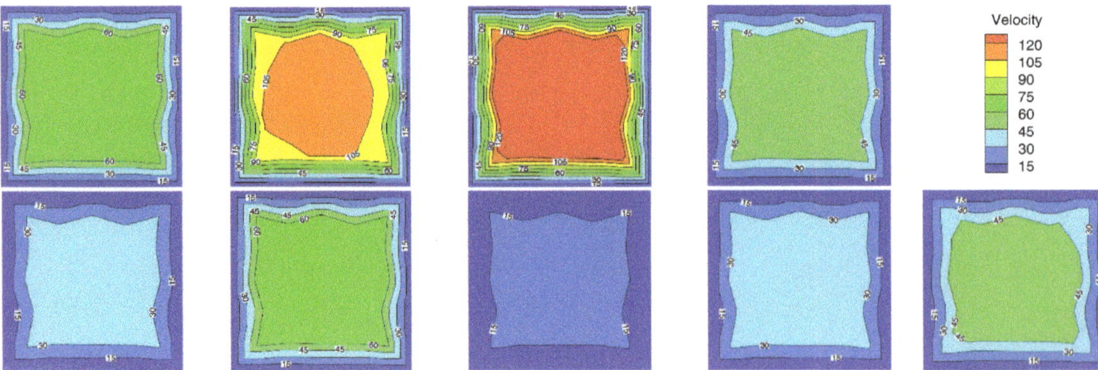

Figure 7. Air velocity of Inlet 1 of the intake airway.

As can be seen from the contour map in Figures 6 and 7, all inlet boundaries are reversed in each test, and the air velocity of Inlet 2 is larger than Inlet 1. Therefore, the influence of the gas outburst accident on the return air system is greater than that on the inlet air system. Extracting the maximum air velocity of the section, the airflow reversal degree is calculated, as shown in Table 4. Meanwhile, the variation curve of air velocity and airflow reversal degree is shown in Figure 8.

Table 4. Air velocity and airflow reversal degree of the inlet boundary.

Test	Maximum Air Velocity/m·s^{-1}		Airflow Reversal Degree	
	Inlet 2	Inlet 1	Inlet 2	Inlet 1
L_1	−142.30	−74.23	−48.43	−25.74
L_2	−206.40	−111.39	−69.80	−38.13
L_3	−252.70	−137.31	−85.23	−46.77
L_4	−112.65	−53.79	−38.55	−18.93
L_5	−78.30	−34.47	−27.10	−12.49
L_6	−137.58	−68.94	−46.86	−23.98
L_7	−55.78	−18.66	−19.59	−7.22
L_8	−80.26	−38.06	−27.75	−13.69
L_9	−98.28	−48.59	−33.76	−17.20

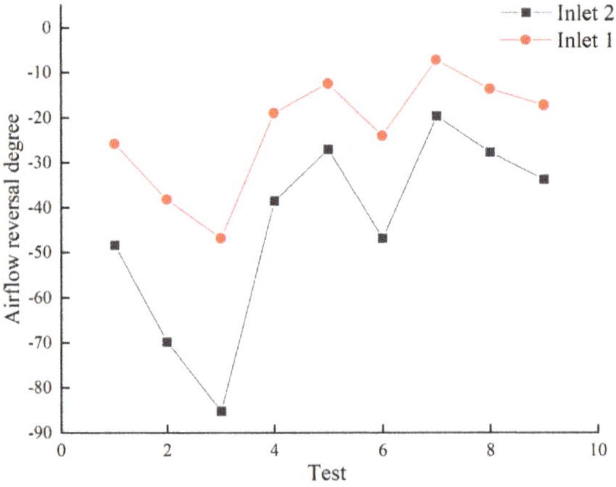

Figure 8. Variation of air velocity and airflow reversal degree.

From Table 4 and Figure 8, the airflow reversal degree of Inlet 2 is larger than Inlet 1, that is, because the air pressure distribution of the ventilation system can restrain the outburst gas, and the air pressure variation of the intake airway is smaller than the return airway.

The numerical simulation results show that the airflow reversal degree of 200 m roadway is greater than that of 1000 m roadway and 2000 m roadway. Therefore, the airflow reversal degree decreases with the increase of the roadway length. Furthermore, the outburst air flow pressure of L_3, L_6, and L_9 is 0.3 MPa, and the absolute value of their airflow reversal degree is greater than the outburst pressure of 0.1 MPa and 0.2 MPa in the same length of the excavation roadway. Therefore, the airflow reversal degree increases with the increase of the outburst pressure and the outburst energy.

5. Comprehensive Evaluation of Airflow Disorder

5.1. Fuzzy Comprehensive Optimization Theory

According to the main factors that affect the safety and stability of the ventilation system as well as the basic data of orthogonal experiments, the air quantity of the auxiliary fan, the air pressure difference of the air door, and the air flow reversal degree of Inlet 1 and Inlet 2 are used to establish the evaluation model of the airflow disorder on the basis of variable fuzzy theory.

Assuming $F = \{F_{ij}\}$, $i = 1, \ldots, m$, $j = 1, \ldots, n$ indicates that the parameter of the m test model corresponds to the set of evaluation factors of n. Here, $m = 4$. The method of relative membership is adopted to deal with the initial data. There are three situations,

(a) The larger the better

$$Fr_{ij} = \frac{F_{ij} - \min_{1 \le i \le m} \{F_{ij}\}}{\max_{1 \le i \le m} \{F_{ij}\} - \min_{1 \le i \le m} \{F_{ij}\}} \tag{2}$$

where Fr_{ij} is the relative membership of the set F_{ij} and its range is [0, 1]; max {} indicates the maximum value of the set; min {} indicates the minimum value of the set.

(b) The smaller the better

$$Fr_{ij} = \frac{\max_{1 \le i \le m} \{F_{ij}\} - F_{ij}}{\max_{1 \le i \le m} \{F_{ij}\} - \min_{1 \le i \le m} \{F_{ij}\}} \tag{3}$$

(c) The value is equal to 1

$$Fr_{ij} = 1, \left(F_{1j} = F_{2j} =, \ldots, = F_{mj}\right) \tag{4}$$

The fuzzy partition matrix is defined as follows,

$$U = \begin{bmatrix} u_{11} & u_{12} & \cdots & u_{1m} \\ u_{21} & u_{22} & \cdots & u_{2m} \end{bmatrix} = \begin{bmatrix} u_{1i} \\ u_{2i} \end{bmatrix} \tag{5}$$

where u_{1i} is subordinate to the superior index and u_{2i} is the inferior index, $\sum_{k=1}^{2} u_{ki} = 1$, $u_{ki} \in [0, 1]$.

The objective function is the minimal sum of the weight difference squared of the m models.

$$\min f(u_{1i}) = \sum_{i=1}^{m} \left(\left(u_{1i} \sqrt{\sum_{j=1}^{4} [\omega_j (gr_j - Fr_{ij})]^2} \right)^2 + \left(u_{2i} \sqrt{\sum_{j=1}^{4} [\omega_j (Fr_{ij} - br_j)]^2} \right)^2 \right) \tag{6}$$

where ω_j is the weight of each factor; gr_j represents the standard superior membership, which is the optimal value of evaluation factor j; br_j represents the standard inferior subordinate, which is the worst value of evaluation factor j.

Supposing $\frac{df(u_{1i})}{du_{1i}} = 0$, the optimal fuzzy partition matrix is calculated. Then, the influence of different factors is evaluated by u_{1i}.

5.2. Quantification and Weight Analysis of the Factors

5.2.1. Factors

The pressure difference of the air door has a strong inhibition of the outburst energy, which belongs to situation (a). The greater the pressure difference of the air door, the greater the pressure energy of the air inlet roadway. The air quantity of the auxiliary fan belongs to situation (a). Because it is opposite to the direction of the outburst gas, the resistance to the migration of the outburst gas increases with the larger air quantity of the auxiliary fan. The airflow reversal degree of Inlet 1 and Inlet 2 also belongs to situation (a), the larger the better.

The above factors consist of the F matrix.

5.2.2. Weight

Since the airflow reversal degree is more important than the pressure difference of air door and the air quantity of the auxiliary fan, the weight of the airflow reversal degree is 0.4 and 0.4, respectively, at Inlet 1 and Inlet 2. The weight of the pressure difference of the air door is equal to the air quantity of the auxiliary fan, their value is 0.1, respectively. Then the weight vector $\omega_j = (0.1, 0.1, 0.4, 0.4)^{\mathrm{T}}$.

5.3. Results and Discussion

The fuzzy comprehensive optimization theory is used to evaluate the airflow disorder degree of the mine ventilation network after a gas outburst accident. From Tables 1 and 4, the max $\{F_{ij}\} = (1500, 1200, -19.50, -7.22)^{\mathrm{T}}$, min $\{F_{ij}\} = (500, 240, -85.23, -46.77)^{\mathrm{T}}$. The membership vector of nine groups is calculated by Equations (2)–(6), $u_{1i} = (0.4928, 2.2160, 23.5568, 0.1586, 0.0394, 0.3398, 0, 0.0746, 0.1209)$. The values of each factor and evaluation results are shown in Figure 9.

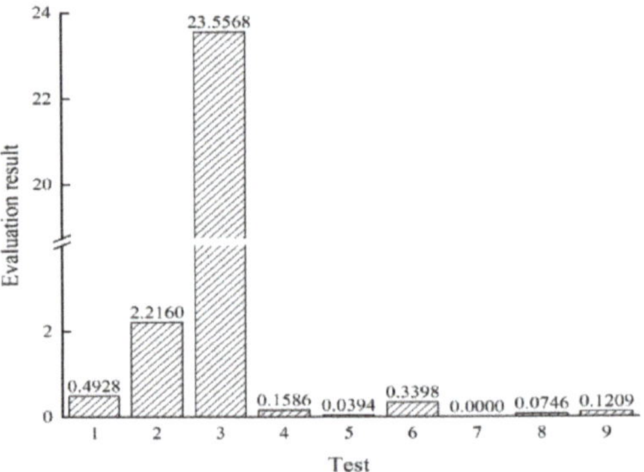

Figure 9. The evaluation results of each test.

From Figure 9, the evaluation result of L_3 is the largest, illustrating that this parameter condition has the strongest influence on the ventilation system, and L_7 is the smallest. In L_3, the length of excavation roadway is 200 m, the outburst pressure is 0.3 MPa, the pressure difference of air door is 1500 Pa, and the air quantity of the auxiliary fan is 600 m^3/min. In L_7, the length of the excavation roadway is 2000 m, the outburst pressure is 0.1 MPa, the pressure difference of air door is 1500 Pa, and the air quantity of the auxiliary fan is 1200 m^3/min.

According to the extreme difference of the range analysis method, the results of the orthogonal experiment are determined. The extreme difference is the difference between the maximum and minimum value of the airflow disorder of each factor. With the increase of the influence range, the disturbance degree of the factor increases. The result is shown in Table 5 and Figure 10. K_i represents the sum of levels of factor i; k_i represents the average value of the levels of factor; R represents the polar difference.

Table 5. Range analysis of the orthogonal experiment.

Factors	A	B	C	D
K_1	26.27	0.53	0.91	0.77
K_2	0.54	2.45	2.38	23.67
K_3	0.20	24.02	23.72	2.56
k_1	8.76	0.18	0.30	0.26
k_2	0.18	0.82	0.79	7.89
k_3	0.07	8.01	7.91	0.85
R	8.69	7.83	7.60	7.63

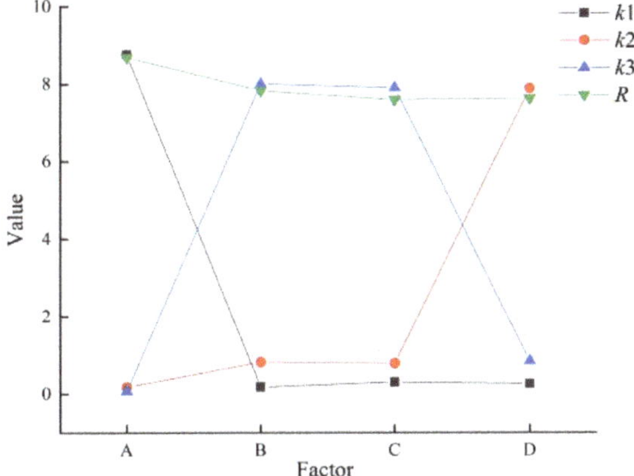

Figure 10. Range analysis of the orthogonal experiment.

From Figure 10, A > B > D > C, the length of the excavation roadway and the outburst pressure are relatively important factors affecting the ventilation system. The airflow reversal degree of the ventilation system increases with the increase of the outburst pressure or the decrease of the length of the excavation roadway.

The energy is continuously attenuated while the outburst gas migrates in the roadway. Adopting the method of reducing the air resistance of the airway or changing the air pressure distribution of the main fans, the initial air pressure in both the roadway and its parallel roadway is easily affected and will be increased in a gas outburst accident. Meanwhile, it will contribute to increase the total air quantity and improve the kinetic energy of the airflow under the condition of the constant capacity of the main fans, so it is harder to reverse the airflow.

6. Conclusions

This paper analyzes the reversal principle of airflow induced by a coal and gas outburst in the excavation roadway and proposes the indexes of the airflow disorder. Through an orthogonal experiment, it is found that all the inlet boundaries are reversed in each test, and the air velocity of Inlet 2 is larger than Inlet 1, indicating that the influence of the gas outburst accident on the return air system is greater than that on the inlet air system. Moreover, the larger air velocity has a greater impact on the ventilation system, especially the air inlet part.

On the basis of the fuzzy comprehensive optimization theory, the evaluation model of the airflow disorder is established. The results show that the order of effect factors for the stability of the mine ventilation system are the length of excavation roadway > outburst pressure > pressure difference of air door > air quantity of auxiliary fan. The most influential result is that the length of the excavation roadway is 200 m, the outburst pressure is 0.3 MPa, the pressure difference of the air door is 1500 Pa, and the air quantity of the auxiliary fan is 600 m^3/min. The airflow reversal degree of the ventilation system increases with the increase of the outburst pressure or decreases with the length of the excavation roadway.

Author Contributions: Methodology, J.S.; software, Y.W.; formal analysis, J.C.; investigation, L.L.; resources, J.S., W.H., L.L. and T.L.; data curation, J.S.; writing—original draft preparation, L.L.; writing—review and editing, J.S., J.C. and Z.L.; supervision, J.C. and Y.W.; project administration, J.S.; funding acquisition, J.S. All authors have read and agreed to the published version of the manuscript.

Funding: This work was supported by "the Fundamental Research Funds for the National Natural Science Foundation of China" (51804120, 52074122), "the National Key Research Project of China" (2018YFC08080306), "the Fundamental Research Funds for the Central Universities" (3142018003) and Science and Technology Research Project of Higher Education in Hebei Province (Z2018004).

Acknowledgments: The authors would like to thank the anonymous referees for their thoughtful comments and suggested edits that have improved the rigor and presentation of this work.

Conflicts of Interest: The authors declare that they have no known competing financial interests or personal relationships that could have appeared to influence the work reported in this paper.

References

1. Zhang, Q.; Yang, C.L.; Li, X.C.; Li, Z.B.; Li, Y. Mechanism and classification of coal and gas outbursts in China. *Adv. Civil Eng.* **2021**, *2021*, 5519853. [CrossRef]
2. Dong, G.F.; Liang, Y.P. Analysis of gas emission from outburst coal and formation conditions of countercurrent. *Coal Sci. Technol.* **2001**, *4*, 46–48.
3. Kazakov, B.P.; Shalimov, A.V.; Semin, M.A. Stability of natural ventilation mode after main fan stoppage. *Int. J. Heat Mass Tran.* **2015**, *86*, 288–293. [CrossRef]
4. Zhou, A.T.; Wang, K. Airflow stabilization in airways induced by gas flows following an outburst. *J. Nat. Gas Sci. Eng.* **2016**, *35*, 720–725. [CrossRef]
5. Bascompta, M.; Sanmiquel, L.; Zhang, H. Airflow stability and diagonal mine ventilation system optimization: A case study. *J. Min. Sci.* **2018**, *54*, 813–820. [CrossRef]
6. Feng, B.; Zhou, A.T.; Xu, X.H.; Ren, Y.Q. Research on the relationship between coal and gas outburst strength and shock wave overpressure. *Min. Eng. Res.* **2011**, *26*, 22–26.
7. Otuonye, F.; Sheng, J. A numerical simulation of gas flow during coal/gas outbursts. *Geotech. Geol. Eng.* **1994**, *12*, 15–34. [CrossRef]
8. Zhang, X.X.; Cheng, J.W.; Shi, C.L.; Xu, X.; Borowski, M.; Wang, Y. Numerical simulation studies on effects of explosion impact load on underground mine seal. *Min. Metall. Explor.* **2020**, *37*, 665–680. [CrossRef]
9. Yang, C.L.; Li, X.C. Statistical analysis of the causes and rules of especially serious gas and coal dust explosion accident in coal mine. *Coal Technol.* **2015**, *34*, 309–311.
10. Cheng, J.W.; Wang, Z.; Li, S.Y.; Song, W.T.; Lu, W.D.; Zhang, Y.J.; Zhao, K. Subsurface strata failure and movement analyses based for improving gas emission control: Model development and application. *Energy Sci. Eng.* **2020**, *8*, 3285–3302. [CrossRef]
11. Xue, S.; Zheng, C.S.; Jiang, B.Y.; Zheng, X.L. Effective potential energy associated with coal and gas outburst during underground coal mining: Case studies for mining safety. *Arab. J. Geosci.* **2021**, *14*, 1–12. [CrossRef]
12. Hong, L.; Gao, D.M.; Wang, J.R.; Zheng, D. The power source for coal and gas outburst. *J. Min. Sci.* **2019**, *55*, 239–246. [CrossRef]
13. Wei, L.J.; Li, S.; Wei, Z.K.; Wang, M.W. Research status and prospect of influence law of coal and gas outburst on ventilation system. *Coal Sci. Technol.* **2021**, *49*, 1–7.
14. Ul'yanova, E.V.; Malinnikova, O.N.; Pashichev, B.N.; Malinnikova, E.V. Microstructure of coal before and after gas-dynamic phenomena. *J. Mining Sci.* **2019**, *55*, 701–707. [CrossRef]
15. Li, C.W.; Yang, W.; Wei, S.Y.; Li, T.; Chi, L.L. Experimental study on influence range of disaster gas after coal and gas outburst. *J. China Coal Soc.* **2014**, *39*, 478–485.
16. Xu, J.; Cheng, L.; Zhou, B.; Peng, S.J.; Yang, X.B.; Yang, W.J. Physical simulation of coal-gas two-phase flow migration during outburst. *Chin. J. Rock Mech. Eng.* **2019**, *38*, 1945–1953.
17. Wu, X.; Xu, J. Numerical simulation of gas ventilation safety in coal and gas outburst process simulation laboratory. *Min. Saf. Environ. Prot.* **2010**, *37*, 5–8, 12.

18. Feng, Y.F.; Wu, Z.Q.; Zhou, A.T. Flow reversals in bypass branch of ascensionally ventilated roadway after coal and gas outburst. *China Coal* **2012**, 83–87. [CrossRef]
19. Yu, X.P.; Huang, S.J. Research on application of downward ventilation to coal mining face in horizontal seam with coal outburst in potential. *Coal Sci. Technol.* **2005**, *6*, 27. [CrossRef]
20. Zhou, A.T.; Wang, K.; Wu, Z.Q. Propagation law of shock waves and gas flow in cross roadway caused by coal and gas outburst. *J. Min. Sci. Technol.* **2014**, *24*, 23–29. [CrossRef]
21. Zhou, A.T.; Wang, K.; Wu, Z.Q.; Li, S. Research on airflow catastrophic law induced by gas pressure in mine. *J. China Univ. Min. Technol.* **2014**, *43*, 1011–1018.
22. Wang, K.; Wu, Z.Q.; Zhou, A.T.; Yu, X.; Shen, S.K. Influence of lateral branch wind resistance on wind reversal induced by gas wind pressure. *J. Basic Sci. Eng.* **2018**, *26*, 538–549.
23. Li, Z.X.; Liu, Y.; Yu, J.X.; Jia, J.Z. Coupling movement simulation of outburst gas flow and mine ventilation system. *J. Chongqing Univ.* **2012**, *35*, 111–116, 135.
24. Li, Z.X.; Li, L.; Yu, J.X.; Han, G. Simulation study on disaster ventilation system of bidirectional burst source in coal face. *J. Nat. Disasters* **2013**, *22*, 185–190.
25. Cheng, W.M.; Wang, G.; Zhou, G.; Chen, L.J. Numerical simulation of blast damper failure after coal and gas outburst. *J. Chongqing Univ.* **2009**, *32*, 314–318.
26. Dai, S.H.; Wang, Z.T.; Xu, Y. Static and dynamic analysis of the structural safety of steel blast gates. *China Saf. Sci. J.* **2017**, *27*, 74–79.
27. Huang, S.J.; Ma, X.X.; Cai, Y. Calculation of influence of outburst gas on ventilation system. *Min. Saf. Environ. Prot.* **2000**, *2*, 16–18.
28. Li, Z.X.; Wang, Y.D.; Gao, G.C. 3D simulation of coal mine disaster ventilation based on active wind network. *China Saf. Sci. J.* **2015**, *25*, 43–49.
29. Wang, K.; Jiang, S.G.; Zhang, W.Q. Destruction mechanism of gas explosion to ventilation facilities and automatic recovery technology. *J. Min. Sci. Technol.* **2012**, *22*, 417–422. [CrossRef]
30. Wu, X.; Cheng, W.Y.; Miu, Y. Research on establishing the indexes system of controlling the coal and gas outburst accident. *Procedia Eng.* **2011**, *26*, 2018–2026.
31. Zhang, C.L.; Wang, E.Y.; Xu, J.; Peng, S.J. A new method for coal and gas outburst prediction and prevention based on the fragmentation of ejected coal. *Fuel* **2021**, *287*, 119493. [CrossRef]
32. Zhai, C.; Xiang, X.W.; Xu, J.Z.; Wu, S.L. The characteristics and main influencing factors affecting coal and gas outbursts in Chinese Pingdingshan mining region. *Nat. Hazards* **2016**, *82*, 507–530. [CrossRef]
33. Yin, S.Y.; Wu, S.P.; Liu, M.B.; You, Z.L. Study on influencing factors of unconfined penetration test based on orthogonal design. *Arab. J. Geosci.* **2021**, *14*, 124. [CrossRef]

Article

Study on Asymmetric Failure and Control Measures of Lining in Deep Large Section Chamber

Yubing Huang [1,*], Bei Jiang [2,*], Yukun Ma [1], Huayong Wei [2], Jincheng Zang [3] and Xiang Gao [3]

1   Research Center of Geotechnical and Structural Engineering, Shandong University, Jinan 250061, China; mayukun@mail.sdu.edu.cn
2   Stake Key Laboratory for Geomechanics and Deep Underground Engineering, China University of Mining and Technology (Beijing), Beijing 100083, China; ZQT2000620163@student.cumtb.edu.cn
3   Yanmei Wanfu Energy Co., Ltd., Heze 274922, China; 201814643@mail.sdu.edu.cn (J.Z.); 201734659@mail.sdu.edu.cn (X.G.)
*   Correspondence: hyb6822@mail.sdu.edu.cn (Y.H.); jiangbei@cumtb.edu.cn (B.J.); Tel.: +86-15668276822 (Y.H.); +86-18253195776 (B.J.)

Abstract: Lining is often used as the last line of defense in deep large section chamber. Under the asymmetric load, it is easy to damage, resulting in the overall repair of the chamber. Aiming at this problem, taking the pump house in Wanfu Coal Mine under construction in China as an engineering example, we analyzed the asymmetric failure of pump house lining caused by construction disturbance, established the lining mechanical model and quantitative evaluation indexes, such as bending moment change rate, bending moment balance rate, displacement change rate and displacement balance rate, studied the influence mechanism of asymmetrical coefficient, section size and lining thickness on the mechanical behavior of lining, and proposed the control measures of deep large section chamber with the core of "strengthening asymmetric support, reducing section size and improving lining strength". The field monitoring shows that the asymmetric deformation of the pump house is effectively controlled, and the maximum displacement is only 7.3 mm, which ensures the long-term stability of the chamber.

Keywords: deep chamber; asymmetric failure; mechanical analysis; control measures

Citation: Huang, Y.; Jiang, B.; Ma, Y.; Wei, H.; Zang, J.; Gao, X. Study on Asymmetric Failure and Control Measures of Lining in Deep Large Section Chamber. *Energies* **2021**, *14*, 4075. https://doi.org/10.3390/en14144075

Academic Editor: Frede Blaabjerg

Received: 9 June 2021
Accepted: 2 July 2021
Published: 6 July 2021

Publisher's Note: MDPI stays neutral with regard to jurisdictional claims in published maps and institutional affiliations.

1. Introduction

With the depletion of shallow coal resources and the improvement of coal mining modernization levels, the mine is developing onto a large scale, and there are more and more large section chambers [1,2]. At present, the combined support method of bolt mesh spray and high strength lining is mostly used in the deep chamber in the large section [3–8]. Especially in the core chamber with long service life, high-strength lining is regarded as the last line of defense for its advantages of high strength, high stiffness, strong sealing and favorable appearance [9,10]. When confronted with the high stress, inclined rock, secondary disturbance and other adverse factors, it is difficult to effectively control the stability of surrounding rock with a bolt mesh spray. As the last line of defense, the lining bears the load of surrounding rock; once damaged, the chamber will face a comprehensive repair. Under complex geological conditions, the loads of surrounding rock are often uneven and asymmetric, and the bearing capacity of the lining is far lower than the theoretical value. Consequently, the phenomenon of local cracking, spalling and even overall instability is very common, which poses a great threat to the safety of mine production [11–13].

Aiming at the deformation and failure of lining under asymmetric load, scholars have carried out a large number of studies [14–24]. Krzysztof Skrzypkowski [25] studied the influence of geological and mining factors on the stability of chamber excavation through roadway anchor support, and determined the key position of chamber failure. Xu Ying, Jing Shenguo et al. [26] established a mechanical model of chamber lining-anchor cable coupling support for a deep large-section pump house, studied the influence law of support parameters on the lining ultimate bearing capacity under non-uniform load, and proposed an effective reinforcement scheme. Wang Shuhong et al. [27] carried out an experimental study of the influence of the position of the cavity behind the wall on the lining failure mode, and summarized the failure modes and failure sequences of different cavity tunnels. Sun Xiaoming, Zhang guofeng, et al. [28] studied the asymmetric deformation and failure of roadway in inclined strata, obtained the key parts of the asymmetric deformation and failure of roadway in deep inclined strata, and proposed the asymmetric coupling control measures. Wang Jiong, Guo Zhibiao, et al. [29] studied the asymmetric deformation and failure characteristics of deep roadway crossings by means of numerical analysis and engineering tests, obtained the key parts of the first failure and put forward the asymmetric coupling support measures of anchor net cables. The above studies mainly focus on the asymmetric failure of lining caused by the cavity behind the lining and the attitude of the surrounding rock. The excavation of adjacent chambers will cause the construction disturbance, bringing the asymmetric load to the stable lining, which leads to deformation and failure, especially for the deep chamber with large section. At present, there are few studies on the mechanical behavior and control method of deep large section chamber lining caused by construction disturbance.

Based on this, this paper takes the permanent pump house of Wanfu Coal Mine in China as the engineering background, establishes the lining mechanical model and quantitative evaluation index based on an asymmetric coefficient, studies the influence mechanism of asymmetric coefficient, section size and lining thickness on lining mechanical behavior, proposes the control method of a deep large section chamber and guides the reinforcement of a permanent pump house in the Wanfu Coal Mine. It has an important guiding significance for engineering under similar conditions.

2. Engineering Background

2.1. Engineering and Geology Background

Wanfu Coal Mine is located at the southernmost point of Juye Coalfield in eastern China. It is a large-scale modern mine under construction, with a designed production capacity of 1.8 Mt/year and a design service life of 67 years. The shaft station is located at the level of −820 m, and the pump house is the largest chamber at this depth. The maximum height of the excavated section of the pump house is 10.4 m, and the maximum width is 7.2 m. The water distribution drift is located on the east side of the pump house, and is the nearest exploitation field to the pump house. The nearest horizontal distance is only 2.5 m, and the nearest vertical distance is only 1.8 m. The mine location and shaft station plan are shown in Figure 1.

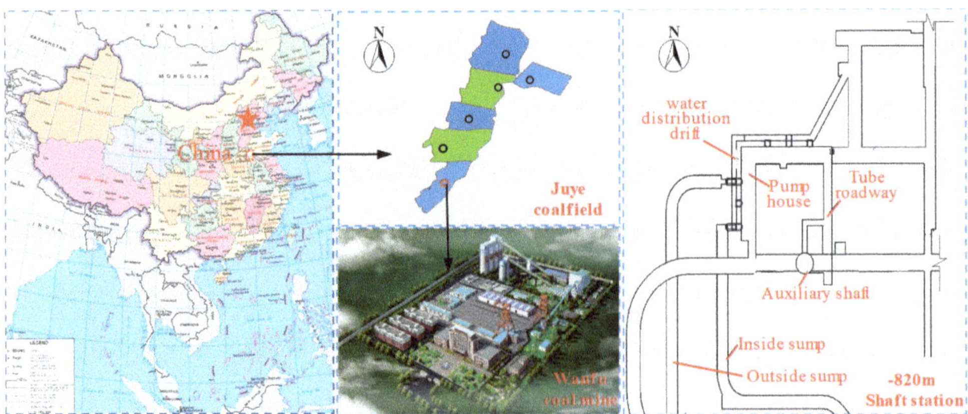

Figure 1. Mine location and −820 m shaft station plan.

The pump house is located in the inclined rock strata with the interaction of mud and sand, and the dip angle of the rock strata is 13°. From the bottom to the top, the lithology is mudstone, siltstone and fine sandstone. The maximum horizontal stress in the area is 37.1 MPa, and the vertical stress is 23.8 MPa. The pump house adopts the support form of anchor net spray + reinforced concrete lining. The parameter of anchor cable is Φ22 × 6200 mm, the inter-row spacing is 1600 × 1600 mm. Bolt parameters are Φ22 × 2500 mm, the inter-row spacing is 800 × 800 mm. Reinforced concrete lining is C40 concrete, 500 mm thick, double reinforced. The support and rock distribution of the pumping house are shown in Figure 2.

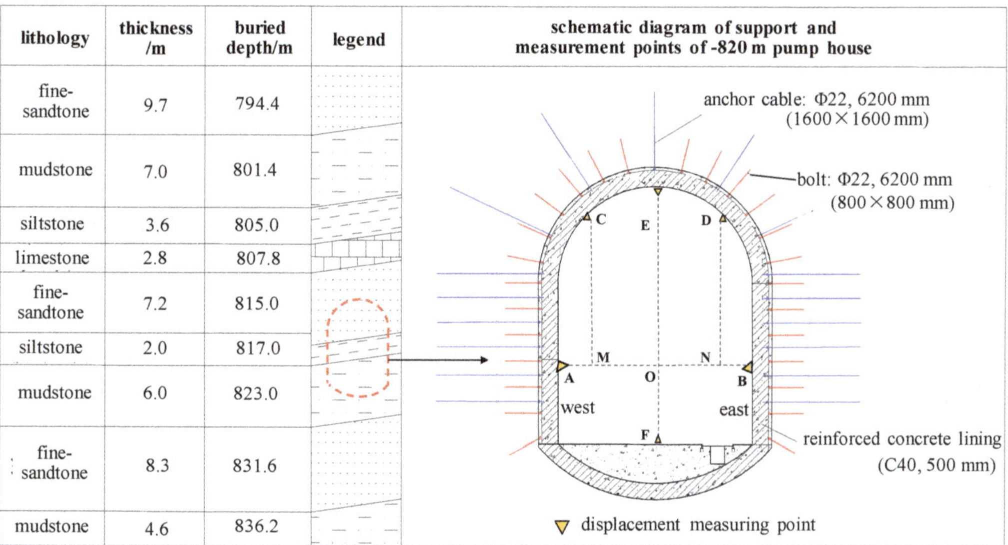

lithology	thickness /m	buried depth/m	legend	schematic diagram of support and measurement points of -820 m pump house
fine-sandtone	9.7	794.4		
mudstone	7.0	801.4		
siltstone	3.6	805.0		
limestone	2.8	807.8		
fine-sandtone	7.2	815.0		
siltstone	2.0	817.0		
mudstone	6.0	823.0		
fine-sandtone	8.3	831.6		
mudstone	4.6	836.2		

Figure 2. The support and rock distribution of the pumping house.

2.2. Field Failure Analysis

After lining work, long-term asymmetric deformation monitoring was carried out, including convergence of two walls (OA and OB), settlement of two arch shoulders (MC and ND) and displacement of roof and floor (OE and OF). The sensor used for convergence measurement is a mechanical convergent ruler, the accuracy is up to 0.1 mm and the monitoring frequency is once every two days by manual. The layout of measuring points is shown in Figure 2, the displacement curve and on-site damage are shown in Figure 3.

The deformation process of lining can be divided into three stages: ① the initial stable stage, as of 328 days after lining working, the pump house is stable as a whole. ② The disturbance stage, during the construction of the inside sump and water distribution drift, the deformation of the pumping house increased rapidly, and the deformation increment is as follows: west straight wall > west arch shoulder > arch top > east straight wall > arch bottom > east arch shoulder. The overall law is that the deformation on the west side is greater than that of the east side. ③ In the slower growth stage, the deformation development ratio slows down, but it is not completely stable.

In conclusion, due to the influence of water distribution drift and inside sump excavation, the original stable lining of the pump house shows asymmetric failure characteristics, and the deformation of the lining near the disturbed side is far greater than that of the other side. After the lining is damaged, the radial restraint effect on the surrounding rock is lost, and the deformation continues to develop slowly. If the reasonable reinforcement measures are not taken, it is difficult to meet the requirements.

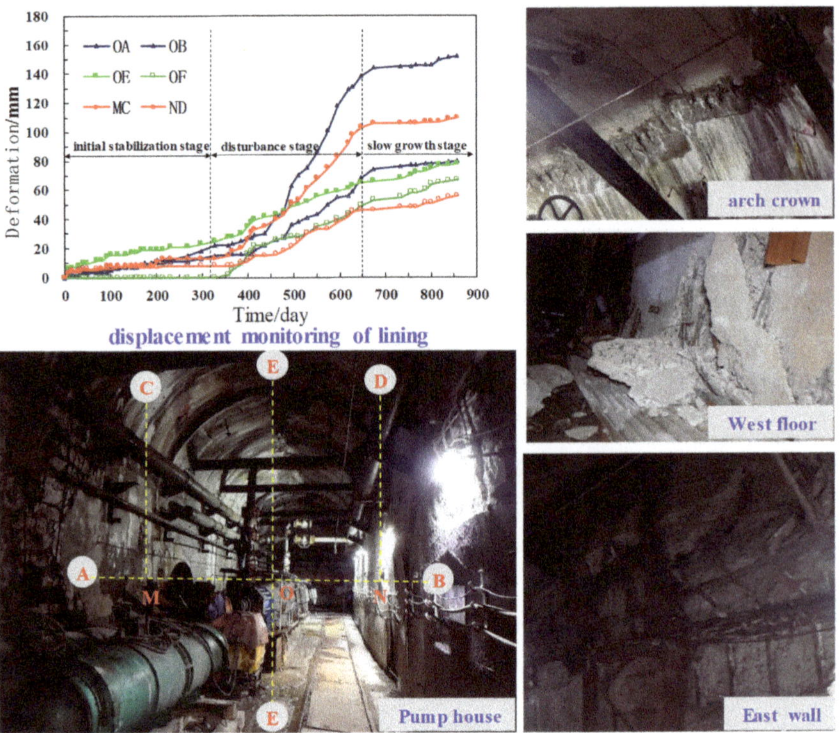

Figure 3. The displacement curve and on-site damage.

3. Asymmetric Mechanical Model and Evaluation Index

In this section, the asymmetric mechanical model and quantitative evaluation index of lining are established to provide a basis for analyzing the deformation and stress rule of the lining under an asymmetric load.

3.1. Asymmetric Mechanical Model

Along the axial direction of the chamber, the lining per unit length is abstracted as a mechanical model, as shown in Figure 4a. In this mechanical model, the arch crown radius is r_1, the arch bottom radius is r_2, and the straight wall height is h. The asymmetrical coefficient of lining stress is λ. The uniform load on the left arch crown is q_1, the uniform load on the left wall is q_2, and the uniform load on the left bottom arch is q_3. The uniform load on the right arch crown, straight wall and bottom arch is λq_1, λq_2 and λq_3, respectively.

Based on the principle of structural mechanics [30], mechanic model of lining under asymmetric conditions is divided into positive symmetry and anti-symmetry, and the "force method" is used to solve them, respectively.

Half of the positive symmetric part is taken for analysis, as shown in Figure 4b, and it is simplified into a quadratic overstatically indeterminate structure. Releasing the support constraint at point A of the vault, and replacing it with the redundant unknown forces X_1 and X_2 and X_1 represents the horizontal support reaction (F_{ax}) at point A, and X_2 represents the constraint bending moment (M_a) at point A, as shown in Figure 4b.

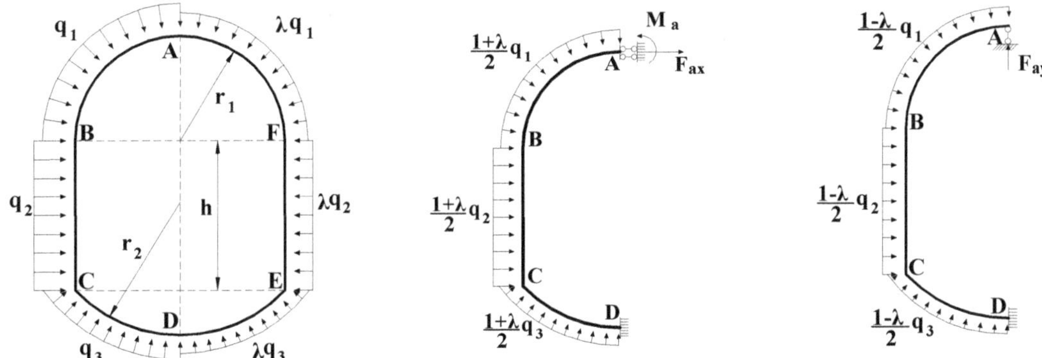

(a) mechanical model   (b) Positive symmetry calculation model   (c) Antisymmetric calculation model

Figure 4. Mechanical calculation model of lining under asymmetric force.

Half of the antisymmetric part is taken for analysis, as shown in Figure 4c, and it is simplified to a statically indeterminate structure. Releasing the support constraint at point A of the vault, and replacing it with the redundant unknown force X_3, which represents the vertical support reaction (F_{ay}) at point A, as shown in Figure 4c.

$$\delta_{11}X_1 + \delta_{12}X_2 + \Delta_{1P} = 0$$
$$\delta_{21}X_1 + \delta_{22}X_2 + \Delta_{2P} = 0 \qquad (1)$$
$$\delta X_3 + \Delta = 0$$

According to Equations (2) and (3), the basic parameters in (1) can be obtained.

$$\delta_{11} = \int \frac{\overline{M_1 M_2}}{EI} ds$$
$$= \frac{9\pi R_1^3 - 24R_1^2 + 12R_1 h + 4h^2}{12EI} + \qquad (2)$$
$$\frac{4R_2\alpha(R_1 h)^2 + R_2^3(2\sin 2\alpha(1-\cos\alpha) - 2\sin\alpha(1-\cos 2\alpha) + 6\alpha)}{4EI}$$

Similarly, δ_{12}, δ_{21}, δ_{22} and δ can be obtained.

$$\Delta_{2P} = \int \frac{\overline{M_2 M_P}}{EI} ds$$
$$= \frac{q_1 R_1 [R_1^2(3\pi - 6) + 6h + 3h^2 + 6R_2 R_1\alpha + 6h\alpha] + q_2 h^2(3\alpha + h)}{6EI} + \qquad (3)$$
$$\frac{R_2^2[q_3 R_2(\alpha - \sin\alpha) + (q_1 R_1 + q_2 h)(\sin\alpha - \cos\alpha) - q_1 R_1(\alpha\sin\alpha + \cos\alpha - 1)]}{EI}$$

Similarly, Δ_{1P} and Δ can be obtained.

For lining structure, the displacement is mainly caused by bending moment, and the displacement caused by axial force and shear force is small. Therefore, when analyzing the mechanical behavior of lining influenced by asymmetric load, only bending stress is considered. According to Equation (4), the bending moment value of any section can be obtained.

$$M_a = M_{pa} + X_1\overline{M_1} + X_2\overline{M_2}$$
$$M_b = M_{pb} + X_3\overline{M} \qquad (4)$$
$$M = M_a + M_b$$

where (1)–(4), δ is the angle from the top of the structure to any section. δ_{ij} represents the displacement of the basic structure along the X_i direction under the single action of the unit force, Δ represents the displacement of the basic structure along the X direction under the action of the load alone, EI is bending stiffness, which refers to the ability of an object to resist its bending deformation. $\overline{M_1}$ and $\overline{M_2}$ represent the bending moment of the unit force in any section of the basic structure, and M_i represents the bending moment of the load in any section of the basic structure [31].

3.2. Analysis Scheme

Mechanical behavior of lining under asymmetric load is most significantly affected by asymmetric coefficient, section size and thickness [32]. Based on the actual parameters of the pump house in Wanfu Coal Mine, different asymmetric coefficients, section size and lining thickness are selected to analyze the influence rule on lining bending moment and displacement, corresponding to the schemes A_i, B_i and C_i, respectively, where $i = 1{\sim}7$. The scheme design is shown in Tables 1–3.

Table 1. Analysis scheme of different asymmetric coefficients.

Scheme No./A_i	Asymmetric Coefficient/λ	Invariant
A_1	1	
A_2	1.1	
A_3	1.2	q_1, q_2, q_3
A_4	1.3	h, r_1, r_2
A_5	1.4	d
A_6	1.5	
A_7	1.6	

Table 2. Analysis scheme of different chamber section size.

Scheme No./B_i	Section Size			Invariant
	Arch Crown Radius/r_1	Wall Height/h	Arch Bottom Radius/r_2	
B_1	4.0	2.3	0.2	
B_2	5.0	3.3	0.3	
B_3	6.0	4.3	0.4	q_1, q_2, q_3
B_4	7.0	5.3	0.5	d
B_5	8.0	6.3	0.6	λ
B_6	9.0	7.3	0.7	
B_7	10.0	8.3	0.8	

Table 3. Analysis scheme of different lining thickness.

Scheme No./C_i	Thickness/m	Invariant
C_1	0.2	
C_2	0.3	
C_3	0.4	q_1, q_2, q_3
C_4	0.5	h, r_1, r_2
C_5	0.6	λ
C_6	0.7	
C_7	0.8	

3.3. Evaluation Indexes

In order to quantitatively analyze the influence rules of different factors on lining bending moment and displacement, evaluation indexes such as bending moment change ratio, bending moment balance ratio, displacement change ratio and displacement balance ratio are established.

- bending moment change ratio

$$\gamma_{xy-z} = \frac{\left| M_{x_n y} - M_{x_1 y} \right|}{\left| M_{x_1 y} \right|} \times 100\% \tag{5}$$

- bending moment balance ratio

$$\delta_{xy} = \frac{2\left|M_{x_ny-l} - M_{x_ny-r}\right|}{\left|M_{x_ny-l}\right| + \left|M_{x_ny-r}\right|} \times 100\% \tag{6}$$

- displacement change ratio

$$\varepsilon_{xy-z} = \frac{\left|D_{x_ny} - D_{x_1y}\right|}{\left|D_{x_1y}\right|} \times 100\% \tag{7}$$

- displacement balance ratio

$$\varphi_{xy} = \frac{2\left|D_{x_ny-l} - D_{x_ny-r}\right|}{\left(D_{x_ny-l} + D_{x_ny-r}\right)} \times 100\% \tag{8}$$

where, M_{x_ny}—In the x_n scheme, the sum of absolute values of crown bending moments on both left and right sides at y distance from the arch bottom. M_{x_ny-l}—In the x_n scheme, the bending moment value at y position of the left. In the x_n scheme, the sum displacement of lining on both left and right sides is at y distance from the arch bottom. In the x_n scheme, the displacement value of the right arch body is from the y position of the arch bottom. Where x is A~C, representing three types of schemes respectively; n is 2~7, representing 2~7 schemes respectively; y is the distance to the arch foot; l and r represent the left and right respectively.

The bending moment change ratio indicates the change of lining bending moment at a certain position. The greater the γ_{xy-z}, the greater the bending moment is affected. The bending moment balance ratio indicates the bending moment difference between the left and the right, and the larger the δ_{xy}, the more unbalanced the lining. The change ratio of displacement indicates the change of lining deformation at a certain position. The larger the ε_{xy-z}, the greater the deformation is influenced. Displacement balance ratio indicates the difference between left and right deformation. The greater the δ_{xy}, the worse the displacement balance.

4. Analysis of Asymmetric Mechanical Behavior of Lining

The section size of the -820 m pump house in Wanfu Coal Mine is 46.5 m² ($R_1 = 3.5$ m, $R_2 = 4.8$ m, h = 5.3 m), the surrounding rock pressure is $q_1 = 100$ kN/m, $q_2 = 150$ kN/m, $q_3 = 50$ kN/m, lining thickness is 0.5 m, and C40 reinforced concrete is adopted. The elastic modulus is 32.5 GPa, and the moment of inertia per unit length lining is 5.3×10^{-3} m⁴. By substituting the specific parameters into the corresponding Equations (1)–(4), the bending moment value M of each position can be calculated. To reduce the computational workload, 42 feature points are taken along the entire lining. On this basis, the influences of asymmetric coefficient, section size of chamber and lining thickness on lining bending moment and displacement are analyzed. In order to quantitatively evaluate the influence of various factors on the lining, the 45° position on both sides of the top arch is taken.

4.1. Asymmetric Coefficient

4.1.1. Bending Moment Analysis

The following result analysis can be obtained from Figure 5.

Under the action of symmetrical force, the bending moment of lining distributes symmetrically. With the increase of the asymmetry coefficient, the bending moment of the left arch crown is larger than the right arch crown, and the distribution of the bending moment of the bottom arch is opposite. The bending moment of the right straight wall is larger than the left.

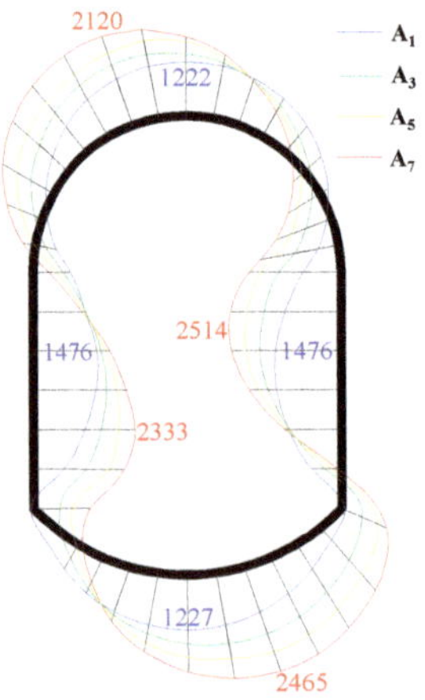

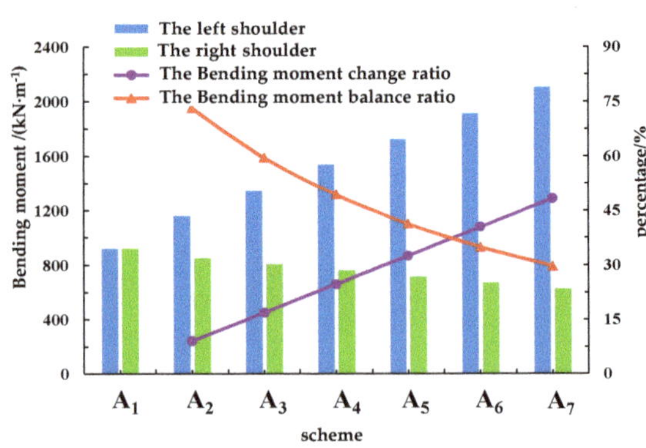

(**a**) Bending moment distribution diagram (kN/m)   (**b**) Bending moment of left and right shoulder

Figure 5. Bending moment analysis with different asymmetric coefficients. (**a**) Bending moment distribution diagram (**b**) Bending moment of left and right shoulder.

Taking the left and right shoulder as an example, the bending moment of the left shoulder increases from 920 kN/m to 2104 kN/m and the bending moment of the right shoulder decreases from 920 kN/m to 625 kN/m, with the increase of the asymmetry coefficient from 1 to 1.6. The change ratio of bending moment of the shoulder is positively correlated with the asymmetric coefficient, which increases from 15% to 48%. The bending moment balance ratio is negatively correlated with the asymmetry coefficient, which decreases from 75% to 30%.

4.1.2. Displacement Analysis

The following result analysis can be obtained from Figure 6.

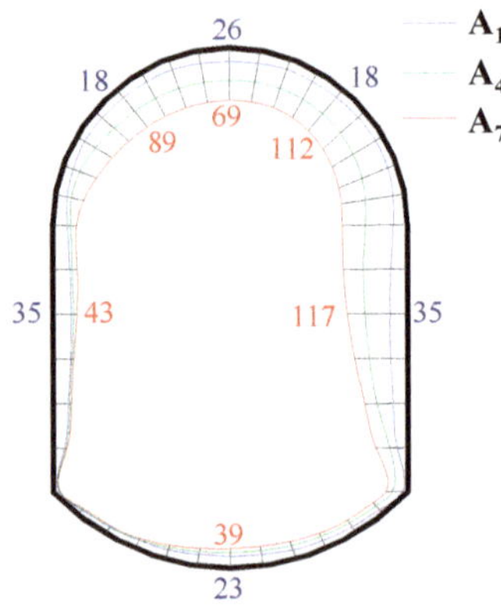

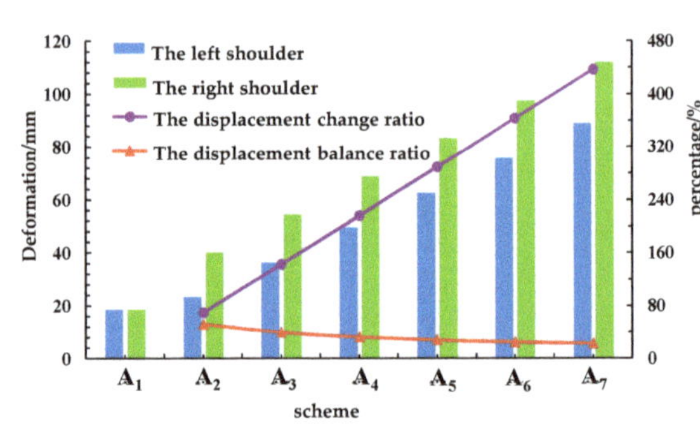

(**a**) Displacement distribution diagram (mm)   (**b**) Displacement of left and right shoulder

Figure 6. Displacement analysis of different asymmetric coefficients. (**a**) Displacement distribution diagram (**b**) Displacement of left and right shoulder.

Under the action of the symmetric force, the displacement is distributed symmetrically. The maximum displacement of the straight wall is 35 mm, and the maximum displacement of the top arch is 26 mm. As the asymmetry coefficient increased, the overall lining displacement increased, and the increase of the right lining displacement (the larger force side) was greater than the left.

Taking the left and right shoulder displacements as an example, the left shoulder displacements increased from 18 mm to 89 mm and the right shoulder displacements increased from 18 mm to 112 mm, with the increase of the asymmetry coefficient from 1 to 1.6. The displacement change ratio of the shoulder is positively correlated with the asymmetry coefficient, which increases from 70% to 437%. The displacement balance ratio is negatively correlated with the asymmetry coefficient and decreased from 60% to 42%.

4.2. Section Size

4.2.1. Bending Moment Analysis

The following result analysis can be obtained from Figure 7.

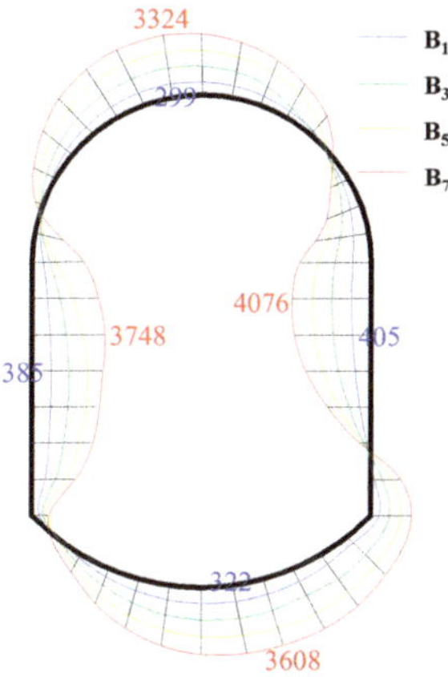

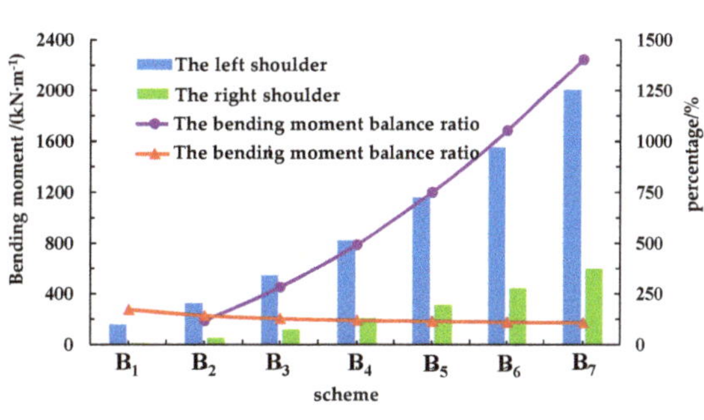

(**a**) Bending moment distribution diagram (kN/m)   (**b**) Bending moment of left and right shoulder

Figure 7. Bending moment analysis of different section sizes. (**a**) Bending moment distribution diagram (**b**) Bending moment of left and right shoulder.

With the increase of the section size of the chamber, the lining bending moment increases, but the position of maximum bending moment does not change. The maximum bending moment of the arch crown is located at the left side of the arch top. The maximum bending moment of the straight wall is located at 2 h/3 to the wall foot. The maximum bending moment of the arch bottom is on the right of the arch bottom middle.

Taking the left and right shoulder as an example, with the section size increasing from 4 m to 10 m, the bending moment of the left shoulder increases from 161 kN/m to 2006 kN/m, and the bending moment of the right shoulder increases from 12 kN/m to 298 kN/m. The bending moment change ratio increases from 118% to 1403% with the increase of section size. The bending moment balance ratio is negatively related with the section size, which decreases from 142% to 108%.

4.2.2. Displacement Analysis

The following result analysis can be obtained from Figure 8.

With the increase in the section size, the displacement of the lining increases rapidly, and the displacement value of the right side increases more obviously than that left, but the maximum displacement position remains unchanged.

Taking the displacement of the left and right shoulder as an example, with the increase of the section width from 4 m to 10 m, the displacement of the left shoulder increases from 1.8 mm to 155.1 mm, and that of the right shoulder increases from 3.1 mm to 220.9 mm.

The displacement change ratio increases from 12% to 810%, and the displacement balance ratio decreases from 51% to 42%.

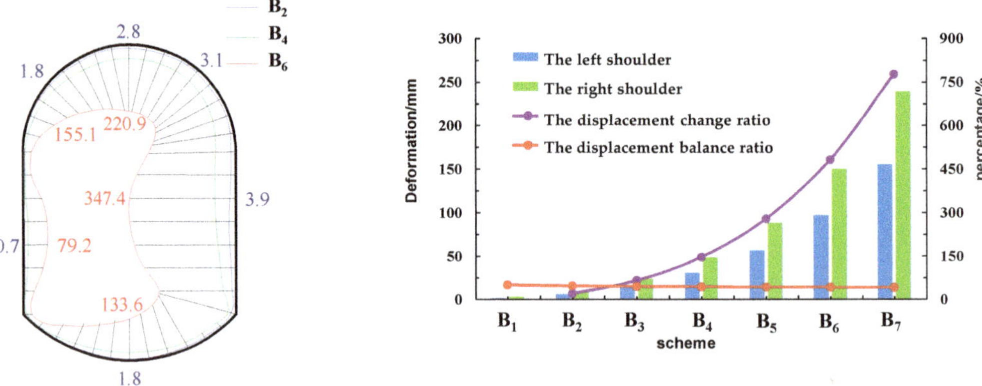

(**a**) Displacement distribution diagram (mm) (**b**) Displacement of left and right shoulder

Figure 8. Displacement analysis of different section sizes. (**a**) Displacement distribution diagram (**b**) Displacement of left and right shoulder.

4.3. Lining Thickness

The following result analysis can be obtained from Figure 9.

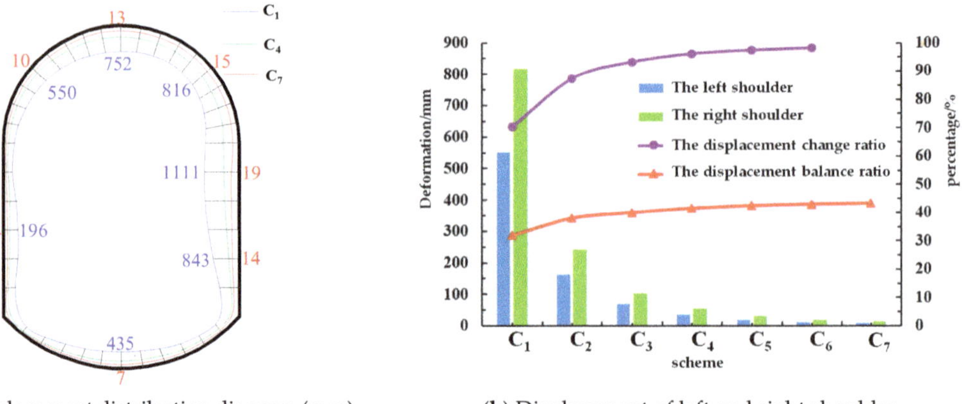

(**a**) Displacement distribution diagram (mm) (**b**) Displacement of left and right shoulder

Figure 9. Displacement analysis of different lining thickness. (**a**) Displacement distribution diagram (**b**) Displacement of left and right shoulder.

With the increase of the lining thickness, the overall displacement decreased, and the displacement of the right side (the side with larger force) decreases more significantly than that of the left, and the maximum displacement position does not change.

Taking the displacement of left and right arches as an example, with the increase of arch thickness from 0.2 m to 0.8 m, the left arch shoulder displacement decreases from 550 mm to 10 mm, and the right arch shoulder displacement decreases from 816 mm to 15 mm. The displacement change ratio increases from 70% to 98%, and the displacement balance ratio increases from 39% to 42%.

When the lining thickness exceeds 0.5 m, both displacement change ratio and displacement balance ratio have little change with the continuous increase of lining thickness.

4.4. Summary and Engineering Recommendations

Under the action of asymmetric load, the bending moment and displacement of the lining show an obvious asymmetry. The larger the asymmetry coefficient is, the more significant the increase of the bending moment and displacement is, and the more serious the failure is. Therefore, for the lining chamber subjected to secondary disturbance, the asymmetric design of bolt mesh spray should be carried out, the support strength of the disturbed side should be strengthened, and the strength of the surrounding rock should be restored through grouting and other measures to reduce the asymmetric load of the lining.

The larger the section size is, the more severely the lining is affected by the asymmetric load, the change rate of bending moment and displacement of the lining increases rapidly, and the symmetry decreases continuously. Therefore, for the lining under asymmetric load, measures such as optimizing equipment selection should be taken to reduce the chamber section size to minimize the size effect.

Increasing the lining thickness can improve the ability of lining to resist asymmetric load, but when the lining thickness exceeds a certain value, the effect of increasing the lining thickness is no longer obvious. Therefore, the section size and the load size of the chamber should be considered comprehensively, and the lining thickness should be reasonably selected by using the evaluation index.

5. Control Measures and Application Effect

5.1. Control Measures

Under the condition of asymmetric load, the core of stability control of the deep large section chamber lining is "strengthening asymmetric support, reducing the section size and improving the lining strength". In view of the asymmetric failure of the permanent pump house in Wanfu Coal Mine, the control method design is shown in Figure 10.

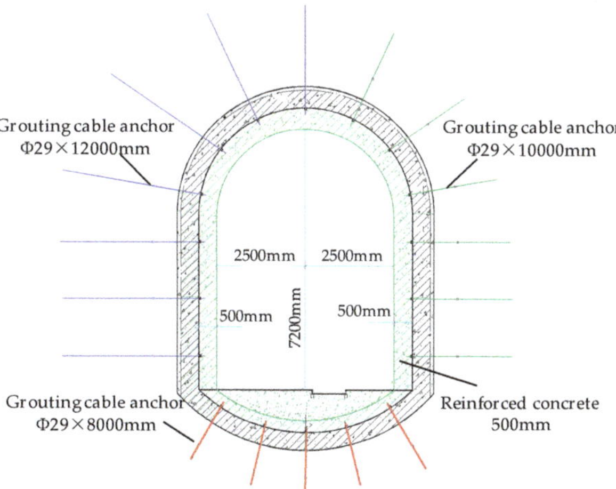

Figure 10. Schematic diagram of reinforcement scheme.

①   After optimization, the net section width of the pump house is 5000 mm and the height is 7200 mm, which avoids the expansion of the lining and reduces the surrounding rock disturbance.

②   The west straight wall and west arch shoulder of the pumping house adopt the Φ29 × 12,000 mm grouting anchor cable, and the other side adopts the Φ29 × 10,000 mm grouting anchor cable. The bottom arch adopts the Φ29 × 8000 mm grouting anchor cables. The diameter of the anchor hole is 38 mm, and two MSZ2870 resin cartridges are used. The cartridges can be cured after being stirred evenly for 5 min. After the end of grouting anchor cable is anchored, 150 kN preload is applied first, and then grouting is carried out one by one to reinforce the broken surrounding rock.

③ On the outside of the existing section, double-layer reinforcement is bound on the whole section, with the diameter of transverse reinforcement of 28 mm and longitudinal reinforcement of 32 mm. C40 concrete is used for pouring, and the lining thickness is 500 mm.

5.2. Application Effect

After the reinforcement following the above scheme, a deformation monitoring section is set, and the position of measuring point is the same as before. Figure 11 shows the displacement change curve of the pumping house within 90 days of reinforcement. It can be obtained that, after the reinforcement of the pumping house chamber, no obvious asymmetrical deformation continues to occur, and the maximum deformation is 7.3 mm, indicating a significant control effect.

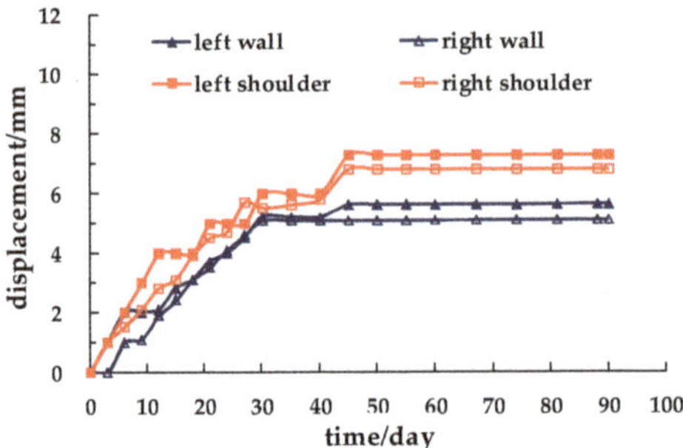

Figure 11. Displacement monitoring of pump house after reinforcement.

For the permanent pump house, continuous deformation is not allowed during the reinforcement phase, and the rigid lining support is adopted to effectively avoid the continuous deformation of the pump house, which meets the support requirements. Figure 12 is the effect drawing of the pump house after reinforcement.

Figure 12. Field effect of pump house after reinforcement.

6. Conclusions

(1) Aiming at the asymmetric failure of lining in Wanfu Coal Mine pump house due to the secondary disturbance, the asymmetric mechanical model of the lining and the quantitative evaluation indexes such as the change rate of bending moment, the balance rate of bending moment, the change rate of displacement and the balance rate of displacement are established, and the influence laws of the asymmetric coefficient,

the section size and the thickness on the bending moment and displacement of the lining are studied.

(2) Under the action of asymmetric load, the bending moment and displacement of lining shows an obvious asymmetry. The larger the asymmetry coefficient is, the more significant the increase of bending moment and displacement will be. The larger the section size is, the more obvious the influence of the asymmetric load is. Increasing the lining thickness is beneficial to improve the lining resistance to the asymmetric load, when the lining thickness exceeds a certain value, the effect of increasing the lining thickness is no longer obvious.

(3) The control measures with the core of "strengthening asymmetric support, reducing section size, improving lining strength" is proposed. The field application shows that the method can effectively restrain the asymmetric deformation of the surrounding rock, and the maximum deformation is only 7.3 mm, which ensures the long-term stability of the chamber. It has an important guiding significance for engineering under similar conditions.

Author Contributions: All the authors contributed to this paper. Conceptualization, Y.H. and B.J.; methodology, H.W.; investigation, Y.H.; resources, J.Z.; data curation, X.G., Y.M.; writing—original draft preparation, Y.H., Y.M. All authors have read and agreed to the published version of the manuscript.

Funding: This work was supported by the National Natural Science Foundation of China (Grant Nos. 52074164 and 42077267); the Natural Science Foundation of Shandong Province, China (Grant Nos. 2019SDZY04 and ZR2020JQ23); the Project of Shandong Province Higher Educational Youth Innovation Science and Technology Program, China (Grant No. 2019KJG013).

Institutional Review Board Statement: Not applicable.

Informed Consent Statement: Not applicable.

Data Availability Statement: The data supporting the findings can be found in the figures, tables and manuscript.

Acknowledgments: This work is supported by Wanfu Energy Co. Ltd. and Yankuang Group Limited, which are gratefully acknowledged.

References

1. He, M.C.; Wang, Q.; Wu, Q.Y. Innovation and future of mining rock mechanics. *J. Rock Mech. Geo Eng.* **2021**, *13*, 1–21. [CrossRef]
2. Xie, H.P. Research review of the state key research development program of China: Deep rock mechanics and mining theory. *J. China Coal Soc.* **2019**, *44*, 1283–1305.
3. Wang, Q.; Gao, H.K.; Jiang, B.; Li, S.C.; He, M.C.; Qin, Q. In-situ test and bolt-grouting design evaluation method of underground engineering based on digital drilling. *Int. J. Rock Mech Min. Sci.* **2021**, *138*, 104575. [CrossRef]
4. Wang, L.G.; Lu, Y.L.; Huang, Y.G.L.; Sun, H.Y. Deep-shallow coupled bolt-grouting support technology for soft rock roadway in deep mine. *J. China Univ. Min. Technol.* **2016**, *45*, 11–18.
5. Li, S.C.; Shao, X.; Jiang, B.; Wang, Q.; Wang, F.Q.; Ren, Y.X.; Wang, D.C.; Ding, L.G. Study of the mechanical characteristics and influencing factors of concrete arch confined by square steel set in deep roadways. *J. China Univ. Min. Technol.* **2015**, *44*, 400–408.
6. Sun, H.B.; Li, S.C.; Wang, Q.; Zhou, L.S.; Jiang, B.; Zhang, X.; Xu, S.; Zhang, H.J. Research and Application of Fabricaed Confined Concrete Construction System for Large Cross Section. *Tunn. Undergr. Space Technol.* **2018**, *31*, 320–327.
7. Wang, Q.; Xin, Z.X.; Jiang, B.; Sun, H.B.; Xiao, Y.C.; Bian, W.H.; Li, L.N. Comparative experimental study on mechanical mechanism of combined arches in large section tunnels. *Tunn. Undergr. Space Technol.* **2020**, *99*, 103386. [CrossRef]
8. Guo, J.; Feng, G.R.; Wang, P.F.; Qi, T.Y.; Zhang, X.R.; Yan, Y.G. Roof Strata Behavior and Support Resistance Determination for Ultra-Thick Longwall Top Coal Caving Panel: A Case Study of the Tashan Coal Mine. *Energies* **2018**, *11*, 1041. [CrossRef]
9. Wang, Q.; Xu, S.; Jiang, B.; Li, S.C.; Xiao, Y.C.; Xin, Z.X.; Liu, B.H. Research progress of confined concrete support theory and technology for underground engineering. *J. China Coal Soc.* **2020**, *45*, 2760–2776.
10. Xu, Y.; Bo, J.B.; Wu, Z.R.; Feng, J.S. Instability Mechanism and Reinforcement Technology of Brickwork Roadway. *J. Min. Saf. Eng.* **2012**, *29*, 790–796.
11. Huang, Y.B.; WANG, Q.; Gao, H.K.; Jiang, Z.H.; Li, S.; Chen, K. Failure mechanism and construction process optimization of deep soft rock chamber group. *J. China Univ. Min. Technol.* **2021**, *50*, 69–78.

12. Sun, K.G.; Li, S.C.; Zhang, Q.S.; Li, S.C.; Li, J.L.; Liu, B. Study on monitoring and simulation of super-long mountain tunnel lining. *Chin. J. Rock Mech. Eng.* **2007**, *26*, 4465–4470.

13. Wang, Q.; Jiang, Z.H.; Jiang, B.; Gao, H.K.; Huang, Y.B.; Zhang, P. Research on an automatic roadway formation method in deep mining areas by roof cutting with high-strength bolt-grouting. *Int. J. Rock Mech. Min. Sci.* **2020**, *128*, 104264. [CrossRef]

14. Wang, Q.; He, M.C.; Jiang, B.; Li, S.C.; Jiang, Z.H.; Wang, Y.J.; Xu, S. Experimental research and application of automatically formed roadway without advance tunneling. *Tunn. Undergr. Space Technol.* **2021**, *114*, 103999. [CrossRef]

15. Ghosh, G.K.; Sivakumar, C. Application of underground microseismic monitoring for ground failure and secure longwall coal mining operation: A case study in an Indian mine. *J. Appl. Geophys.* **2018**, *150*, 21–39. [CrossRef]

16. Saeedi, G.; Shahriar, K.; Rezai, B.; Karpuz, C. Numerical modelling of out-of-seam dilution in longwall retreat mining. *Int. J. Rock Mech. Min. Sci.* **2010**, *47*, 533–543. [CrossRef]

17. Rezaei, M.; Hossaini, M.F.; Majdi, A. Development of a time-dependent energy model to calculate the mining-induced stress over gates and pillars. *J. Rock Mech. Geotech. Eng.* **2015**, *7*, 306–317. [CrossRef]

18. Zhai, X.X.; Huang, G.S.; Chen, C.Y.; Li, R.B. Combined Supporting Technology with Bolt-Grouting and Floor Pressure-Relief for Deep Chamber: An Underground Coal Mine Case Study. *Energies* **2018**, *11*, 67. [CrossRef]

19. Waldemar, K.; Krzysztof, S.; Łukasz, H. Metody badania obudowy kotwowej w Katedrze Górnictwa Podziemnego AGH. *Cuprum* **2015**, *3*, 49–60.

20. Wang, Q.; Qin, Q.; Jiang, B.; Xu, S.; Zeng, Z.N.; Luan, Y.C.; Liu, B.H.; Zhang, H.J. Mechanized construction of fabricated arches for large-diameter tunnels. *Automat. Constr.* **2021**, *124*, 103583. [CrossRef]

21. Chang, B.L.; Li, J.B.; Xu, Y. Instability mechanism and control technology of Surrounding rock of brickwork roadway. *Saf. Coal Min.* **2015**, *46*, 100–103.

22. Zhao, M.Z.; Meng, W.F.; Lu, D.H.; Huang, C.F. Damage of upper arch walling support roadway with lower working face mined. *J. Min. Saf. Eng.* **2011**, *36*, 23–27.

23. Wang, Q.; He, M.C.; Xu, S. Mechanical properties and engineering application of bolt made of new constant resistance energy absorbing materials. *J. China Coal Soc.* **2021**. [CrossRef]

24. Liu, Q.S.; Shi, K.; Kang, Y.S.; Huang, X. Monitoring analysis of secondary lining structure of central pump house in deep coal mine. *Chin. J. Rock Mech. Eng.* **2011**, *30*, 1596–1603.

25. Skrzypkowski, K. Case Studies of Rock Bolt Support Loads and Rock Mass Monitoring for the Room and Pillar Method in the Legnica-Gogów Copper District in Poland. *Energies* **2020**, *13*, 2998. [CrossRef]

26. Jing, S.G.; Su, Z.L.; Wang, X.K. Research and application on the coupling mechanism of cable and masonry for chamber with large-section. *J. Min. Saf. Eng.* **2018**, *35*, 1158–1163.

27. Wang, S.H.; Wang, P.Y.; Liu, Y.; Zhu, B.Q. Experimental Study on failure mode of tunnel model containing cavity in different locations. *J. Northeast. Univ.* **2020**, *41*, 863–869.

28. Sun, X.M.; Zhang, G.F.; Cai, F.; Yu, S.B. Asymmetric deformation mechanism within inclined rock strata induced by excavation in deep roadway and its con-trolling countermeasures. *J. Rock Mech. Eng.* **2009**, *28*, 1137–1143.

29. Wang, J.; Guo, Z.B.; Cai, F.; Hao, Y.X.; Liu, X. Study on the asymmetric deformation mechanism and control countermeasures of deep layers roadway. *J. Min. Saf. Eng.* **2014**, *31*, 28–31.

30. Long, Y.Q.; Bao, S.H. *Course of Structural Mechanics*; Higher Education Press: Beijing, China, 2001; pp. 79–91.

31. Fan, Z.; Shi, S.; Li, Z.B.; Wang, Q.Q.; Zhao, C.J.; Kou, L.; Shi, J.W. Research on large diameter concrete filled tubular column and H-section beam connection. *J. Build. Struct.* **2016**, *37*, 1–12.

32. Zhu, W.S.; He, M.C. *Stability and Dynamic Construction Mechanics of Surrounding Rock under Complex Condition*; Beijing Science Press: Beijing, China, 1995; pp. 155–182.

Article

Analysis of Natural Hydraulic Fracture Risk of Mudstone Cap Rocks in XD Block of Central Depression in Yinggehai Basin, South China Sea

Ru Jia [1,2], Caiwei Fan [3], Bo Liu [1,2,*], Xiaofei Fu [2] and Yejun Jin [1,2]

[1] Institute of Unconventional Oil & Gas, Northeast Petroleum University, Daqing 163318, China; jiaru_dq@163.com (R.J.); Yejun.jin@nepu.edu.cn (Y.J.)
[2] Key Laboratory of Continental Mudstone Hydrocarbon Accumulation and Efficient Development, Ministry of Education, Northeast Petroleum University, Daqing 163318, China; fuxiaofei2008@sohu.com
[3] Hainan Branch of China National Offshore Oil Corporation Ltd., Haikou 570100, China; fancw@cnooc.com.cn
*   Correspondence: liubo@nepu.edu.cn

Citation: Jia, R.; Fan, C.; Liu, B.; Fu, X.; Jin, Y. Analysis of Natural Hydraulic Fracture Risk of Mudstone Cap Rocks in XD Block of Central Depression in Yinggehai Basin, South China Sea. *Energies* **2021**, *14*, 4085. https://doi.org/10.3390/en14144085

Academic Editors: Sheng-Qi Yang, Min Wang, Qi Wang, Wen Zhang, Kun Du and Chun Zhu

Received: 18 May 2021
Accepted: 2 July 2021
Published: 6 July 2021

Abstract: The Yinggehai Basin is an important Cenozoic gas bearing basin in the South China Sea. With the gradual improvement of gas exploration and over-development in shallow layers, deep overpressured layers have become the main target for natural gas exploration. There are no large-scale faults in the strata above the Meishan Formation in the central depression, and hydraulic fracturing caused by overpressure in mudstone cap rocks is the key factor for the vertical differential distribution of gas. In this paper, based on the leak-off data, pore fluid pressure, and rock mechanics parameters, the Fault Analysis Seal Technology (FAST) method is used to analyze the hydraulic fracture risk of the main mudstones in the central depression. The results show that the blocks in the diapir zone have been subjected to hydraulic fracturing in the Huangliu cap rocks during the whole geological history, and the blocks in the slope zone which is a little distant from the diapirs has a lower overall risk of hydraulic fracture than the diapir zone. In geological history, the cap rocks in slope zone remained closed for a longer time than in diapir zone and being characterized by the hydraulic fracture risk decreases with the distance from the diapirs. These evaluation results are consistent with enrichment of natural gas, which accumulated in both the Yinggehai Formation and Huangliu Formation of the diapir zone, but it only accumulated in the the Huangliu Formations of the slope zone. The most reasonable explanation for the difference of the gas reservoir distribution is that the diapirs promote the development of hydraulic fractures: (1) diapirism transfers deep overpressure to shallow layers; (2) the small fault and fractures induced by diapir activities weakened the cap rock and reduced the critical condition for the natural hydraulic fractures. These effects make the diapir zone more prone to hydraulic fracturing, which are the fundamental reasons for the difference in gas enrichment between the diapir zone and the slope zone.

Keywords: Yinggehai; overpressure; hydraulic fracture; mudstone; fluid pressure

1. Introduction

Hydraulic fracturing, also known as natural hydraulic fracturing and hydraulic extensional fracturing, is a common kind of geological phenomenon which refers to fractures caused by the increase of pore fluid pressure [1–5]. It covers the hydraulic effects on intact rocks and fractured rocks, which play an important guiding role in fluid migration, oil and gas preservation, and the safe exploitation of oil and gas fields. Research on hydraulic fracturing began in the late 1960s [1]. In a study on the law of joint development, Secor proposed the mechanism of natural tensile fractures and pointed out that the fracture caused by fluid overpressure can open original fractures in the formation while also playing an important role in the migration of groundwater, oil and gas, and ore-forming fluids [6,7]. After that, Phillips formally proposed the concept of hydraulic fracturing in his study of the formation mechanism of mineralized normal faults in Wales, the UK, which he described as the process of fracture expansion caused by the increase of pore fluid pressure within the fracture [2]. In 1990, Sibson analyzed the migration process of abnormally high pressure fluid near the fault, described the dynamic process of fault hydraulic rupture, and pointed out that this process is periodic, intermittent, and pulsating [8]. Subsequently, the research on hydraulic fracture was deepened, many overpressure basins around the world were analyzed and studied, and it was clearly pointed out that hydraulic fracture is a potential risk leading to oil and gas leakage [9]. For example, in the research of Timor Sea, they pointed out that the fault reactivity caused by hydraulic fracturing is the main cause of oil and gas leakage [10].

Through the study of the development of hydraulic fracture, the hydraulic fracture phenomenon is similar to the fault valve mechanism [11]. It has the characteristics of episodic eduction, especially in the overpressure basins. When the pore fluid pressure exceeds the rock fracturing pressure, rocks will begin fracturing leading to highly permeable channels for fluid to flow into which carries a large quantity of minerals through the fractures. In the process, pore fluid pressure is released and gradually reduced. The mineral cementation, precipitation, and pressolution gradually decreases the width of the fractures which finally forms a closed hydraulic fracture until the fluid pressure rises to enter the next "broken-closed" cycle.

Studies have also shown that these hydraulic fractures have specific geological characteristics both macroscopically and microscopically. For example, due to the periodic activities of hydraulic fracturing, different periods of hydraulic fracturing and fluid migration lead to changes in the formation structure and in 3D seismic imaging. Furthermore, the amplitude intensity is usually increased or decreased locally along the reflection layer and it forms the vertical continuous fuzzy zone and pipe features [12,13]. In the field of outcropping, it is mainly manifested as extensional fractures extending perpendicular to the direction of minimum principal stress and filled by quartz, calcite, and gypsum veins [14–16]. The multi-stage filling of fractures by fluid minerals can also be observed under a microscope.

How to effectively and accurately determine the critical condition of hydraulic fracturing is particularly important to quantitatively evaluate the periodicity of hydraulic fracturing. Through a lot of experience, many scholars regard 85% of the overburden lithostatic pressure as the critical pressure for hydraulic fracturing to occur [6,17–20]. Similarly, a study on the Norwegian Snorre field found that when the pore fluid pressure at the top of the reservoir reaches 82% of the static pressure, approximately equal to the minimum horizontal principal stress, hydraulic fracturing increases the permeability of mudstone and leads to the escape of oil and gas from the reservoir [21]. Some investigators propose that the critical pore fluid pressure for hydraulic fractures is the sum of the principal of minimum stress and the tensile strength of cap rock [1,22–25]. Gaarenstroom defined the difference between the minimum horizontal principal stress and pore fluid pressure as the retention capacity in his study on the risk of hydraulic fracture in the cap rock of the North Sea Basin. The larger the pore fluid pressure is, the retention capacity will be smaller, and the risk of hydraulic fracture in the cap rock will be greater. He found that when the retention capacity is lower than 7 MPa, the risk of hydraulic fracture will significantly increase [24]. Most of these evaluation methods only consider the risk of hydraulic tensile fracture of rock and ignore the tensile strength of rock. However, the actual cap rock is often heterogeneous, especially when there is strong and weak interbedding in the cap rock (sandstone and mudstone are both distributed), and the cap rock may form shear fracture or mixed-mode fracture failure. Therefore, this simple evaluation method is obviously not applicable to evaluate the hydraulic fracture risk of deep overpressure cap rock such as Yinggehai Basin. At present, we mainly use the Fault Analysis Seal Technology (FAST) for the hydraulic fracture evaluation [10,26]. The Fast evaluation method is a comprehensive method proposed by Mildren et al. in 2002 based on the principle of rock rupture. This method can be used to evaluate the conditions and types of faults or fractures formed by rock rupture and the conditions under which the faults reactivity or maintain stability. The basic principle is whether the formation fluid pressure under the current stress and pressure conditions can cause the structure to reactivate or remain stable. The quantitative evaluation results of the FAST method are in good agreement with the actual situation, and it has been widely used in oil and gas field production and gas storage [10,26]. The data required by the FAST method mainly include the characteristics of present stress field, fluid pressure, friction strength, and cohesion of the rock. In this study, although there is a lack of large-scale fault development in the study area, a large number of fractures are still developed in the cap rock. Therefore, we can still use the FAST method to evaluate whether hydraulic fracture will be induced by fluid pressure in the cap rocks. It can be used to evaluate not only the risk of complete cap fracture, but also the risk of the occurrence of pre-existing extensional fracture and shear fracture before movement takes place, which is an effective method for studying the integrity of cap rocks.

In recent years, many scholars have done a lot of research on the effectiveness of cap rocks. We keep focusing on the study of hydraulic fracturing on the effectiveness of cap rock and the risk of oil and gas leakage and have used this method to analyze hydraulic fractures of cap rocks in several blocks in different basins [27–32] which have achieved some good results. Especially in the study area, a lot of preliminary research has actually been done on the cap rocks, such as the macro- and micro-development characteristics and sealing mechanism and sealing ability of the cap rocks. Some scholars have made preliminary studies on the migration and preservation conditions of natural gas caused by hydraulic fracturing, but most of them focus on theoretical and qualitative analysis. In order to further study the influence of hydraulic fracturing on natural gas enrichment in different structural locations, the same FAST evaluation method was used in this study. Based on the in-situ stress characteristics, mudstone mechanics and pore fluid pressure are used to analyze the potential of mudstone hydraulic fracturing in present conditions, and combine with the paleo-pressure to analyze the hydraulic fracture risk in geological history to discuss its influence on gas accumulation and provide theoretical guidance for oil and gas exploration in basins with the same geological background.

2. Geological Setting

The Yinggehai Basin is a typical Cenozoic sedimentary basin on the northwestern continental shelf of the northern South China Sea [33–38]. The northeast side of the basin adjoins the southern uplift of the Beibu Gulf Basin and the Hainan Uplift. The west side is connected to the Kunsong Uplift and it extends to the Hanoi Depression in Vietnam. This basin is 750 km long and 200 km wide, with an area of about 11.3×104 km^2. Yinggehai Basin is located in the convergence zone of the Eurasian, Pacific, and Indo-Australian plates [38]. Under the background of the expansion of the South China Sea, it is restricted by the strike-slip activity of the Red River Fault and the compression of the plate [39]. The basin is distributed in a rhomboid shape along the NW-SE direction, and composed of four first-order structural units: The Central Depression, The Yingdong Slope Zone, The Yingxi Slope Zone, and The Lingao Uplift (Figure 1).

The basin has experienced multi-stage regional tectonic movements, which can be divided into four stages: Eocene—Early Oligocene rift period, Late Oligocene fault—Depression period, Early—Middle Miocene post-rift thermal subsidence period, and Late Miocene—Quaternary accelerated subsidence period. Drilling data shows that the strata drilled from bottom to the top of the basin are as follows: The Lingtou Formation, The Yacheng Formation, The Lingshui Formation, The Sanya Formation, The Meishan Formation, The Huangliu Formation, The Yinggehai Formation, and The Ledong Formation (Figure 2).

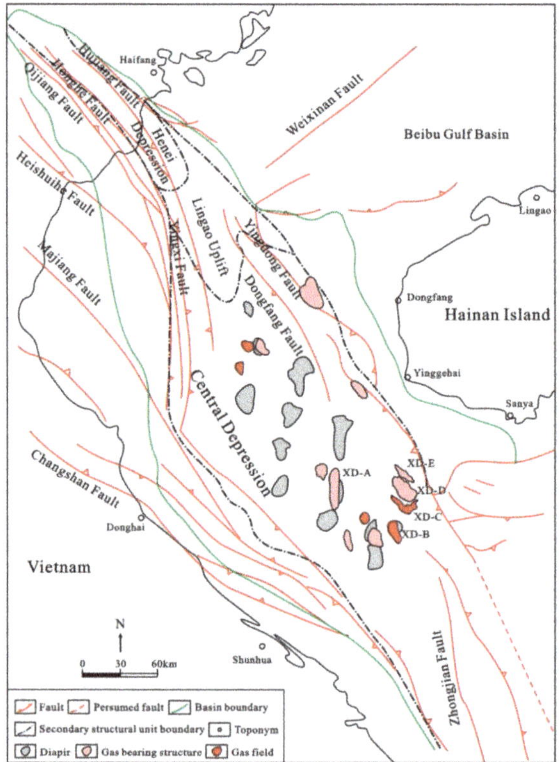

Figure 1. The structure outline map of Yinggehai Basin.

The Central Depression developed a series of nearly N–S orientations diapir structures, which is distributed in an echelon arrangement [35,37]. The research area is located in the southeast of the central depression in the Yinggehai Basin (Figure 1), which is characterized by a lack of fracture, rapid settlement, high temperature and high pressure, and it has more mudstone and less sand. The Yinggehai Basin experienced rapid subsidence in the late structural period, with a deposition rate of 500–1400/Ma, and the Cenozoic maximum sedimentary thickness is over 17 km [34]. Test data of the both the drill pipe and the cable indicates that deep formation pressures below 4000 m exceed 100 MPa, with a maximum pressure factor of 2.30. The numerical simulation of different overpressure formation mechanisms shows that due to the limited abundance of organic matter and the weak contribution of hydrocarbon generation and compression, the abnormally high pressure of the formation comes from the under-compaction caused by the rapid deposition of mudstone, and the conduction effect of the diapir structure acting on the deep overpressure [40,41].

The Ledong area has developed mainly two types of traps, tectonic lithologic trap under the diapir background and lithologic trap group developed within the background context of gravity flow deposition. The Sanya Formation and the Meishan Formation in the Miocene are shallow marine facies mudstones which is the main source rocks in study area. The mudstones in Huangliu Formation and the Yinggehai Formation in upper Miocene and Quaternary Formation are the main cap rocks for nature gas. In addition, these strata are the main reservoir-forming assemblage in the study area (Figure 2). The longitudinal enrichment layers of natural gas are quite obviously different in various structural positions. In the diapir zone, oil and gas are mainly enriched in the shallow strata (Yinggehai and Ledong Formation), such as XD-A area and XD-B area, but natural gas is generally enriched in deeper layers (the Meishan and the Huangliu Formation) and in the slope zones which are far from the diapirs, such as XD-C, XD-D, XD-E area (Figure 3). As a whole, the farther from the diapir, the deeper the gas enrichment. Published literature indicates that there are signs of large-scale fluid migration in the Yinggehai Basin (e.g., gas chimney, pockmark, and pipe) the origins of which many scholars attribute to hydraulic fracturing by overpressure fluid [36]. We speculate that the difference of vertical distribution of natural gas may be caused by the difference in hydraulic fracture strength in different structural positions. Therefore, based on the in-situ stress in the study area, this paper will use the fluid pressure in mudstone and the mechanical parameters of mudstone to analyze the risk of hydraulic fracture in the diapir area and the slope area in the long-term geological historical course as well as the current period. Furthermore, we clarify the main reasons for the natural gas differentially accumulated in the vertical between the different structural zones.

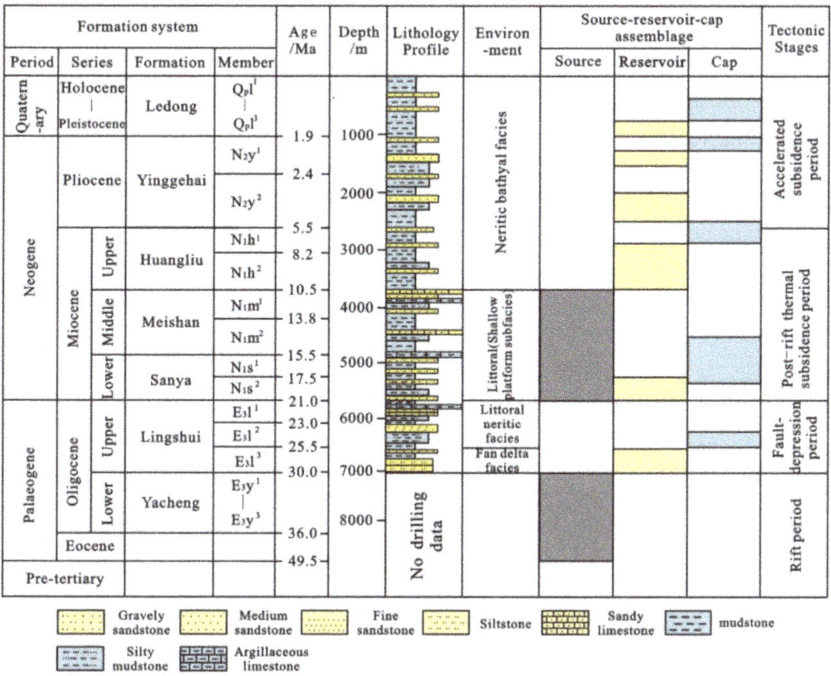

Figure 2. Comprehensive strata log diagram of the Yinggehai Basin with the source-reservoir-cap assemblage and main tectonic event.

Figure 3. Vertical distribution of gas in diapir zone and slope zone in Yinggehai Basin.

3. Methods

3.1. Principle of Cap Rock Hydraulic Fracture Evaluation

There are two kinds of hydraulic fracture modes of cap rock under the dominance of fluid, namely, the fracture of intact cap rock and the re-opening of pre-existing fractures. The constant increase of pore fluid pressure leads to the gradual decrease of the effective stress on the cap rocks, and eventually, pressure is relieved through hydraulic fractures [4].

The fracture mode on cap rocks is controlled by two factors, the tensile strength of cap rock (T) and its differential stress ($S_1 - S_3$) [1,42,43]. When the $S_1 - S_3$ of the stratum is less than 4 times the tensile strength, extensional fractures are formed; when the $S_1 - S_3$ is more than 6 times its tensile strength, it results in compressional shear fractures; the tensile strength and shear mixing fractures occur in relation to the above two conditions when the $S_1 - S_3$ is more than 4 times and less than 6 times the tensile strength (Table 1) [26]. For the intact cap rocks, in the case of extensional fractures, extensional shear fractures, and shear fractures, the fracture pressure (P) of cap rock corresponds to A_1C_1, A_2C_2, and A_3C_3, respectively (see points in Figure 4) [42]. However, for cap rocks with pre-existing fractures, hydraulic fractures are always re-opening along the weak surface. We used the incoherency of envelope to characterize fracture conditions, fracture pressure (P) for extensional fractures, extensional shear fractures, and shear fractures which respectively correspond to Figure 4 B_1C_1, B_2C_2, and B_3C_3. Analysis shows that the existence of early fractures largely increases the risk of hydraulic cracks in the cap rocks. Therefore, in order to determine the hydraulic fracture pressure, the fracture mode of the cap rock must be determined first.

Table 1. The fracture mode and criterion of rocks in different differential stress [26].

Fracture Mode	Fracture Criterion	Differential Stress Condition
Extensional fracture	$P = S_3 + T$	$S_1 - S_3 < 4T$
Extensional shear fracture	$P = S_n + (4T^2 - \tau^2)/4T$	$4T < S_1 - S_3 < 6T$
Shear fracture	$P = S_n + (C - \tau)/\mu$	$S_1 - S_3 > 6T$

Where P is fracture pressure (MPa), S_1 is maximum principal stress (MPa), S_3 is minimum principal stress (MPa), $S_1 - S_3$ is differential stress (MPa), S_n is normal stress (MPa), T is tensile strength of rock (MPa), C is cohesion (MPa), τ is shear stress (MPa), μ is coefficient of friction.

3.2. Parameters of Cap Rock Hydraulic Fracture Evaluation

Based on the above description, the development of hydraulic fracture depends on the pore fluid pressure, the characteristics of the in-situ stress field, and the mechanical properties of the cap rock. Among them, pore fluid pressure and mechanical parameters of cap rocks can be obtained through formation tests and laboratory tests. Therefore, in order to be able to quantitatively evaluate the hydraulic fracture risk of cap rock, it is necessary to define the in-situ stress characteristics of the study area.

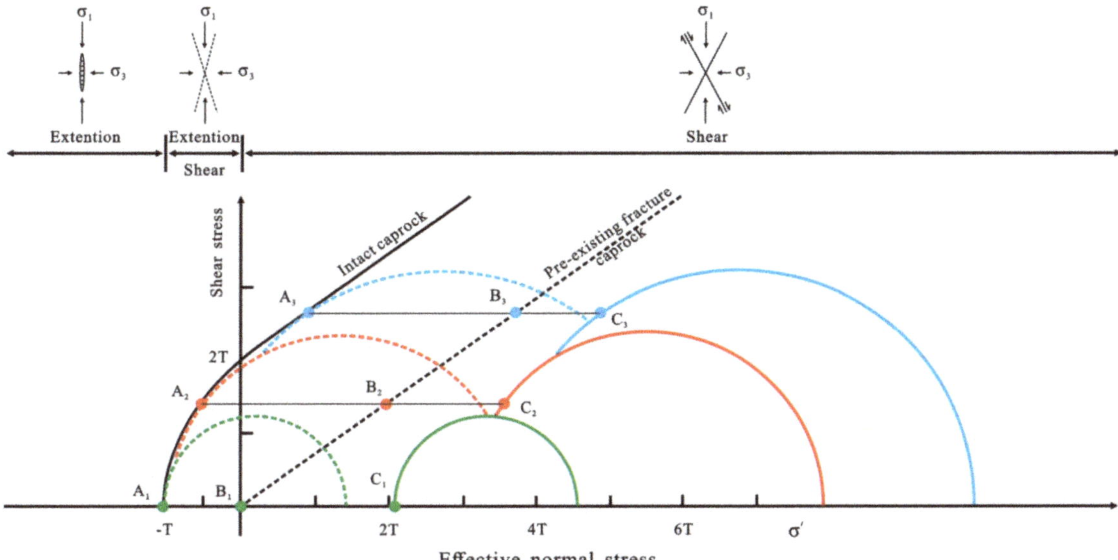

Figure 4. Hydraulic fracture pressure for the cap rock in different stress and mechanical properties. The Mohr circles, from large to small, indicate the critical conditions of shear, extension-shear and extension fracture of intact cap rocks and fractured cap rocks, and the corresponding maximum sustainable fluid pressures are A_1C_1, A_2C_2, A_3C_3 and B_1C_1, B_2C_2, B_3C_3(modified by Sibson(1996)).

3.2.1. Vertical Principal Stress

The vertical principal stress mainly comes from the rock gravity of the overburden strata. Therefore, it can be obtained by integrating density log data. Because the Yinggehai Basin is in the South China Sea, it is necessary to consider the effect of seawater on the vertical principal stress [44] (Equation (1)).

$$S_v = \rho_w g h_w + \int_{h_w}^{h} \rho_c(h) g dh,$$ (1)

where S_v is the vertical principal stress (MPa), ρ_w is the density of water(1.03 g/cm^3 for Yinggehai Basin), h_w is the water depth(average depth is 65 m in the study area), ρ_c is the density of overburden layers (g/cm^3), g is the gravitational acceleration constant (N/kg), h is the burial depth below sea level (m). However, there are always some problems when calculating vertical principal stress. For example, the density logging usually does not start from the sea bottom or surface. The missing density value can be converted from the check-shot velocity data to the density data through the Nafe-Drake transformation formula [45].

3.2.2. Horizontal Principal Stress

Horizontal principal stress is usually the most difficult to interpret. At present, a relatively reliable method is extracted from hydraulic pressure tests [46]. However, these tests have not been widely used on drilling. In this study, we have chosen to use pumping pressure tests data to analyze horizontal principal stress which can be classified as FITs (Formation Integrity Tests), LOTs (Leak-Off Tests), and XLOTs (Extended Leak-Off Tests) [47]. They are curves of bottom hole pressure and pumped mud volume/time, and can be distinguished by the difference between the number of pumping cycles and the point at which pumping is ceased [47] (Figure 5a). LOTs are performed in the formation below the casing shoe, and its primary purpose is to determine the maximum mud weight that could be allowed without drilling risk (formation loss) [24]. Whereas LOTs are not pressurized after the loss pressure is reached to avoid further formation damage, XLOTs are pressurized after the loss pressure is reached, even with multiple pressure cycles performed to obtain more information. Engineering practice shows that a classic single-cycle XLOT curve includes Leak-off Pressure (LOP), Formation Break-down Pressure (FBP), Fracture Propagation pressure (FPP), Instantaneous Shut in Pressure (ISIP), Fracture Closure Pressure (FCP) (Figure 5b–d). Of these points on the curve, LOP is the first point that deviates from the linear relationship between pressure and cumulative mud volume, which can be considered to be the point where the fluid begins to penetrate into the formation, and the rock begins to deform inelastically. Thus, the lower envelope of all LOP data can be used to represent horizontal minimum principal stress [24].

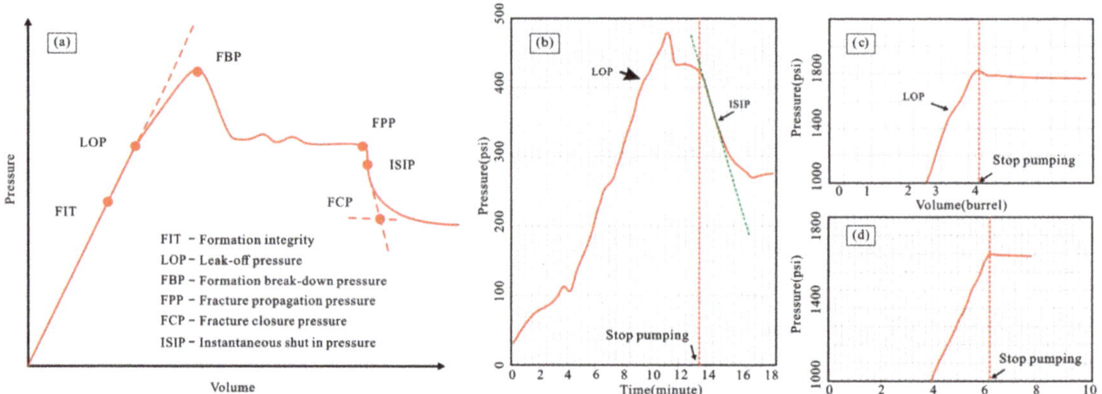

Figure 5. (**a**) Classic XLOT curve in the drilling [24]; (**b**–**d**) The different types XLOT curves in study area.

During the XLOT test, the failure pattern of the wellbore wall is very similar to the hydraulic fracturing in cap rock. The difference is that hydraulic fracturing detects the wellbore conditions through technology and then selects the target interval without fractures. The presence of fractures causes the formation break-down pressure to be close to the fracture reopening pressure, and the tensile strength of wellbore wall is approximately zero [48,49]. Thus, the Hubbert–Willis equation [48] can be simplified to

$$S_{Hmax} = 3S_{hmin} - FRP - P_p, \tag{2}$$

where S_{Hmax} and S_{hmin} are the maximum and minimum horizontal principal stress, respectively (MPa); FRP is the fracture reopening pressure (MPa); and P_p is the pore pressure (MPa).

3.2.3. Pore Fluid Pressure and Mechanical Parameters of Cap Rocks

The current pore fluid pressure is derived from formation Drill Stem Test (DST) and Modular Formation Dynamics Test (MDT) data and drilling fluid weight. The palaeo-pressure data was derived from CNOOC (China National Offshore Oil Corp., Beijing, China).

The mechanical parameters of cap rocks can be tested by direct tensile test and the Brazilian tensile test. However, these two methods have high requirements for sample preparation and low success rate. Considering the limited core samples, conventional triaxial compression test was chosen for this paper. Specifically, we used the GCTS Rapid Triaxial Rock Test System, and the instrument model is RTR-2000. Through tests, we measured the stress-strain curves of rock under different confining pressures and assumed that the tensile strength envelope in the tensile region conforms to the Griffith fracture criterion; the tensile strength of rock can be estimated by its relation to the cohesion, which is usually approximately twice the tensile strength.

4. Results

4.1. In-Situ Stress and Mechanical Properties of the Cap Rock

4.1.1. Pore Fluid Pressure

Drill pipe or wireline test data obtained from DST and MDT show that most of the wells in the Yinggehai Basin have encountered abnormal high fluid pressure formations in the middle and deep layers. In the study area, gas reservoirs are distributed in both overpressure and normal pressure layers, and the Huangliu Formation contains the main regional cap rocks for overpressure gas reservoirs in the deep layer. The maximum fluid pressure can be reached 96 MPa and the fluid pressure coefficient is close to 2.3. The fluid pressure increases with the burial depth longitudinally, and the magnitude of overpressure in the diapir zone is stronger than the slope zone at the same depth. This overpressure distribution may be due to the fact that overpressure can be transferred from deep to shallow layers in diapir zone.

4.1.2. Vertical Principal Stress

In this study, the density–depth relationship of five Wells in the three blocks of the study area (Figure 6a, Equation (2)) was used to integrate to obtain the magnitude of vertical principal stress in the study area (Figure 6b, Equation (3)). The vertical stress gradient increases with depth and is close to 22.8 MPa/km in Huangliu Formation.

$$\rho = 1158 \times h^{0.0949}, \tag{3}$$

$$Sv = 0.010576h^{1.0949} - 0.607 \tag{4}$$

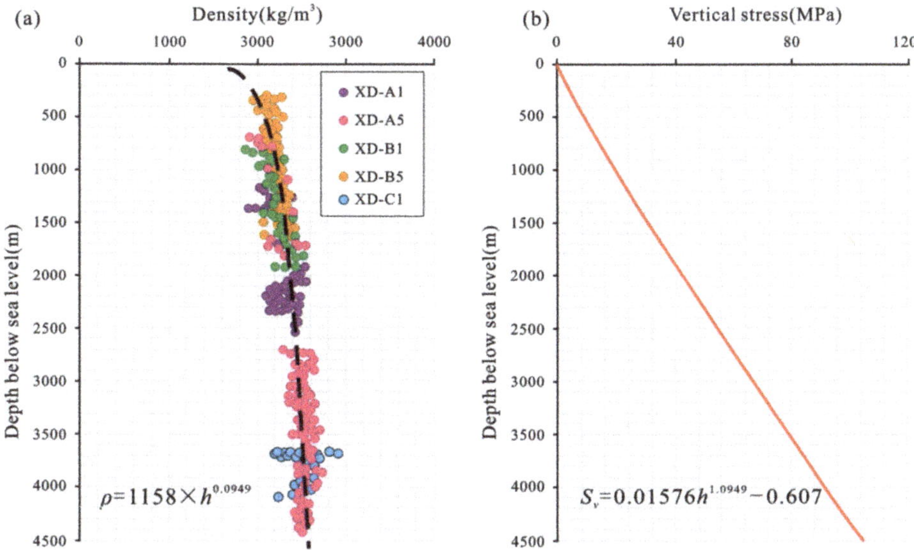

Figure 6. (**a**) The density measurements of 5 Wells in the three blocks of the study area. The relationship between density and depth follows a power function. (**b**) Vertical principal stress obtained by the density log from the 5 wells. The vertical stress gradient is shown to increase with depth and close to 22.8 MPa/km in Huangliu Formation.

4.1.3. Horizontal Principal Stress

Minimum Horizontal Principal Stress

According to the statistics of the LOTs and XLOTs data from the study area, they can be divided into three types. The first type is the XLOTs, although only one pressure cycle was performed, we can observe the FBP, FPP, and ISIP clearly (Figure 5b). In the second category, the well was shut in immediately after LOP was observed, with no significant pressure drop and no formation fracture (Figure 5c). The third category of data is that the formation did not fracture, and no significant losses were observed (Figure 5d). We cannot determine whether the shut-in was performed before or after the LOP was reached, but it can be inferred from the data of the offset wells.

For the first category of date, ISIP can better reflect S_{hmin} than LOP, so ISIP is chosen to represent S_{hmin}. For the second category, S_{hmin} was represented by LOP. As for the third category of data, it is not possible to determine whether the maximum pressure is between FIT and LOP or between LOP and FBP, since no LOP formation was observed and no fracture occurred. However, compared to the offset well, the later well date can be used as the upper limit of S_{hmin}. Synthesizing three categories of data analysis in the study area, the S_{hmin} can be described by ISIP and LOP data in depths (Figure 7).

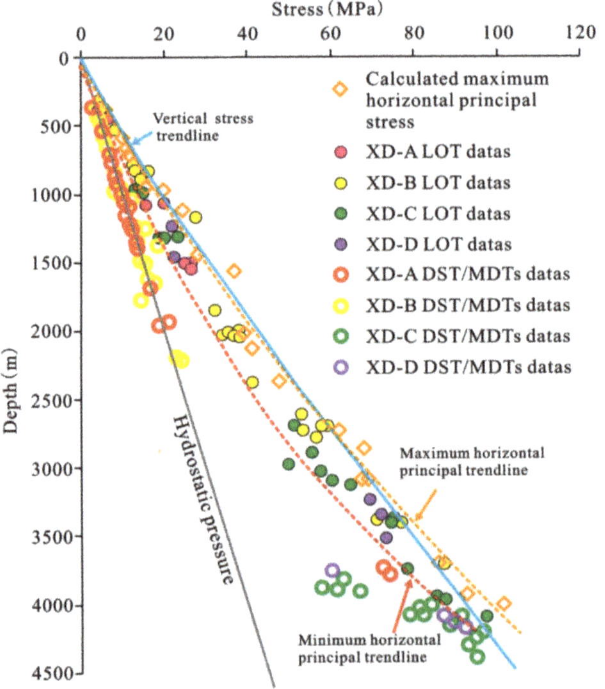

Figure 7. Horizontal principal stress in study area. S_{Hmax} is calculated by the Hubbert–Willis equation [50] under the assumption of zero tensile strength. S_{Hmax} is interpreted from LOTs and XLOTs. The data show the characteristics of stress pressure coupling, that is, the horizontal principal stress increases with the pore fluid pressure increasing [51–53].

Maximum Horizontal Principal Stress

According to the XLOT curve of the study area, Equation (2) is used to calculate the maximum horizontal principal stress. The tensile strength is approximated by the difference between the fracture pressure and the loss pressure. The results show that the in-situ stress state is $S_v > S_{Hmax} > S_{hmin}$ when the depth is above 2700 m and corresponds to the normal faulting stress regime. The in-situ stress state is $S_{Hmax} > S_v > S_{hmin}$ when the depth is below 2700 m and corresponds to the strike-slip faulting stress regime (Figure 7). The horizontal principal stress is obviously affected by the pressure-stress coupling, that is, the horizontal principal stress shows a trend of significant increase with the increase of fluid pressure.

4.1.4. Geomechanical Properties of the Cap Rock

The Huangliu Formation cap rock in the Yinggehai Basin is a shallow-marine-facies deposit composed of mudstone which is a regional cap layer widely developed in the study area with a maximum thickness of up to 400 m. The mechanical properties of the cap rock directly control the fracture conditions and the required fracture pressure. As mentioned above, we used the GCTS Rapid Triaxial Rock Test System to test the mechanical parameters of cap rocks. The laboratory temperature was 25 °C, the humidity was 50%, the strain rate was 0.03 mm/min, and the sample was a cylinder with a diameter of 25 mm and a height of 50 mm taken from cap rock in the Huangliu Formation's well cores. According to the real geological background, the pore fluid pressure was set to 55 MPa. Through data processing and calculation, the friction coefficient (μ) and cohesion (C) were 0.426 and 17 MPa. The tensile strength of rock is 8.5 MPa (Figure 8). Furthermore, the well logging data can be used to calculate the tensile strength of the strata by Equation (5) [54]. The prediction of tensile strength is shown in Figure 9.

$$T = [0.0045E_d (1 - V_{sh}) + 0.008E_d V_{sh}]/K, \tag{5}$$

where T is tensile strength (MPa), E_d is dynamic modulus of elasticity (MPa), V_{sh} is mudstone content (%), and K is the correction coefficient obtained by fitting the experimental test data and logging calculation data, which is approximately 16 (dimensionless).

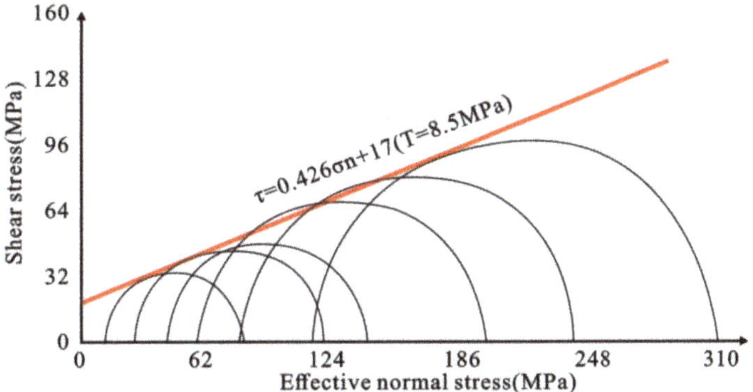

Figure 8. Mohr diagram plotting shear stress against effective normal stress.

4.2. Risk of Hydraulic Rupture of Mudstone Cap Rock

Hydraulic fracture is characterized by episodic opening, which not only provides channels for fluid migration, but also damages the integrity of cap rock, leading to the loss of oil and gas. It is, therefore, of great significance for evaluating the effectiveness of oil and gas traps and finding favorable exploration targets to determine whether hydraulic fracturing is occurring or about to occur in the cap rock.

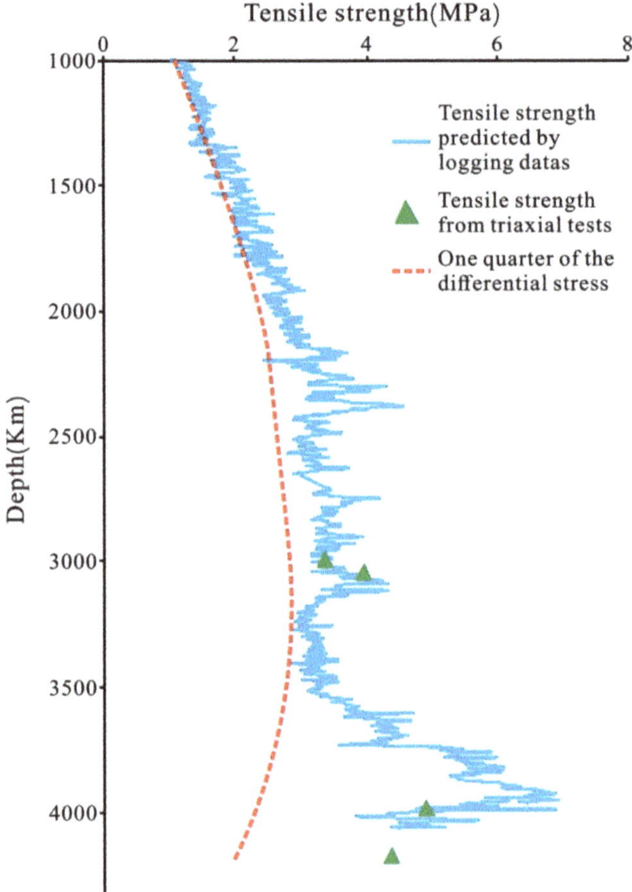

Figure 9. Prediction of tensile strength from the well XD-A1. The one quarter of the differential stress is always less than the tensile strength in the whole depth, it means that extensional fracture always occurs for the intact cap rock.

4.2.1. Risk of Hydraulic Fracture of Cap Rock in Present Conditions

The development of hydraulic fractures depends on the pore fluid pressure, the characteristics of in-situ stress field, and the mechanical properties of the cap rock. As mentioned above, when the differential stress of the cap rock is less than four times the tensile strength, the cap rock will suffer extensional fracture; when the difference stress is more than six times the tensile strength, the cap rock will suffer shear fracture; and when the difference stress is between the two conditions, the cap rock will suffer tensile shear mixed fracture (Table 1). The comparison of tensile strength and differential stress shows that the differential stress of strata at different depths in the study area is generally less than four times the tensile strength, and the difference between differential stress and four times the tensile strength gradually increases with the increase of depth, which means that extensional fracture is always developed in the strata. Due to the different fracture conditions of the cap rocks with early fracture development and those without fracture development, the existence of fractures increases the risk of hydraulic fracture of the cap rocks to a large extent and makes the cap rocks more prone to hydraulic fracture. Considering the geological background of early strike-slip fault activity and diapir structure, subsequent hydraulic fracturing will occur along the pre-existing fracture weak surface. In this paper, the minimum horizontal principal stress is used to represent the minimum fracture pressure required for hydraulic fracturing of cap rock, and the ratio of fluid pressure to the minimum fracture pressure is defined as the hydraulic fracture risk factor of cap rock (R_f) (Equation (6)). When its value is greater than 1, it indicates that hydraulic fracture has occurred in fractured cap rock.

$$R_f = P_{HF}/P_p \, , \tag{6}$$

where R_f is the hydraulic fracture risk factor of cap rock, P_{HF} is the minimum hydraulic fracture pressure (MPa), equal to the minimum horizontal principal stress (MPa), P_p is the pore fluid pressure (MPa).

Five blocks with relatively complete data in the diapir zone and the slope zone in Ledong area were selected for analysis and evaluation. We interpolated the R_f values from the 19 wells to generate a contour map of R_f (Figure 10). The results show that at the top of XD-A and XD-B reservoirs have developed strong overpressure so that the R_f ratio is greater than 1, means that it is still in the period of hydraulic leakage. Blocks XD-C, XD-D, and XD-E maintain hydraulic seal in Huangliu Formation. It can be concluded from the plane variation of the risk coefficient that the closer to the center of the diapir, the higher the fracture risk, which reflects the control effect of the burial depth of the cap rock and the degree of overpressure development on the gas reservoir distribution.

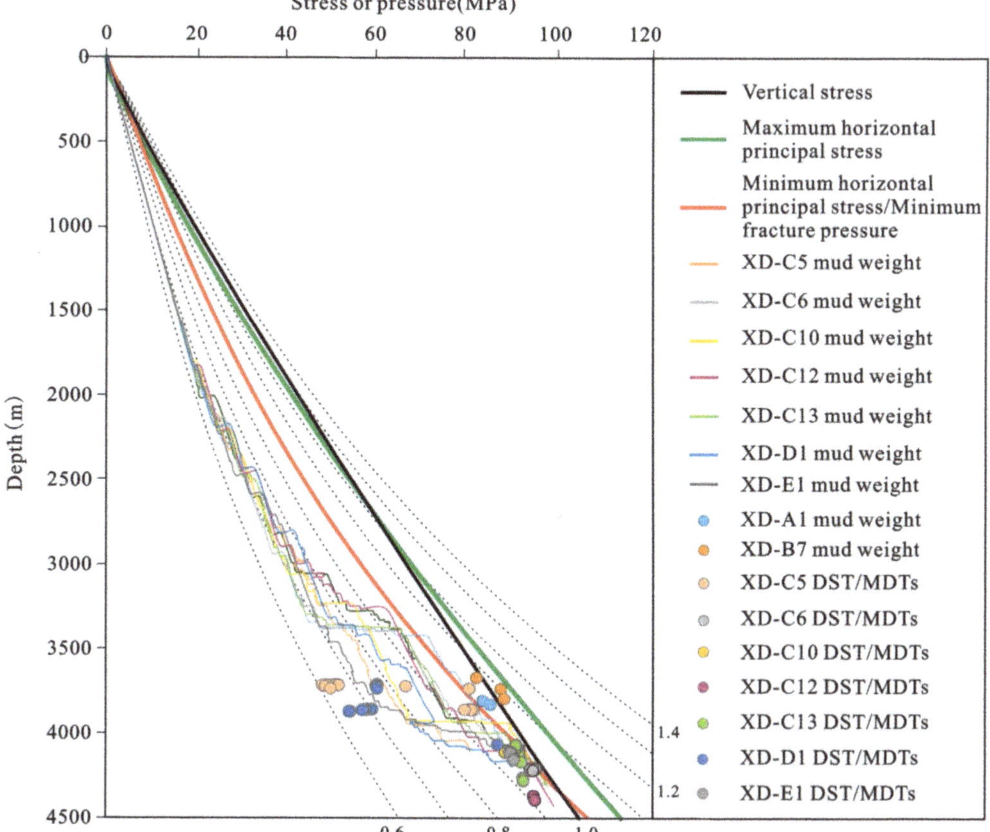

Figure 10. Hydraulic fracture risk of Huangliu Formation in the present stress and pressure condition.

4.2.2. The Evolution Process of Hydraulic Fracture in Cap Rock

At present, the natural gas in the Yinggehai Basin is enriched in multiple reservoirs, and the enrichment of shallow oil and gas reflects the fracture of cap rock during the historical period, which leads to the migration of natural gas from deep to shallow layers. The risk of cap rock hydraulic fracture in geological history mainly depends on the paleo-fluid pressure, paleo-in-situ stress, and the strength of cap rock (cap rock fracture pressure) in geological history. Compared with the in-situ stress in present conditions, it is very difficult to recover the palaeo-stress which is controlled by many factors (e.g., depth, tectonic activity, and palaeo-pressure) [55]. Due to the limitation of the original data, the subsequent research of this paper assumes that the tectonic environment is stable after the middle Miocene, and the burial history is combined with the current stress data to analyze the magnitude of the palaeo-stress. Based on that assumption, the vertical principal stress at the same depth varies little in geological history, but the magnitude of horizontal principal stress is mainly composed of tectonic stress and the Poisson effect of overburden, so the strength of tectonic activity has a great influence on it. As stated earlier, the Yinggehai Basin experienced multiphase tectonic evolution, in middle-late Miocene into a period of rapid subsidence and basin tectonic activity is weak overall. The strata of Meishan Formation and above formations generally not develop large faults, and the faults in deep layers mostly have stop activities. The Ledong area is in a relatively stable tectonic environment. The key horizon of this study is the cap layer of the Huangliu Formation and Meishan, which are overpressured strata developed during the period of structural stability. Therefore, the present in-situ stress (Figure 7) combined with burial history can be roughly used in this paper to predict the magnitude of paleo-stress under a stable tectonic setting (Figure 11). Although there are some uncertainties in the prediction of paleo-stress, it can still provide some reference for the evaluation of whether hydraulic fracture occurs in the cap rock. In addition, compaction is the main factor to control the variation of rock tensile-strength with similar mineral content, and compaction is closely related to the stress on the stratum. Therefore, this paper also assumes that the tensile strength of mudstone cap rocks in different geological periods has little difference under the same burial depth. The relationship between differential stress and tensile strength in the study area should also be referred to when studying whether hydraulic fracture occurs in the cap rock in historical periods. As shown in Figure 9, the differential stress from shallow to deep layers is always less than four times tensile strength, which indicates that even in geological historical periods, the hydraulic fracture of intact rock is dominated by tensile fracture. Therefore, approximate vertical fractures developed in the diapir structural zone confirm the conclusion of extensional fracture in the rock [41]. We calculated hydraulic fracture risk factor (R_f), the ratio of paleo-fluid pressure, and rock fracture minimum pressure in different periods in cap rock to evaluate whether hydraulic fracturing occurred in the historical period.

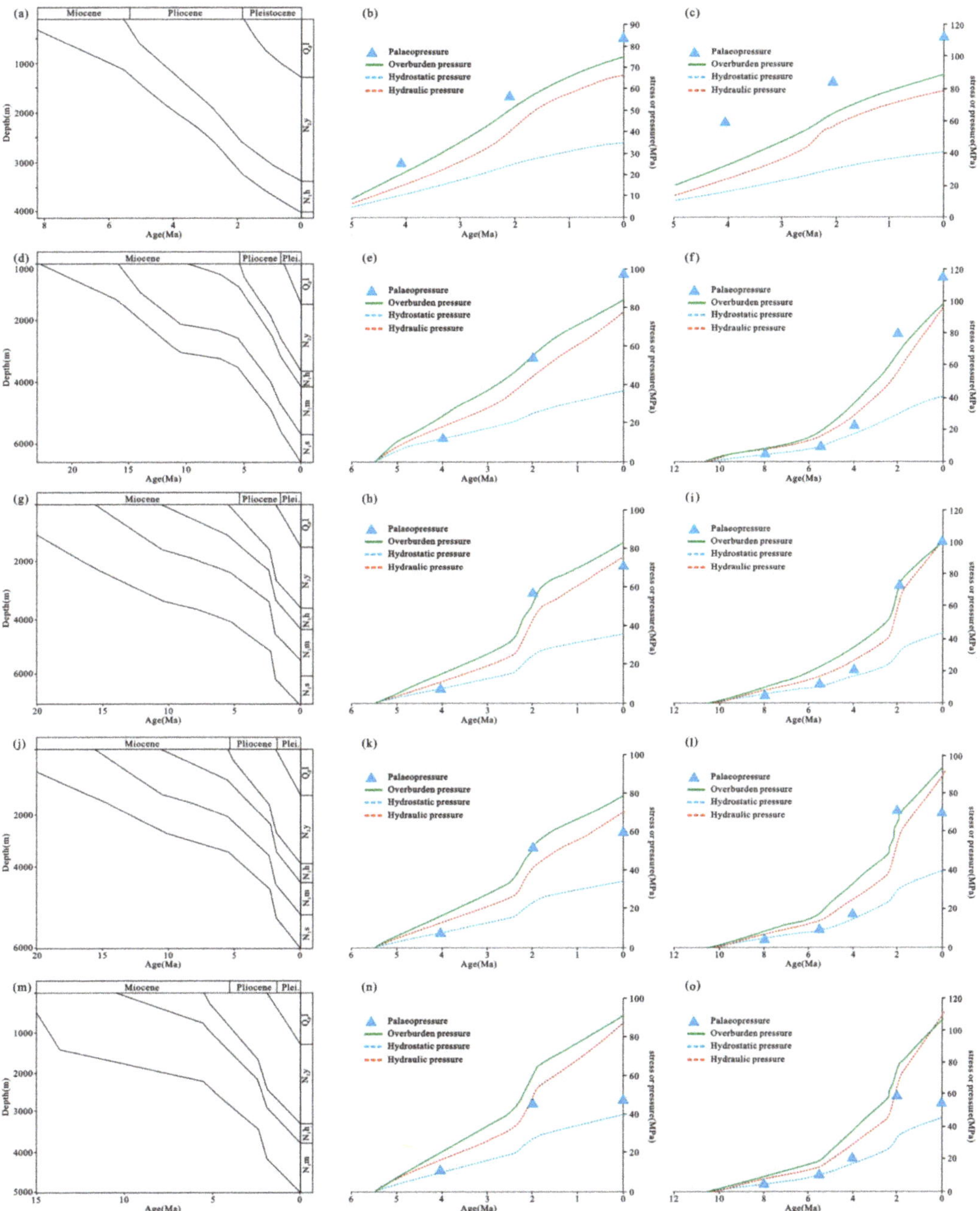

Figure 11. Burial history of the five blocks in study area and the hydraulic risk analysis of Huangliu and Meishan Formation in geological time. (**a–c**) is the XD-A, the palaeopressure was larger than the hydraulic, indicating continuous leakage in whole geological time in both Huangliu and Meishan Formation. The Xd-B show in (**d–f**) is similar to the XD-A, the leakage time was later than XD-A, and still leaking today. (**g–i**) and (**j–l**) is present XD-C and XD-D block, hydraulic fracture occurred about 2 Ma. (**m–o**) is the XD-E which is farthest from the diapir, has not reached the hydraulic fracture pressure in the historical period until now.

The paleo-pressure data of hydraulic fracture evaluation are mainly from CNOOC. We evaluated gas reservoirs in the diapir zone and slope zone. Taking XD-A area as an example in the diapir zone, it can be seen that the paleo-fluid pressure is consistently higher than the hydraulic fracturing pressure of rock strata, indicating that with the passage of geological time, the hydraulic fracturing of cap rock continues to occur in Huangliu and Meishan Formation, providing an effective channel for oil and gas migration, enabling natural gas to enter the shallow Ledong Formation and Yinggehai Formation from the deep source rock and finally to form reservoirs (Figure 11b,c). The evolution process of hydraulic fracture of XD-B block is similar to that of the XD-A block. Although the hydraulic fracture started later than XD-A block, both the cap rocks of Huangliu and Meishan Formations were in the stage of hydraulic fracturing after 2 Ma (Figure 11e,f) and finally formed shallow gas reservoirs or showing gas-bearing properties (Figure 11e,f). In relatively distant slope zones, the XD-C and XD-D blocks, in the historical period, hydraulic fracture occurred about 2 Ma in Huangliu and Meishan Formation, but later, as the pressure was released, hydraulic fractures began closing again; only a small amount of oil and gas shows but failed to form oil and gas accumulation (Figure 11h,i,k,l). The XD-E block, which is farthest from the diapir, has not reached the hydraulic fracture pressure in the historical period until now, although the pressure has accumulated to a certain extent (Figure 11n,o). The cap rocks have always been sealed, and the oil and gas have not migrated or diffused through the layers, and finally accumulated in the sandstone of the Meishan Formation (Figure 3). From the whole evaluation results, the evolution of hydraulic fracture of cap rock has obvious characteristics. It can be divided into three categories, "continuous fracturing" in diapir zones, "continuous sealing" in slope zones, and "sealed-fractured-sealed" in between.

5. Discussion

5.1. The Uncertainty of Calculation in Hydraulic Fracture Risk Factor

The evaluation method of this study is the FAST method which can be used to evaluate both tensile and shear hydraulic fracture. We first determined the hydraulic fracture mode of cap rock, and then selected the appropriate evaluation method for evaluation. In the evaluation process, we fully applied the formation test data of the oil field to analyze the hydraulic fracture risk of the cap rock under real geological conditions as far as possible. In the measurement of rock mechanics parameters, we also restored the real geological conditions to the greatest extent within the scope of instrument testing.

In spite of our efforts, there is still some uncertainty in the determination of R_f, mainly because it is controlled by many factors, including the state of paleo-stress, mechanical properties of rock in historical period, and paleo-fluid pressure. The stress field can be divided into three scales [56]: the first-order stress field mainly comes from the plate driving force; the second-order stress field is derived from major intraplate stress, such as isostatic compensation, glaciation retreat, and lithospheric bend; and the third-order stress field is determined by stress and strain such as faults activity, local intrusion, horizontal detachment, and density inversion. The influences of the various factors described above bring great difficulties to the evaluation of the current stress field, especially the determination of the maximum horizontal principal stress. In our study, the in-situ stress is determined based on two assumptions. Firstly, it is assumed that the in-situ stress in the study area is homogeneous as is the magnitude of the in-situ stress in different locations. Secondly, it is assumed that the influence intensity of various factors on the in-situ stress in the geological history period is the same as that of today, so that the magnitude of the ancient in-situ stress can be obtained from the burial-depth. In addition, the study in this paper ignored the heterogeneity of stratigraphic sedimentation in the same geological history period and assumed a consistent tensile strength of rocks, which was obviously inconsistent with reality. However, due to the limited core samples, we cannot test drill cores at different locations, so more tensile strength data could not be obtained. In theory, at some locations in the cap rock, due to the influence of the rock's mineral composition, it is possible to have lower tensile strength, leading to other fracture modes and lower hydraulic fracture pressure. Fortunately, the data show that the differential stress is always less than the four times the tensile strength (Figure 9) and the fracture mode of cap rock is mainly extensional fracture.

5.2. Control Effect of Hydraulic Fracturing on Gas Accumulation

The preservation condition of the cap rock is very important to the hydrocarbon accumulation in the trap, which is mainly reflected in the microscopic sealing ability of the cap rock itself and whether the cap rock is damaged. Scholars have conducted a lot of studies on the cap rocks in Yinggehai Basin [33–38]. The mathematical model and the test method show that the cap rock of Yinggehai Basin has strong displacement pressure, which is enough to seal up the hydrocarbon column of a certain height. However, there are still some traps that fail to be explored due to inadequate preservation conditions. The primary reason of this phenomenon is that the cap rocks have been damaged. In accordance with the process of the hydraulic fracture of mudstone, it is not hard to see that hydrocarbons leak episodically at the geological timescale. In particular, some scholars have found evidence of the existence of hydraulic fractures in the three-dimensional seismic reflection and cores of Yinggehai Basin [36], and the cemented fractures also show multi-phase characteristics, which all indicate that fractures are an important channel for fluid migration. In addition, other geochemical parameters in the study area, such as fluid inclusion homogenization temperature, natural gas isotope difference, and quantitative reservoir fluorescence analysis, can also reflect the multi-stage fluid charging caused by the fracturing of mudstone cap rocks in the past.

These phenomena mean that gas accumulation in the study area is actually a process of dynamic equilibrium of accumulation and dispersion. Although oil and gas will be leaked at a certain stage in geological history, as long as the rate of oil and gas charging is greater than the rate of loss, oil and gas can accumulate in traps and even form large and medium-sized oil and gas fields. In the future, if accurate in-situ stress and mechanical properties of cap rock can be obtained based on the analysis of gas generation capacity and dominant migration path, the hydraulic fracture pressure of cap rock can be effectively predicted, which will be an important method to reduce the risk of gas exploration in deep overpressured reservoirs.

In our present study, the hydraulic fracture status of cap rock in different periods is given, but the current data do not completely cover the entire historical period, it is difficult to construct a complete and accurate hydraulic fracture process of mudstone with existing data and means. Further and comprehensive work may be carried out in the future.

It is clear to see from the above formula for calculating hydraulic fracture pressure that the effectiveness of overpressure cap rock is controlled by three types of factors: (1) the in-situ stress factor, (2) the fluid in the formation, and (3) the mechanical properties of the cap rocks. In this paper, we only discuss the effect of overpressure in lithologic traps on the preservation conditions of traps and its effect on gas accumulation in Yinggehai Basin. As we all know, there are many diapir structures developed in Yinggehai Basin [57]. Diapirism formed a series of associated high-angle faults; these faults and diapirs can conduct deep pressure to shallow layers. Under the action of strong overpressure, the periodic rupture and healing of the cap rocks at the top of trap can regulate the fluid pressure in the formation and control the accumulation and loss of natural gas. Moreover, the faults in conjunction with hydraulic fracturin allow fluid to migrate more easily. The seismic reflection fuzzy zone above the Meishan Formation is the evidence of a large amount of hydraulic fracturing. By compacting the hydraulic fracture processes of gas reservoirs or gas-bearing structures in the different structural zones, it can be seen that the diapir zone has a more obvious hydraulic fracture response and higher risk than the slope zone. On the plane, the closer to the diapir structure, the greater the risk of hydraulic fracture. For example, XD-C and XD-D are more affected by diapir activities and more likely to fracture than XD-E. Under the background of the overall lack of large faults in the study area, the tectonic activities of the diapir promote the occurrence of hydraulic fracture, which is mainly reflected in two aspects: (1) diapir structure has conduction effect on deep overpressure; (2) the diapir structure activity will deform the overburden strata, accompanied by a many structural fractures (small faults and fractures), making mudstones more prone to hydraulic fracture. This is the main reason why hydraulic fracturing is more likely to occur in the diapir zone, and it also explains why only in the diapir zone oil and gas can accumulate in shallow layers, while the natural gas in the slope zone cannot break through to the shallow layers.

6. Conclusions

In this paper, the hydraulic fracture of the main cap rock in the study area is analyzed based on the in-situ stress, pore fluid pressure, and rock mechanical parameter in the geological historical period and present time. The difference in the evolution process and risk of hydraulic fracture is the fundamental reason for the obvious difference in gas distribution between diapir zone and slope zone.

The research of hydraulic fracturing shows that gas leakage has occurred in cap rocks in the diapir zone throughout geological time and remains in the leakage period like XD-A and XD-B; as a result, the natural gas accumulated in both deep and shallow reservoirs at this locations. However, in the slope zone, except XD-E block, gas leakage has occurred in the Huangliu cap rock and Meishan cap rock at a certain geological time in XD-C and XD-D block. Therefore, gas accumulated only in the Meishan Formation in XD-E, but both in the Huangliu and Meishan reservoirs in XD-C and XD-D.

The tectonic activity of the diapirs weakened the strength of the strata and made them more prone to hydraulic fracturing, thus providing the vertical channel for fluid migration, resulting in differential distribution of natural gas in diapir zone and slope zone. This general phenomenon has two main causes: a. diapirism transfers deep overpressure to shallow layers; b. the activity of diapir cause strata to deform and form structural fractures (small faults and fractures), reduce critical stress condition, and make the cap rock more susceptible to hydraulic fracturing. The slope zones may have better preservation conditions than the diapir zones.

Author Contributions: Conceptualization, R.J.; methodology, C.F.; software, Y.J.; investigation, R.J. and C.F.; data curation, Y.J.; writing—original draft preparation, R.J. and B.L.; writing—review and editing, B.L. and X.F.; supervision, X.F.; funding acquisition, R.J. All authors have read and agreed to the published version of the manuscript.

Funding: This research was funded by National Natural Science Foundation of China, grant number 42002152; Science and Technology Project of Heilongjiang Province, grant number 2020ZX05A01.

Institutional Review Board Statement: Not applicable.

Informed Consent Statement: Not applicable.

Data Availability Statement: Not applicable.

Nomenclature

FAST	Fault Analysis Seal Technology
XD-A, XD-B, XD-C, XD-D, XD-E	Typical block code in the study area
P	Fracture pressure, MPa
P_p	Pore pressure, MPa
P_{HF}	Minimum hydraulic fracture pressure, MPa
S_1	Maximum principal stress, MPa
S_3	Maximum principal stress, MPa
$S1 - S3$	Differential stress, MPa
S_n	Normal stress, MPa
S_v	Vertical stress, MPa
S_{hmin}	Horizontal minimum principal stress, MPa
S_{Hmax}	Horizontal maximum principal stress, MPa
τ	Shear stress, MPa
T	Tensile strength of rock, MPa
C	Cohesion, MPa
μ	Coefficient of friction
ρ_w	Density of water, g/cm^3
Pc	Density of overburden layers, g/cm^3
g	Gravitational acceleration constant, N/kg
h_w	Water depth, m
h	Burial depth below sea level, m
FIT	Formation Integrity Tests
LOT	Leak-Off Test
XLOT	Extended Leak-Off Test
LOP	Leak-off Pressure, MPa
FBP	Formation Break-down Pressure, MPa
FPP	Fracture Propagation pressure, MPa
ISIP	Instantaneous Shut in Pressure, MPa
FCP	Fracture Closure Pressure, MPa
FRP	fracture reopening pressure, MPa
DST	Drill Stem Testing
MDT	Modular Formation Dynamics Test
CNOOC	China National Offshore Oil Corp.
GCTS	Geotechnical Consulting and Testing Systems
E_d	Dynamic modulus of elasticity, MPa
V_{sh}	Mudstone content, %
K	Correction coefficient, dimensionless
R_f	Hydraulic fracture risk factor of cap rock

References

1. Secor, D.T. Role of fluid pressure in jointing. *Am. J. Sci.* **1965**, *263*, 633–646. [CrossRef]
2. Phillips, W.J. Hydraulic fracturing and mineralization. *J. Geol. Soc.* **1972**, *128*, 337–359. [CrossRef]
3. Ozkaya, I. Analysis of natural hydraulic fracturing of mudstones during sedimentataion. *SPE Prod. Eng.* **1986**, *1*, 191–194. [CrossRef]
4. Watts, N.L. Theoretical aspects of cap-rock and fault seals for single- and two-phase hydrocarbon columns. *Mar. Pet. Geol.* **1987**, *4*, 274–307. [CrossRef]
5. Miller, T.W. New insights on natural hydraulic fractures induced byabnormally high pore pressures. *AAPG Bull.* **1995**, *79*, 1005–1081.
6. Roberts, S.J.; Nunn, J.A. Episodic fluid expulsion from geopressured sediments. *Mar. Pet. Geol.* **1995**, *12*, 195–202. [CrossRef]
7. Løseth, H.; Gading, M.; Wensaas, L. Hydrocarbon leakage interpreted on seismic data. *Mar. Pet. Geol.* **2009**, *26*, 1304–1319. [CrossRef]
8. Sibson, R.H. Conditions for fault-valve behavior. *Geol. Soc. Lond. Spec. Publ.* **1990**, *541*, 15–28. [CrossRef]
9. Ingram, G.M.; Urai, J.L. Top-seal leakage through faults and fractures: The role of mudrock properties. *Geol. Soc. Lond. Spec. Publ.* **1999**, *158*, 125–135. [CrossRef]
10. Mildren, S.D.; Hillis, R.R.; Dewhurst, D.N.; Lyon, P.J.; Meyer, J.J.; Boult, P.J. FAST: A new approach to risking fault reactivation and related seal breach of fault seals. *Am. Assoc. Petrol. Geol. Bull.* **2005**, *2*, 73–85.

11. Sibson, R.H. Fluid flow accompanying faulting: Field evidence and models. *Earthq. Predict.* **1981**, *4*, 593–603.
12. Cartwright, J.; Huuse, M.; Aplin, A. Seal bypass systems. *AAPG Bull.* **2007**, *91*, 1141–1166. [CrossRef]
13. Zuhlsdorff, L.; Spiess, S.V. Three-dimensional seismic characterization of a venting site reveals compelling indications of natural hydraulic fracturing. *Geology* **2004**, *32*, 101–104. [CrossRef]
14. Cosgrove, J.W. Hydraulic fracturing during the formation and deformation of a basin: A factor in the dewatering of low-permeability sediments. *AAPG Bull.* **2001**, *85*, 737–748.
15. Meng, Q.; Hooker, J.; Cartwright, J. Genesis of natural hydraulic fractures as an indicator of basin inversion. *J. Struct. Geol.* **2017**, *102*, 1–20. [CrossRef]
16. Petrie, E.S.; Evans, J.P.; Baure, S.J. Failure of cap-rock seals as determined from mechanical stratigraphy, stress history, and tensile-failure analysis of exhumed analogsStress History and Tensile Failure Analysis of Exhumed cap rock Seal Analogs. *AAPG Bull.* **2014**, *98*, 2365–2389. [CrossRef]
17. Fang, H.A.O.; Jian-Zhang, L.I.U.; Hua-Yao, Z.O.U.; Peng-Peng, L.I. Mechanisms of natural gas accumulation and leakage in the overpressured sequences in the Yinggehai and Qiongdongnan basins, offshore South China Sea. *Earth Sci. Front.* **2015**, *22*, 169–180.
18. Ungerer, P.; Burrus, J.; Doligez, B.; Chenet, P.Y.; Bessis, F. Basin evaluation by integrated two-dimensional modeling of heat transfer, fluid flow, hydrocarbon generation, and migration. *AAPG Bull.* **1990**, *74*, 309–335.
19. Wang, C.; Xie, X. Hydrofracturing and episodic fluid flow in mudstone-rich basins-a numerical study. *AAPG Bull.* **1998**, *82*, 1857–1869.
20. Xie, X.N.; Li, S.T.; Wang, Q.Y. Hydraulic fracturing and episodic compaction of mudstones in sedimentary basin. *Chin. Sci. Bull.* **1997**, *42*, 2193–2195.
21. Caillet, G. The cap rock of the Snorre Field, Norway: A possible leakage by hydraulic fracturing. *Mar. Pet. Geol.* **1993**, *10*, 42–50. [CrossRef]
22. Du Rouchet, J. Stress fields, a key to oil migration. *AAPG Bull.* **1981**, *65*, 74–85.
23. Li, T.W.; Meng, L.D.; Feng, D.; Zhou, X.N.; Xia, N.; Wei, Z.P. Hydraulic fracturing mechanism and its application in stability evaluation of caprock and fault. *Fault-Block Oil Gas Field* **2015**, *22*, 47–52.
24. Gaarenstroom, L.; Tromp, R.; Brandenburg, A.M. Overpressures in the Central North Sea: Implications for trap integrity and drilling safety. In *Proceedings of the Geological Society, London, Petroleum Geology Conference Series*; Geological Society of London: London, UK, 1993; Volume 4, pp. 1305–1313.
25. Konstantinovskaya, E.; Malo, M.; Castillo, D.A. Present-day stress analysis of the St. Lawrence Lowlands sedimentary basin (Canada) and implications for cap rock integrity during CO2 injection operations. *Tectonophysics* **2012**, *518–521*, 119–137. [CrossRef]
26. Mildren, S.D.; Hills, R.R.; Lyon, P.J.; Meyer, J.J.; Dewhurst, D.N.; Boult, P.J. FAST: A new technique for geomechanical assessment of the risk of reactivation-related breach of fault seals. *AAPG Hedberg Ser.* **2005**, *2*, 73–85.
27. Li, T.W. Hydraulic Fracturing Mechanism and Quantitative Evaluation Method of Cap Rock and Fault. Ph.D. Thesis, Northeast Petroleum University, Daqing, China, 2015.
28. Jia, R. The Integrity of Cap Rocks and Gas Accumulation in Yingqiong Basin. Ph.D. Thesis, Northeast Petroleum University, Daqing, China, 2018.
29. Jin, Y.J.; Fan, C.W.; Fu, X.F.; Meng, L.; Liu, B.; Jia, R.; Cao, S.; Du, R.; Li, H.; Tan, J. Risk analysis of natural hydraulic fracturing in an overpressured basin with mud diapirs: A case study from the Yinggehai Basin, South China Sea. *J. Pet. Sci. Eng.* **2021**, *196*, 107621. [CrossRef]
30. Liu, B.; He, S.; Meng, L.D.; Fu, X.; Gong, L.; Wang, H. Sealing mechanisms in volcanic faulted reservoirs in Xujiaweizi extension, Northern Songliao Basin, Northeastern China. *AAPG Bull.* **2020**. ahead of print.
31. Liu, B.; Sun, J.; Zhang, Y.; Junling, H.E.; Xiaofei, F.U.; Liang, Y.A.N.G.; Jilin, X.I.N.G.; Xiaoqing, Z.H.A.O. Reservoir space and enrichment model of shale oil in the first member of Cretaceous Qingshankou Formation in the Changling sag, southern Songliao Basin, NE China. *Pet. Explor. Dev.* **2021**, *48*, 608–624. [CrossRef]
32. Gong, L.; Wang, J.; Gao, S.; Fu, X.; Liu, B.; Miao, F.; Zhou, X.; Meng, Q. Characterization, controlling factors and evolution of fracture effectiveness in shale oil reservoirs. *J. Pet. Sci. Eng.* **2021**, *203*, 108655. [CrossRef]
33. He, J.X.; Xia, B.; Wang, Z.X.; Liu, B.M.; Sun, D.S. The regularity of oil and gas migration and accumulation and direction of exploration of northern south China sea continental shelf of the western. *Acta Pet. Sin.* **2006**, *27*, 12–18.
34. Xie, Y.H. *Sequence Stratigraphic Analysis and Hydrocarbon Accumulation Models in Tectonically Active Basins: Case Study on the Yinggehai Basin*; Geological Publishing House: Beijing, China, 2009.
35. Xie, Y.H.; Zhang, Y.C.; Li, X.S.; Zhu, J.C.; Tong, C.X.; Zhong, Z.H.; Zhou, J.X.; He, S.L. Main controlling factors and formation models of natural gas reservoirs with high-temperature and overpressure in Yinggehai Basin. *Acta Pet. Sin.* **2012**, *33*, 601–609.
36. Zhao, B.F.; Chen, H.H.; Kong, L.T.; Wang, Q.R.; Liu, R. Vertical migration system and its control on natural gas accumulation in Yinggehai Basin. *Earth Sci. J. China Univ. Geosci.* **2014**, *39*, 1324–1332.
37. Hao, F.; Li, S.T.; Gong, Z.S.; Yang, J. Mechanism of diapirism and episodic fluid injections in the Yinggehai Basin. *Sci. China Ser. D* **2001**, *31*, 471–476.
38. Fan, C.W. The identification and characteristics of migration system induced by high pressure, and its hydrocarbon accumulation process in the Yingqiong Basin. *Oil Gas Geol.* **2018**, *39*, 254–267.

39. Ren, J.Y.; Lei, C. Tectonic stratigraphic framework of Yinggehai-Qiongdongnan Basins and its implication for tectonic province division in South China Sea. *Chin. J. Geophys.* **2011**, *54*, 3303–3314. [CrossRef]
40. Luo, X.; Dong, W.; Yang, J.; Yang, W. Overpressuring mechanisms in the Yinggehai Basin, South China Sea. *AAPG Bull.* **2003**, *87*, 629–642. [CrossRef]
41. Lei, C.; Ren, J.; Clift, P.D.; Wang, Z.; Li, X.; Tong, C. The structure and formation of diapirs in the Yinggehai–Song Hong Basin, South China Sea. *Mar. Pet. Geol.* **2011**, *28*, 980–991. [CrossRef]
42. Sibsion, R.H. Structural permeability of fluid-driven fault-fracture meshes. *J. Struct. Geol.* **1996**, *18*, 1031–1042. [CrossRef]
43. Sibsion, R.H. Tensile overpressure compartments on low-angle thrust faults. *Earth Planets Space* **2017**, *69*, 1–15. [CrossRef]
44. Mcgarr, A.; Gay, N.C. State of stress in the earth's crust. *Annu. Rev. Earth Planet. Sci.* **1978**, *6*, 405. [CrossRef]
45. Ludwing, W.J.; Nafe, J.E.; Drake, C.L. Seismic refraction. In *The Sea*; Maxwell, A.E., Ed.; Wiley-Interscience: New York, NY, USA, 1971; Volume 4, pp. 53–84.
46. Engelder, T. *Stress Regimes in the Lithosphere*; Princeton University Press: Kandy, Sri Lanka, 1993; Volume 102, p. 457.
47. White, A.J.; Traugott, M.O.; Swarbrick, R.E. The use of leak-off tests as means of predicting minimum in-situ stress. *Pet. Geosci.* **2002**, *8*, 189–193. [CrossRef]
48. Brudy, M.; Zoback, M.D. Drilling-induced tensile wall-fractures: Implications for determination of in-situ stress orientation and magnitude. *Int. J. Rock Mech. Min. Sci.* **1999**, *36*, 191–215. [CrossRef]
49. Zoback, M.D.; Barton, C.A.; Brudy, M.; Castillo, D.A.; Finkbeiner, T.; Grollimund, B.R.; Moos, D.B.; Peska, P.; Ward, C.D.; Wiprut, D.J. Determination of stress orientation and magnitude in deep wells. *Int. J. Rock Mech. Min. Sci.* **2003**, *40*, 1049–1076. [CrossRef]
50. Hubbert, M.K.; Aime, M.; Willis, D.G.; Aime, J.M. Mechanics of hydraulic fracturing. *Pet. Trans. AIME* **1957**, *210*, 153–168. [CrossRef]
51. Hills, R. Pore pressure/stress coupling and its implications for seismicity. *Explor. Geophys.* **2000**, *31*, 448–454. [CrossRef]
52. Tingay, M.R.P.; Hills, R.R.; Morley, C.K.; Swarbrick, R.E.; Okpere, E.C. Pore pressure/stress coupling in Brunei Darussalam—implications for mudstone injection. *Geol. Soc. Lond. Spec. Publ.* **2003**, *216*, 369–379. [CrossRef]
53. Liu, R.; Liu, J.; Zhu, W.; Hao, F.; Xie, Y.; Wang, Z.; Wang, L. In Situ stress analysis in the Yinggehai Basin, northwestern South China Sea: Implication for the pore pressure-stress coupling process. *Mar. Pet. Geol.* **2016**, *77*, 341–352. [CrossRef]
54. Jin, Y.; Yuan, J.; Chen, M.; Chen, K.P.; Lu, Y.; Wang, H. Determination of rock fracture toughness K IIC and its relationship with tensile strength. *Rock Mech. Rock Eng.* **2011**, *44*, 621. [CrossRef]
55. Morris, A.P.; Ferrill, D.A.; Mcginnis, R.N. Using fault displacement and slip tendency to estimate stress states. *J. Struct. Geol.* **2016**, *83*, 60–72. [CrossRef]
56. Heidbach, O.; Reinecker, J.; Tingay, M.; Müller, B.; Sperner, B.; Fuchs, K.; Wenzel, F. Plate boundary forces are not enough: Second-and third-order stress patterns highlighted in the World Stress Map database. *Tectonics* **2007**, *26*, TC6014. [CrossRef]
57. Gong, Z.S.; Li, S.T. *Continental Margin Basin Analysis and Hydrocarbon Accumulation of the Northern South China Sea*; China Science Press: Beijing, China, 1997.

Article

Creep Rupture and Permeability Evolution in High Temperature Heat-Treated Sandstone Containing Pre-Existing Twin Flaws

Sheng-Qi Yang [1,*], Jin-Zhou Tang [1,2,3] and Derek Elsworth [2]

[1] State Key Laboratory for Geomechanics and Deep Underground Engineering, School of Mechanics and Civil Engineering, China University of Mining and Technology, Xuzhou 221116, China; tangjinzhou@cumt.edu.cn

[2] Department of Energy and Mineral Engineering, EMS Energy Institute and G3 Center, Pennsylvania State University, State College, PA 16802, USA; elsworth@psu.edu

[3] State Key Laboratory of Mining Response and Disaster Prevention and Control in Deep Coal Mines, Anhui University of Science and Technology, Huainan 232001, China

* Correspondence: yangsqi@hotmail.com; Tel.: +86-516-8399-5856

Citation: Yang, S.-Q.; Tang, J.-Z.; Elsworth, D. Creep Rupture and Permeability Evolution in High Temperature Heat-Treated Sandstone Containing Pre-Existing Twin Flaws. *Energies* **2021**, *14*, 6362. https://doi.org/10.3390/en14196362

Academic Editor: Nikolaos Koukouzas

Received: 13 August 2021
Accepted: 21 September 2021
Published: 5 October 2021

Abstract: Utilizing underground coal gasification cavities for carbon capture and sequestration provides a potentially economic and sustainable solution to a vexing environmental and energy problem. The thermal influence on creep properties and long-term permeability evolution around the underground gasification chamber is a key issue in UCG-CCS operation in containing fugitive emissions. We complete multi-step loading and unloading creep tests with permeability measurement at confining stresses of 30 MPa on pre-cracked sandstone specimens thermally heat-treated to 250, 500, 750 and 1000 °C. Observations indicate a critical threshold temperature of 500 °C required to initiate thermally-induced cracks with subsequent strength reduction occurring at 750 °C. Comparison of histories of creep, visco-elastic and visco-plastic strains highlight the existence of a strain jump at a certain deviatoric stress level—where the intervening rock bridge between the twin starter-cracks is eliminated. As the deviatoric stress level increases, the visco-plastic strains make up an important composition of total creep strain, especially for specimens pre-treated at higher temperatures, and the development of the visco-plastic strain leads to the time-dependent failure of the rock. The thermal pre-treatment produces thermal cracks with their closure resulting in increased instantaneous elastic strains and instantaneous plastic strains. With increasing stress ratio, the steady-state creep rates increase slowly before the failure stress ratio but rise suddenly over the final stress ratio to failure. However, the pre-treatment temperature has no clear and apparent influence on steady creep strain rates. Rock specimens subject to higher pre-treatment temperatures exhibit higher permeabilities. The pre-existing cracks close under compression with a coplanar shear crack propagating from the starter-cracks and ultimately linking these formerly separate cracks. In addition, it is clear that the specimens pre-treated at higher temperatures accommodate greater damage.

Keywords: red sandstone; pre-existing cracks; creep behavior; temperature; long-term permeability

1. Introduction

Underground coal gasification (UCG) is a technology that converts hydrocarbons into syngas in-situ [1,2]. Crucially, UCG combined with carbon capture and sequestration (CCS) may have a special promise [3]. While UCG-CCS has potential promise, some key environmental and engineering issues require resolution. We focus on key rock mechanics problems including excessive subsidence and fugitive gases—each critically impacted by the response of strength and permeability to the imposed high temperatures. Similar to long-wall mining, excavation of coal from depth may cause subsidence, especially over long-term operations and under high crustal stress. It ought to be possible to inject CO_2 into former UCG voids which are currently at a depth of more than 800 m [4]. The UCG process is intensely exothermic with the temperatures occasionally exceeding 900 °C in the combustion zone. Prior observations indicate that high temperatures significantly transform the physical and chemical properties and resulting mechanical behavior of rocks (e.g., heating and quenching the first caprock may result in cracking or other behavior that compromises the integrity of the storage [5]). In addition, the high permeability of the UCG goaf for geological CCS projects, CO_2 storage in artificially developed UCG goaf with high permeability appears to be an attractive proposition [4], provided storage security of the caprocks can be guaranteed.

Heat transfer mechanisms in rocks are complex and nonlinear [6–11]. UCG production boreholes typically experience extreme temperatures of the order of ~900 °C which significantly exceeds the critical temperature (*Tc*~400–500 °C) that induces changes in the mechanical and permeability behavior of typical sandstones [11]. Temperatures from ambient to 400 °C correspond to vaporization then escape of adhered water, combined water and structural water. Between 400–600 °C, the minerals in the sandstone thermally react and transform, resulting in porosity increase, reduction in permeability and diffusivity, and changes in heat capacity [12]. Reservoir rheology is also sensitive to temperature [6,13,14]. An increase in temperature accelerates creep strains and consequently reduces the time-to-failure [15]. The total axial strain increases with increasing temperature, with instantaneous elastic strain and instantaneous plastic strain increasing slightly as the temperature increases from 25 to 700 °C but then increasing substantially as the temperature reaches 1000 °C [16]. However, the response of fractured rocks is usually different from that of intact rock—since stress concentrations are present at the ends of pre-existing cracks—promoting the initiation of new cracks and the reduction in rock mass strength [17,18]. Previous investigations [19–22] addressed crack initiation and coalescence for jointed rock masses under short-term loading with key mechanisms of creep contributing to time-dependent deformation and crack propagation [23,24]. Investigation of time-dependent response of jointed rock [21,25–28] have defined key failure mechanisms but few characterizations study the comprehensive impact of thermal effects on time-dependent interaction of thermal cracking across the scales.

The high-temperature treatment of the rock mass during underground coal gasification leads to the formation of fractures and changes in the fluid transmission characteristics of the strata, which is a direct result of the combustion process [29]. Thermal cracks result and increase permeability as the temperature increases from 400–600 °C [10]. Prior observations have promoted our understanding of permeability evolution due to short-term heating and loading demonstrating a linear correlation with volumetric strain [24,30]. However, the timescales of permeability evolution of the goaf in UCG-CCS operations should also be noted. During creep loading the permeability will typically decrease before increasing with the onset of failure at late deformation stages [28,30]. Some observations reveal that the temperature effects are on timescales of permeability evolution and that during the multi-step loading and unloading cycles process the permeability first decreases slightly as temperature increases from 25 to 300 °C and then increases with increasing temperature [28]. Heating rate significantly affects thermal damage as evident in that permeability is exponentially correlated with the number of AE events [24].

Consequences of the combined presence of pre-existing fissures, high confining pressures and high temperatures on the evolution of strength and permeability for UCG-CCS operations has not been previously considered. Correspondingly, the following focuses on defining the influence of thermal damage on creep properties and permeability evolution of pre-cracked specimens of red sandstone from micro-scale to laboratory scale. The primary focus is to identify creep parameters and overall permeability evolution in the pre-cracked rock in response to different pre-treatment temperatures in defining characteristics of the evolution of transport properties and mechanical strength.

2. Experimental Materials and Procedures

2.1. Sandstone Specimens and Heating Procedure

The red sandstone used for these experiments was collected from Rizhao City in Shandong Province, China. The dry density is 2402 kg/m^3, and effective porosity is ~6.26–6.55% [28]. The sandstone mainly comprises quartz, feldspar, dolomite, hematite and clay minerals. The tested specimens were cylindrical with 50 mm in diameter and 100 mm in height. Then, two fissures of near-identical length ($2a$ = 12 mm) and width (d = 2 mm) were cut in the specimens by diamond wire saw. The inclination angle of two coplanar cracks was 45° with an intervening rock bridge of 2b = 12 mm. The samples were dried for ~24 h and then heated to 250, 500, 750 and 1000 °C at a rate of 5 °C/min. The specimens were held at their target temperature for two hours before cooling naturally to room temperature (Figure 1). The color of the specimens changed as a result of the heat-treatment changing from drab-red (250 °C), progressively redder (500 and 750 °C) before becoming brown at 1000 °C.

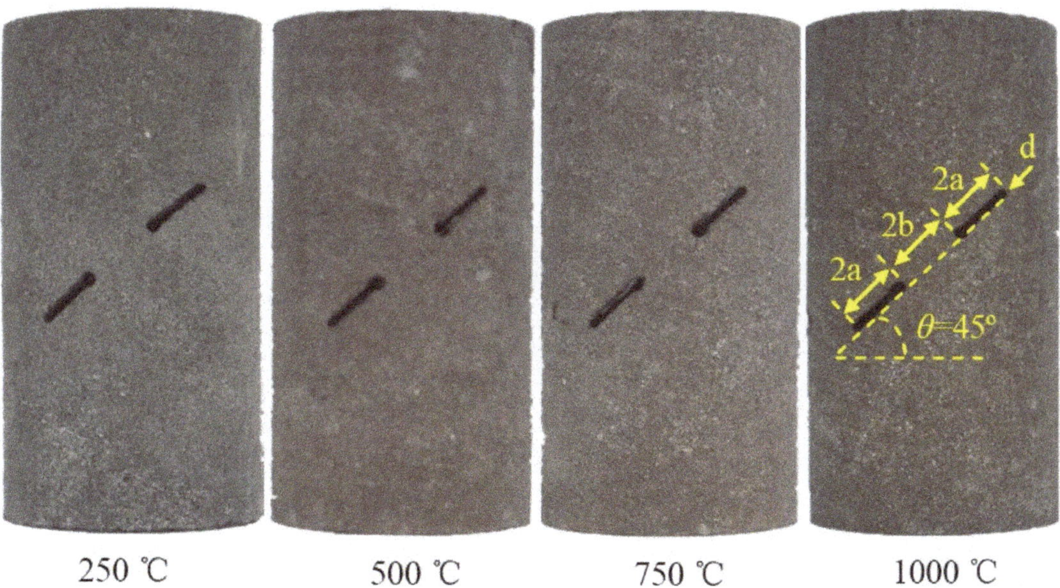

Figure 1. Pre-cracked red sandstone specimens after exposure to different pre-treatment temperature.

Figure 2 shows SEM images of the red sandstone after different pre-treatment temperatures. At 250 °C (Figure 2a), minor micro-cracks appear, and these develop into boundary cracks at the edges of grains during thermal cooling and differential expansion/contraction (Figure 2b,c). In addition, some transgranular cracks appear across grains. When the temperature reaches 1000 °C (Figure 2d) the density of boundary cracks increases and the crack apertures between grains also increases. The number of transgranular cracks also increases significantly. The results indicate a threshold temperature to produce thermally induced cracks in this study is 500 °C.

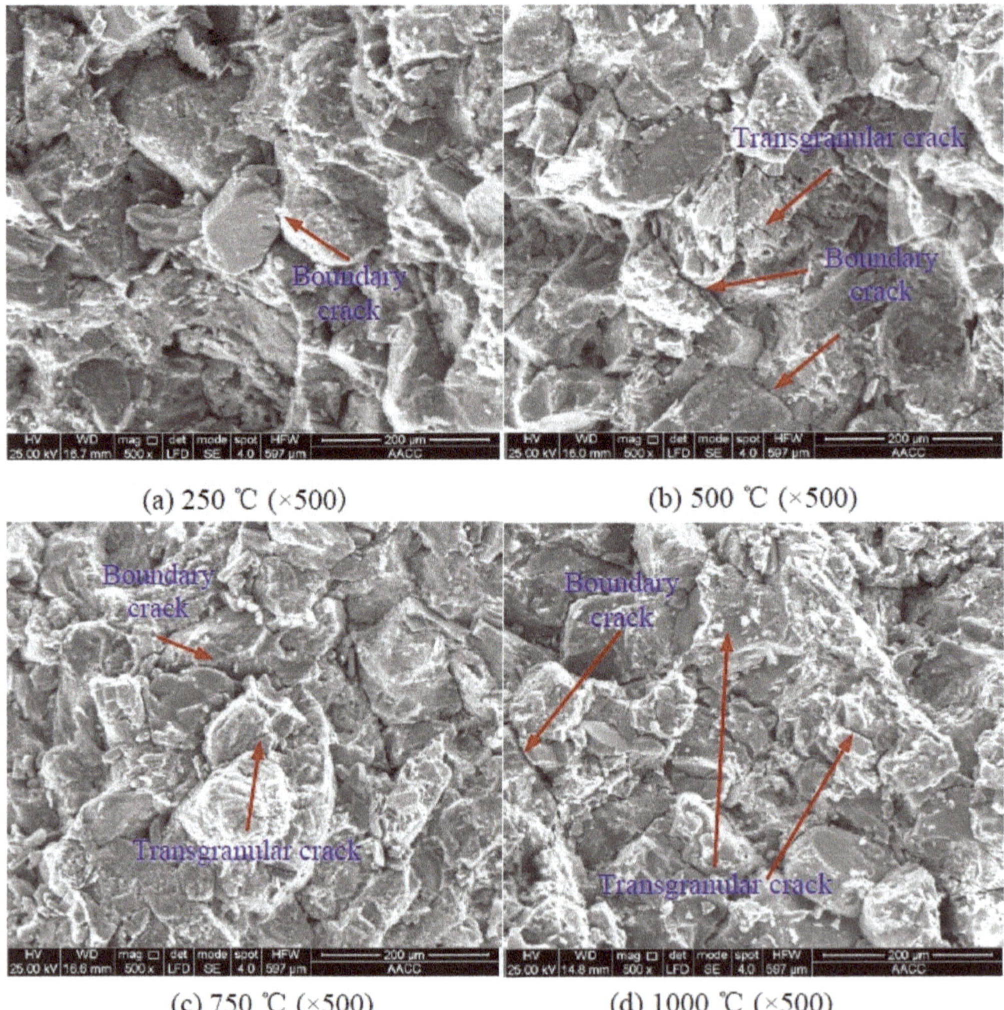

(a) 250 ℃ (×500)

(b) 500 ℃ (×500)

(c) 750 ℃ (×500)

(d) 1000 ℃ (×500)

Figure 2. Sampling location, granite specimens, SEM photograph and Optical microscopy. (**a**) $T = 250$ °C, (**b**) $T = 500$ °C, (**c**) $T = 750$ °C, (**d**) $T = 1000$ °C.

2.2. Testing Procedure

Conventional triaxial compression and multi-step loading- and unloading-cycle creep tests were completed under a confining pressure of 30 MPa (Figure 3a). The test system consists of three servo-hydraulic control units applying: axial, confining pressure and pore pressure loads. The maximum axial loading capacity of testing system is 800 kN, and the maximum confining pressure and pore pressure can reach 60 MPa. The axial deformation of the specimen was measured with a pair of linear variable displacement transducers (LVDTs), and the circumferential deformation was measured with a circumferential extensometer (Figure 3b).

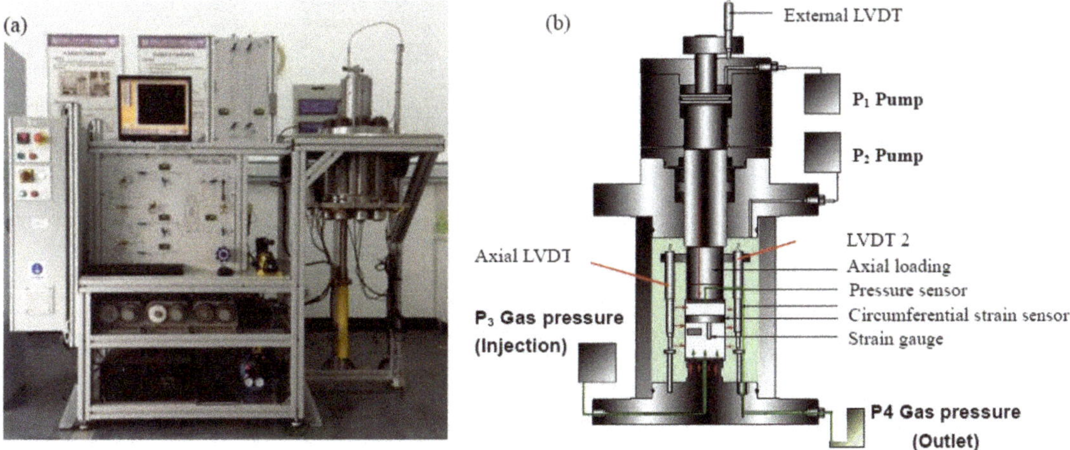

Figure 3. Rock tri-axial rheological testing apparatus (**a**) Testing equipment; (**b**) Specimen installation diagram.

The conventional triaxial compression tests were first completed to obtain the peak strength of the double pre-cracked sandstone specimens after exposure different temperature treatments. The test procedure is as follows: (i) at a loading rate of 4 MPa/min, 30 MPa hydrostatic pressure was applied to the specimen; (ii) then the specimen was axial loaded at a rate of 0.04 mm/min until it reached the residual strength. The stress-strain curves and triaxial compressive strengths (TCS) were thereby obtained.

Multi-step loading- and unloading-cycle creep tests were subsequently completed to study the creep characteristics of the pre-cracked sandstone. Stress levels for the creep tests were set according to the TCS. The test procedure was as follows: (i) the specimens were subjected to 30 MPa hydrostatic pressure at a rate of 4 MPa/min; (ii) a pre-determined first stress level was applied under pressure-controlled condition of 5 MPa/min; (iii) the specimen was then allowed to deform for approximately 48 h under constant stress; (iv) the axial deviatoric stress was then unloaded to 0 MPa under pressure-controlled conditions of 5 MPa/min; (v) these unloading conditions were retained for 24 h. Next, these loading and unloading cycles to creep were repeated until creep failure occurred. During the test, the axial and lateral strains of the pre-cracked red sandstone specimens were recorded continuously. The basic physical parameters of the specimens are listed in Table 1 with each of the deviatoric stress levels shown in Table 2.

With inert nitrogen as the flow medium, the permeability of pre- cracked specimens was measured by steady-state flow method during the creep test. The gas pressures at upstream and downstream were 3 MPa and 0.1 MPa, respectively. The permeability was calculated from Equation (1).

$$k = \frac{Q \cdot \mu \cdot L \cdot 2p_d}{A \cdot (p_u^2 - p_d^2)} \tag{1}$$

where Q represents volumetric gas flow rate (m³/s), μ represents the viscosity of the nitrogen (Pa·s); L and A represent the length (m) and the cross-sectional area (m²) of the rock specimen, respectively; P_u and P_d represent the inlet and outlet gas pressures (MPa), respectively. We observe that as the temperature increases the elastic modulus decreases and the Poisson's ratio overall increases (Table 1).

Table 1. Basic physical properties of sandstone specimens.

Specimen	*L*/mm	*D*/mm	*M*/g	ρ/kg/m^3	*T*/ °C	σ_3/MPa	σ_P/MPa	*E*/GPa	*v*	**Testing Design**
EDI-45-2	101.22	50.00	473.1	2380.44	250	30	193.28	25.89	0.12	Triaxial compression
EDI-45-3	101.25	50.28	472.8	2351.81	500	30	210.77	25.33	0.06	Triaxial compression
EDI-45-4	100.29	51.03	472.0	2301.14	750	30	238.06	20.17	0.13	Triaxial compression
EDI-45-5	102.66	51.46	475.2	2225.59	1000	30	222.29	12.29	0.21	Triaxial compression
EDI-45-8	100.54	49.97	479.2	2430.35	250	30	-			Creep compression
EDI-45-9	100.72	50.04	473.0	2387.93	500	30	-			Creep compression
EDI-45-10	100.79	50.18	468.9	2352.40	750	30	-			Creep compression
EDI-45-11	102.65	51.10	472.8	2245.88	1000	30	-			Creep compression

Table 2. Applied stress levels in creep tests.

Specimen	1st ($0.6\sigma_P$)	2nd ($0.7\sigma_P$)	3rd ($0.8\sigma_P$)	4th ($0.9\sigma_P$)	5th ($1.0\sigma_P$)
EDI-45-8	115.97	135.296	154.62	173.95	193.28
EDI-45-9	126.46	147.54	168.62	189.69	210.77
EDI-45-10	142.83	166.64	190.44	214.25	238.06
EDI-45-11	133.37	155.60	177.83	200.06	222.29

3. Experimental Results and Discussion

3.1. Creep Test Results

The axial deviatoric stress-strain curves of the pre-cracked sandstone specimens under conventional triaxial compression and cyclic creep compression are presented in Figure 4. The peak strength obtained from the cyclic creep compression is lower than that obtained from conventional compression for the different pre-treated temperatures. The creep curves plateau at each stress level, representing creep deformations under each constant stress. Each loading produced plastic deformation, with the plastic deformation increasing with an increase in the deviatoric stress. In addition, as with the increases in stress levels, the length of the stress-plateau also gradually increases, except a certain threshold stress level where coalescence of two coplanar cracks produced a larger plastic deformation, such as Figure 4a–c. A stress drop occurs after reaching the first peak strength due to the closure of the pre-existing cracks. This is because the ensemble closure of the pre-existing cracks and newly formed fractures increases the load capacity of the specimen. The resulting second peak strength was always larger than the first peak strength due to this closure of the pre-existing cracks. It should be noted that this double peak strength phenomenon is different from that for intact specimens that show only a single peak strength because intact specimens without pre-existing cracks the plastic hardening is not obvious [28,30].

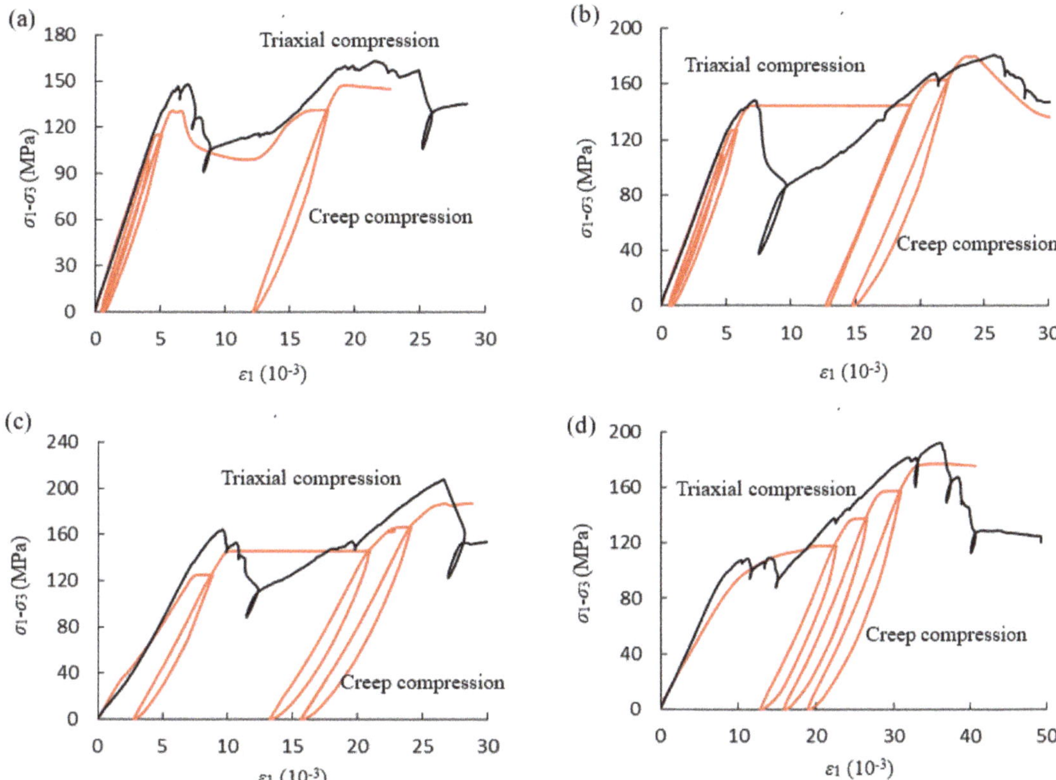

Figure 4. Axial stress-strain curves of pre-cracked sandstone under different loading paths. (**a**) T = 250 °C, (**b**) T = 500 °C, (**c**) T = 750 °C, (**d**) T = 1000 °C.

Figure 5 illustrates the variations in the axial and lateral strains over time for the pre-cracked specimens for different pre-treatment temperatures subject to multi-level creep loading and unloading. Instantaneous axial and lateral strains are produced as the deviatoric stress is applied with the elastic strains recovered as the deviatoric stress is removed. The increasing number of loading and unloading cycles will lead to more unrecoverable axial deformation, especially at a certain level of deviatoric stress that causes the closure of pre-existing cracks and produces large irrecoverable deformations, such as the bule dotted oval in Figure 5a–c. When the target deviatoric stress was reached, the deviatoric stress remained constant for 48 h. When the stress level is relatively low, deceleration deformation and steady creep deformation are the main components of axial deformation. As the deviatoric stress increases, the duration of the primary creep tends to increase, and when the deviatoric stress exceeds a threshold stress level, a fracture propagated between the two colinear cracks, producing a large plastic deformation. For instance, the specimens pre-treated to 250 and 500 °C (Figure 5a,b) fractured during the 3rd stress level while the specimen exposed to 750 °C (Figure 5c) fractured during only the 2nd stress level. Whereas primary, secondary and tertiary creep occurred under creep failure stress levels, as shown in the red dotted oval. The unloading deviatoric stress was 0 MPa since the elastic hysteretic behavior of rock maintained unloading conditions for 24 h.

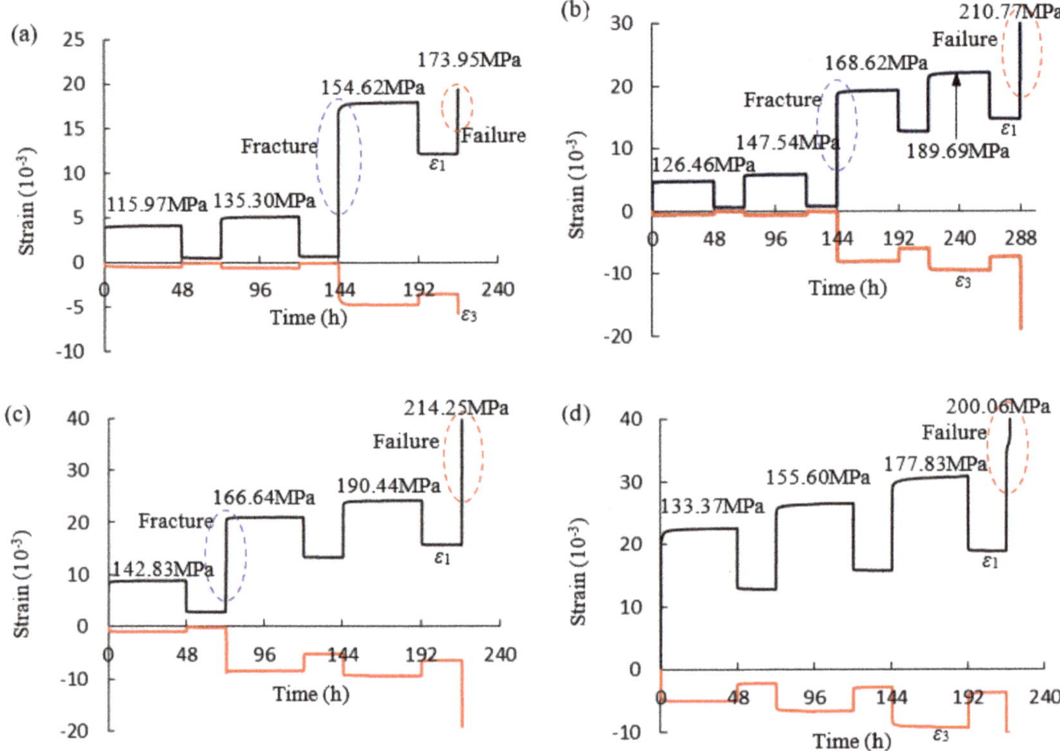

Figure 5. Axial and lateral strains of pre-cracked red sandstone specimen after undergoing different temperature treatments. (**a**) T = 250 °C, (**b**) T = 500 °C, (**c**) T = 750 °C, (**d**) T = 1000 °C.

3.2. Strain Separation

During creep experiments, the specific strains are separated by cyclic loading and unloading, including components of both recoverable strain and irrecoverable strain. The recoverable strain consists of the instantaneous elastic strain (ε_{me}) and visco-elastic strain (ε_{ve}), whereas the irrecoverable strain is composed of the instantaneous plastic strain (ε_{mp}) and visco-plastic strain (ε_{vp}), as shown in Figure 6. Moreover, total strain (ε) can also be divided into instantaneous (ε_m) and creep (ε_c) strains, which can be expressed as

$$\varepsilon = \varepsilon_m + \varepsilon_c \tag{2}$$

The instantaneous strain ε_m is the measured strain when the targeted deviatoric stress level is reached and the creep strain ε_c refers to the increase of strain measured with time at the stage of constant loading. The instantaneous strain ε_m is composed of two parts, i.e., the instantaneous elastic strain ε_{me} and the instantaneous plastic strain ε_{mp}:

$$\varepsilon_m = \varepsilon_{me} + \varepsilon_{mp} \tag{3}$$

The creep strain ε_c also consists of two parts, the recoverable visco-elastic strain ε_{ve} and the unrecoverable visco-plastic strain ε_{vp}:

$$\varepsilon_c = \varepsilon_{ve} + \varepsilon_{vp} \tag{4}$$

The duration of the creep loading is often longer than the duration of unloading—for instance, in this present study, the duration of the loading stage is 48 h, while the duration of unloading is 24 h. Hence, no visco-elastic recovery data are available from unloading for only 24 h. To obtain the visco-plastic strains under different stress levels, the visco-elastic strains based on experimental results were fitted by Equation (5). Then, the difference between the creep strain and the fitted visco-elastic strain may be evaluated as the visco-plastic strain under the loading conditions [16].

$$\varepsilon_{ve} = A(1 - \exp(-Bt^{m}))$$

(5)

where A, B and m are fitting parameters ($m > 0$).

Due to accelerated creep failure, the visco-elastic strain and visco-plastic strain in the third creep cannot be separated. and the tertiary creep should be represented by coupling visco-plasticity to damage [31–33]. In this study, the deformation after failure of the specimen is considered irreversible and the entire creep strain is taken as the visco-plastic strain.

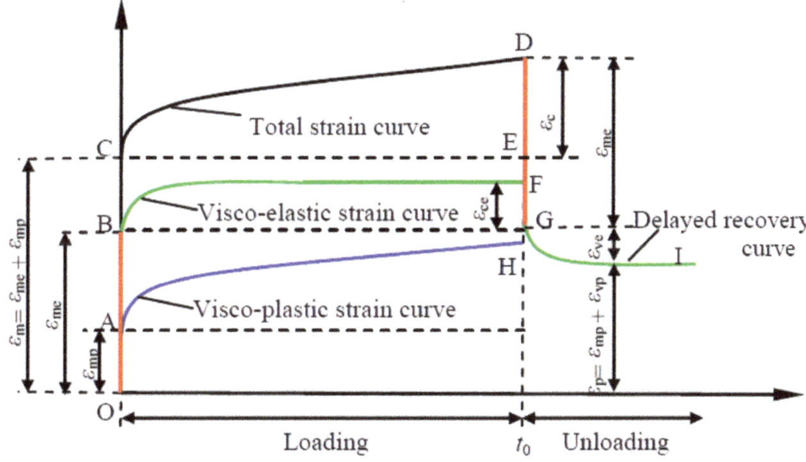

Figure 6. Representative strain-time curve defining different types of strains [28].

The creep strain histories of the thermally-treated pre-cracked specimens are separated into visco-elastic and visco-plastic strains according to the prior separation method. The variation of the total creep strains, visco-elastic strains and visco-plastic strains with time for four specimens are plotted in Figure 7 (the subfigure in dashed box is the enlarged version of the dashed part in main figure). For the specimen pre-treated at 250 °C (Figure 7a), there are 4 levels of deviatoric stress in total. The creep and visco-plastic strains can be characterized by primary creep and steady-state creep during the 1st and 2nd deviatoric stress levels. At the 3rd deviatoric stress level, as a result of crack growth, the creep strain and visco-plastic strain rates gradually increases until 95.597 h when the rock bridge between the two cracks fractures. At that point, the strain rate decreases, and begins steady-state creep at 95.61 h. At the 4th deviatoric stress level, because the ultimate failure of the specimen, the creep strain cannot be separated—therefore, it is considered that the creep strain is equal to the visco-plastic strain. The creep strain sequence can be characterized into primary, steady-state and tertiary creep. The visco-elastic strain can be characterized as primary and steady-state creep over the 1st stress interval and as primary creep from the 2nd deviatoric stress level. It should be noted that, the creep strain curves of the 3rd and 4th deviatoric stress levels are different: the creep rate under the 3rd deviatoric stress level first increases, and then decreases; however, the creep rate under 4th deviatoric stress level is opposite and is characterized by a typical creep curve.

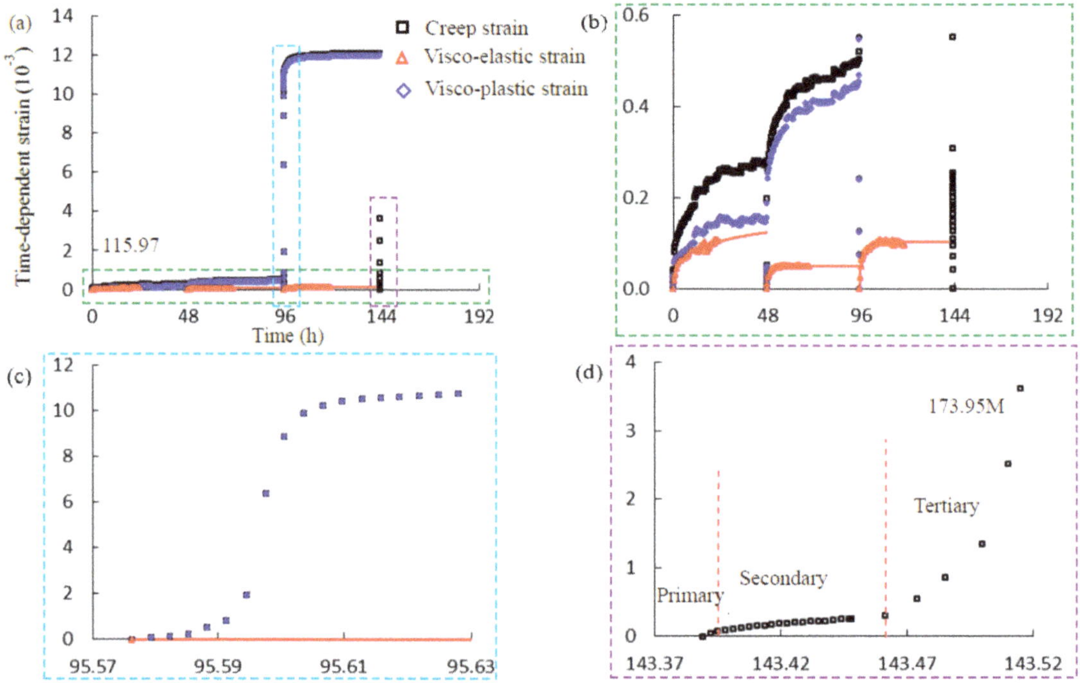

Figure 7. Creep strain, visco-elastic strain and visco-plastic strain of the pre-cracked specimens after different temperature treatment. (**a**) 250 °C; (**b**) 500 °C; (**c**) 750 °C; (**d**) 1000 °C.

The separated strains for the specimen pre-heated to 500 °C are shown in Figure 7b. The results are similar to that for the 250 °C heated specimen as the creep and visco-plastic strains show both primary and steady-state creep at the 1st to 2nd deviatoric stress levels with a strain jump at the 3rd deviatoric stress level. At the 4th deviatoric stress level, the creep and visco-plastic strains are larger than those at the 1st and 2nd deviatoric stress level but smaller than at the 3rd deviatoric stress level that exhibits primary and steady-state creep. As the deviatoric stress is increased, the primary creep, steady-state creep and tertiary creep clearly exhibit at the 5th deviatoric stress level when the specimen finally fails during tertiary creep.

Figure 7c shows the separated strains for the 750 °C treatment. The creep strain curve and the visco-plastic strain curve are characterized by primary and steady-state creep at the 1st deviatoric stress levels but with very small strain rates of steady-state creep. The visco-elastic strains are lower than the visco-plastic strains at the 1st deviatoric stress level. At the 2nd deviatoric stress level, in the initial loading stage, the creep and visco-plastic strain increases rapidly, but the rate of increase decreases until 47.5 h when the creep and visco-plastic strain jump from 0.85×10^{-3}, 0.83×10^{-3} to 10.37×10^{-3}, 10.33×10^{-3}, respectively, but the visco-elastic strain only increases from 0.027×10^{-3} to 0.042×10^{-3}, before establishing a steady-state. The specimen ultimately fails at the 4th deviatoric stress level after 0.11 h.

The separated strains for the 1000 °C specimen are different from the previously noted specimen behaviors (Figure 7d). First, the creep and visco-plastic strains are larger than those for specimens pre-heated to 250, 500 and 750 °C. Secondly, as the creep progresses, the creep and visco-plastic strains increase smoothly and progressively with no strain jump occurring until final failure. This results from the damage produced by the intense thermal cracking—during creep loading, closure of these cracks consumes most of the strain energy with insufficient remainder to fracture the rock.

Comparison of histories of creep, visco-elastic and visco-plastic strains highlight the existence of a strain jump at a certain deviatoric stress level—where the intervening rock bridge between the two starter-cracks fractures. However, the joining of these pre-existing cracks does not result in complete failure, but merely produces a larger irreversible visco-plastic strain with reduced ultimate strength. As the deviatoric stress level further increases, the visco-plastic strain accounts for an important proportion of total creep strain, especially for the high pre-treatment temperature specimens, with the development of the visco-plastic strain leading to the time-dependent failure of the rock.

3.3. Pre-Treatment Temperature Effects

These specimens exhibit dual peak strengths in compression as a result of the twin coplanar flaws (Figure 4), as illustrated by the corresponding axial strain versus pre-treatment temperature in Figure 8. Figure 8a shows the two peak strengths obtained from both the triaxial and creep tests both increase as the temperature increases from 250–750 °C, but then decreases substantially as temperatures reach 1000 °C. The principal reason for this is the vaporization and escape of adhered water, combined water and structural water between temperatures of 250–750 °C, which leads to an increase in the coefficient of internal friction [11] and increase in strength. When the temperature reaches 1000 °C, boundary and transgranular cracks develop and intergranular cracks widen—again resulting in a reduction in strength. In addition to its influence on ultimate strength, thermal treatment also has a influence on the deformation characteristics. Figure 8b shows changes in the peak strains with pre-treatment temperature. The axial strain at both peak strengths increase with pre-treatment temperature as it increases from 250 to 1000 °C.

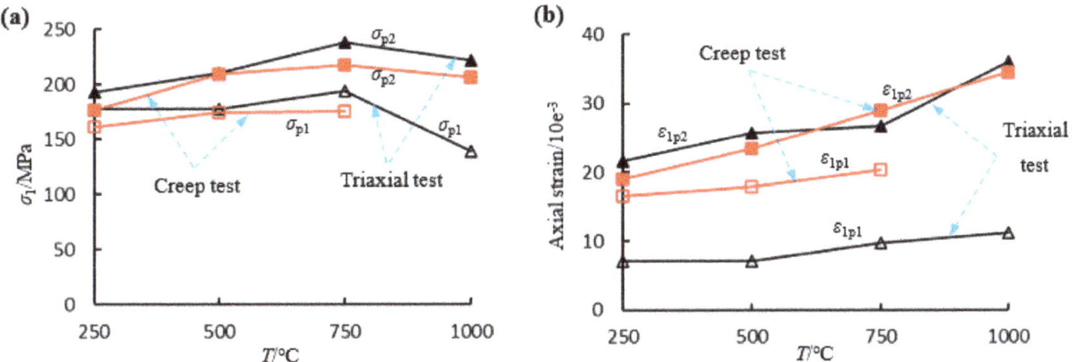

Figure 8. The temperature effects on strength and axial strain. (**a**) Strength; (**b**) axial strain.

To better characterize the effects of temperature on different separated strains, those strains are plotted against stress ratio for different pre-treatment temperatures Figures 9 and 10. Figure 9 shows the variation in the instantaneous elastic strains and the instantaneous plastic strains with stress ratio for the four pre-treatment temperatures. It is apparent from Figure 9a that the instantaneous elastic strains of the pre-cracked specimens increase linearly with an increase in the stress ratio. The instantaneous elastic strains increase with pre-treatment temperature, for example, the difference in value *a* between 250 °C and 500 °C is less than the difference in value *b* between 500 °C and 750 °C, and both are significantly less than the difference in value *c* between 750 °C and 1000 °C. Figure 9b illustrates the variation of instantaneous plastic strains with stress ratio showing that between 250~750 °C there is no obvious change with the initial stress ratio. However, when reaching the stress level for fracture, the instantaneous strain increases significantly due to the closure of pre-existing cracks. The instantaneous plastic strain of the specimen under a treatment temperature of 1000 °C is relatively large due to the closure of thermally-induced cracks with the instantaneous plastic strain increasing with an increase in the stress ratio.

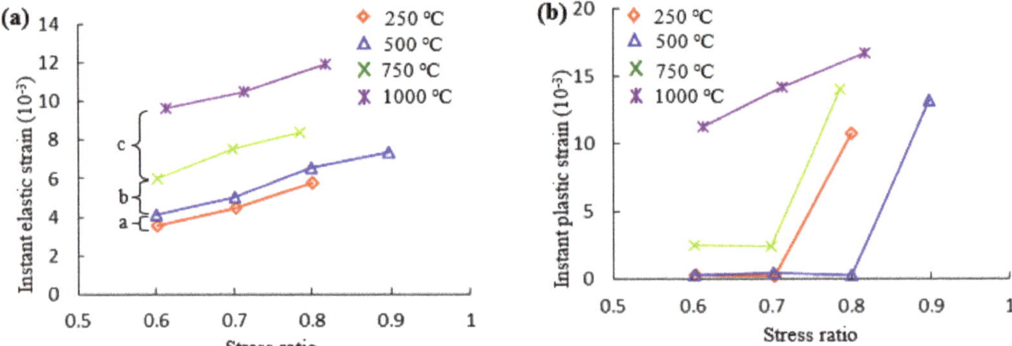

Figure 9. Relationships between stress ratio with instantaneous elastic strain and plastic strain after different temperature treatment. (**a**) Relationship between stress ratio and instantaneous elastic strain; (**b**) Relationship between stress ratio and instantaneous plastic strain.

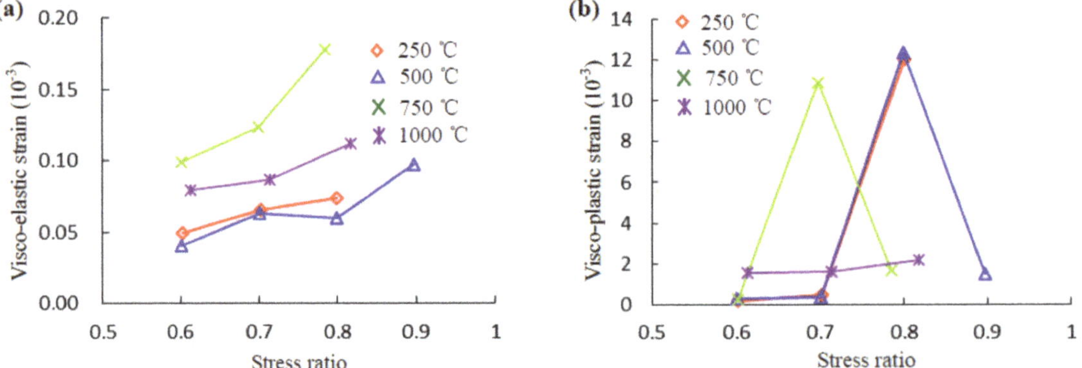

Figure 10. Relationships between stress ratio with visco-elastic strain and visco-plastic strain after different temperature treatment. (**a**) Visco-elastic strain; (**b**) visco-plastic strain.

From the above analysis, the instantaneous elastic deformation is usually larger than the instantaneous plastic deformation, but the closure of pre-existing cracks and the fracture of rock bridges produces a large plastic deformation that presents unconventional response. The instantaneous plastic strain exceeds any instantaneous elastic strain when closure and fracture occur. The thermal pre-treatment produces thermal cracks with their closure resulting in increased instantaneous elastic strains and instantaneous plastic strains.

Figure 10 shows the evolution of the visco-elastic and visco-plastic strains with stress ratio. The visco-elastic strains increase nonlinearly with an increase in the stress ratio and exhibit a complex relationship with temperature. The visco-elastic strains of the pre-cracked specimen after pre-treatment at 250 and 500 °C are almost equal, but smaller than that of the specimen pre-treated at 750 °C. However, as the pre-treatment temperature increases to 1000 °C, the visco-elastic strain is less than that for the specimen pre-treated at 750 °C because a larger irreversible plastic deformation is produced by the closure of thermally-induced cracks of the specimen pre-treated at 1000 °C. The visco-plastic strain apparent in the 1000 °C specimen (Figure 10b) also supports this—the visco-plastic strain of this hot specimen is larger than that for lower temperatures, except at certain stress ratios where the closure of pre-cracks and fracturing of the rock bridge causes a significant plastic deformation.

The creep strain rates during steady creep can be obtained by taking the derivative of steady creep curves, considered as of constant gradient. Figure 11 shows the steady-state creep strain rate versus stress ratio for all tested specimens. The results show that the steady-state creep rates increase slowly with stress ratio before the failure stress ratio but rise suddenly over the final stress ratio. The steady-state creep strain rate over the last stage is usually several orders of magnitude larger than that of the previous stages—requiring the use of logarithmic rates. The pre-heated temperature has no clear nor apparent influence on steady creep strain rates.

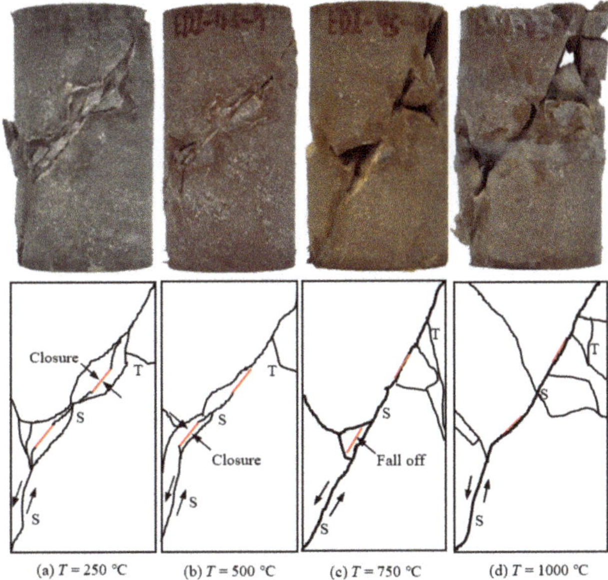

Figure 11. Relationships between stress ratio with steady-state strain ratio and damage after different temperature treatment.

The degree of material deterioration under changing stress states can be quantitatively described by the damage variable (D) [34]. Usually, D can be calculated using the elastic modulus, ultrasonic wave velocity, density and indexed to energy adsorption, strain and acoustic emission [35]. However, in cyclic loading-unloading creep tests, instantaneous elastic and visco-elastic strains are recovered after unloading, but instantaneous plastic and visco-plastic strains are not. Hence, the damage variable may be calculated using the ratio of non-elastic strain (instant plastic strain and visco-plastic strain) to total strains. Finally, D can be calculated from the instantaneous elastic and visco-elastic strains as [28]:

$$D = 1 - \frac{\varepsilon_e}{\varepsilon} = 1 - \frac{\varepsilon_{me} + \varepsilon_{cve}}{\varepsilon_m + \varepsilon_c} = \frac{\varepsilon_{mp} + \varepsilon_{cvp}}{\varepsilon_m + \varepsilon_c} \tag{6}$$

where ε_e and ε are the elastic and the total strains, respectively; ε_{me} and ε_{cve} represent the instantaneous elastic and the visco-elastic strains, respectively; ε_{mp} and ε_{cvp} represent the instantaneous plastic and the visco-plastic strains, respectively; and ε_m and ε_c represent the instantaneous and the creep strains, respectively.

Figure 12 plots D against the different stress ratios (*SRs*). As *SR* increases, D also increases—the higher the stress level the greater the damage. This is especially true when the *SR* increases from 0.7 to 0.8 for the specimens pre-treated at 250 and 500 °C, and as *SR* increases from 0.6 to 0.7 for the specimen pre-treated at 750 °C where D increases sharply due to the closure of pre-existing cracks. For identical *SRs*, the specimen with the higher temperature pre-treatment usually accumulates the greater damage. For example, under a stress ratio of 0.6, the D of the specimen subjected to 1000 °C is 0.56 and is larger than the 0.31 for the 750 °C, and much larger than the 0.11 and 0.13 for the 250 and 500 °C specimens.

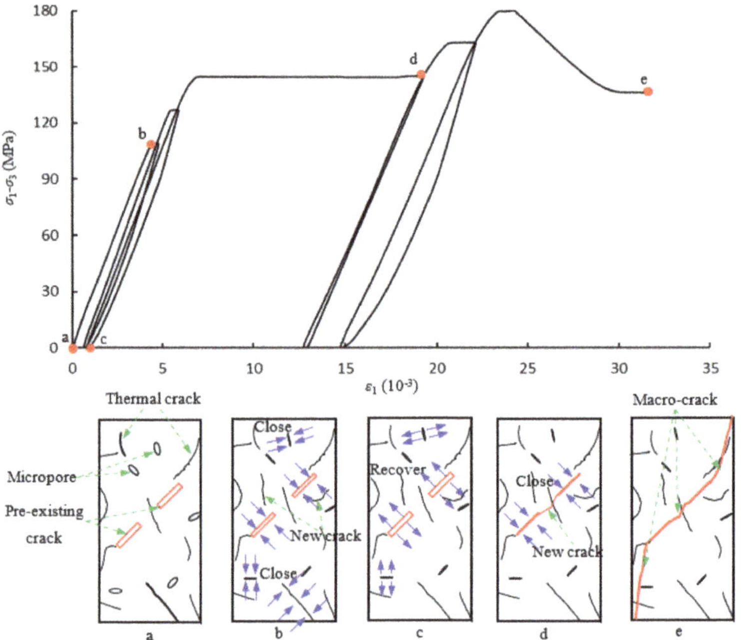

Figure 12. Relationships between stress ratio and steady-state strain ratio and damage after different temperature treatment.

3.4. Permeability Evolution

The evolution of volumetric creep and gas permeability for pre-treatment temperatures of 250 and 500 °C is presented in Figure 13. Equation (1) is adopted to calculate the coefficient of permeability. Permeability is closely related to volumetric strain during the creep process. During cyclic loading and unloading creep tests (Figure 13a) the permeability of sandstone remains near constant at low deviatoric stress (first two stress) before slightly increasing with unloading. The volumetric strain during the 3rd deviatoric stress level sharply increases and then decreases, resulting in the permeability also first increasing before subsequently decreasing. This results from the closure of pre-existing cracks and the production of some new cracks, although the newly produced cracks subsequently close under compression. At the final deviatoric stress level, the volume of the sandstone specimen initially decreases but then dilates before failure—the permeability shows an inverse trend to that expected from the volume strain signal. When the specimen fails, macro-cracks are generated and the permeability increases. Figure 13b shows the relationship between volumetric strain and permeability for the specimen after pre-heating to 500 °C. The permeability of the pre-cracked sandstone decreases with loading and increases with unloading of the deviatoric stress (Figure 13b). This results from the re-opening of pre-existing pores and fissures. With an increase in the deviatoric stress from 293 MPa to 247 MPa, the permeability due to each loading and unloading creep cycle becomes progressively lower—this can be explained by the presumed material strengthening of the sandstone under cyclic loading [30]. Over the final deviatoric stress level, the large macroscopic cracks linking the starter cracks have already been formed and this precipitates structural failure of the specimen—the permeability then increases significantly to 45×10^{-18} m^2.

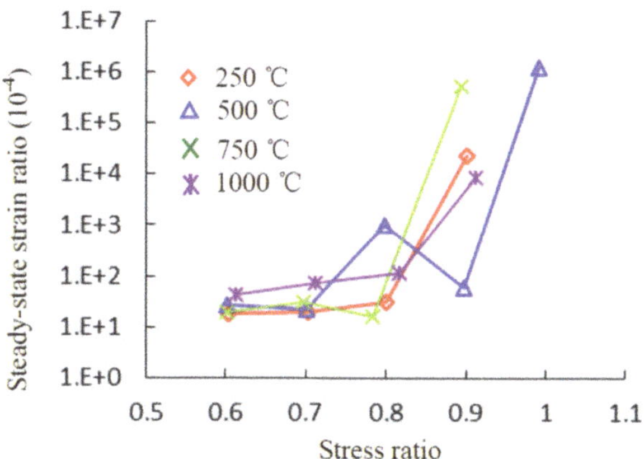

Figure 13. Permeability evolution of pre-cracked red sandstone after heated at different temperature. (**a**) T = 250 °C; (**b**) T = 500 °C.

Rock specimens undergoing hotter pre-treatments exhibit higher permeabilities. The average permeability of specimens pre-heated to 250 °C before failure is 5×10^{-18} m^2 which is almost an order of magnitude lower than that for 500 °C (45×10^{-18} m^2). The higher temperature pre-treatment produces more thermal cracks and provides more connected channels for gas transport.

3.5. Failure Modes

Failure modes are presented in Figure 14 showing the presence of a macro shear crack as indicative of failure with some tensile cracks induced by the shear sliding. The pre-existing cracks are closed under compression with a coplanar shear crack linking these formerly separate cracks. The pre-existing cracks exert a considerable influence on the creep failure mode in each specimen. In addition, it is clear that the specimens pre-treated at higher temperatures accommodate greater damage. For example, the shear fracture for specimens heated to 750 and 1000 °C is clearly wider than that for specimens pre-treated to 250 and 500 °C. In addition, the specimens undergoing higher temperature pre-treatments (Figure 14c,d) accumulated more severe damage than those at lower temperatures (Figure 14a,b)—apparent in the loss of some spalling along the main shear crack (Figure 14c,d).

We choose five stages to describe crack evolution in the loading and unloading creep tests (Figure 15). Due to the interaction of thermal cracks (produced by high temperature) and pre-existing cracks, the crack evolution is more complex than that in isothermal untreated intact specimens. Stage a represents the initial state with the specimen containing two coplanar pre-existing cracks dipping at 45° relative to the direction of the maximum principal stress. We conclude from Figure 2 that the specimen contains some thermal cracks and micropores. Stage b corresponds to the first level of loading. Under compression, the micropores and some thermal cracks close and produce a small number of cracks, but the stress is insufficient to cause wholesale fracture. Stage c is when the second unloading is completed. As the loading stress is removed, the closed micropores and cracks partially recover, elastically. Stage d produces a collapse as a stress concentration at the ends of the pre-existing cracks causes a new shear crack to connect the two pre-existing cracks. A large axial deformation results due to the closure of the pre-existing cracks. Stage e is the ultimate failure of the specimen. A macro-crack forms colinear to the pre-existing cracks, linking them, and which provides a connected channel for gas flow that yields a corresponding jump in permeability.

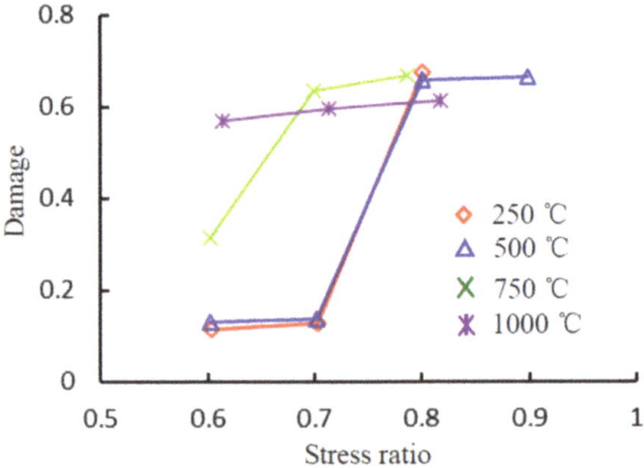

Figure 14. Ultimate failure modes of pre-cracked sandstone specimen after creep compression. (**a**) T = 250 °C, (**b**) T = 500 °C, (**c**) T= 750 °C, (**d**) T = 1000 °C.

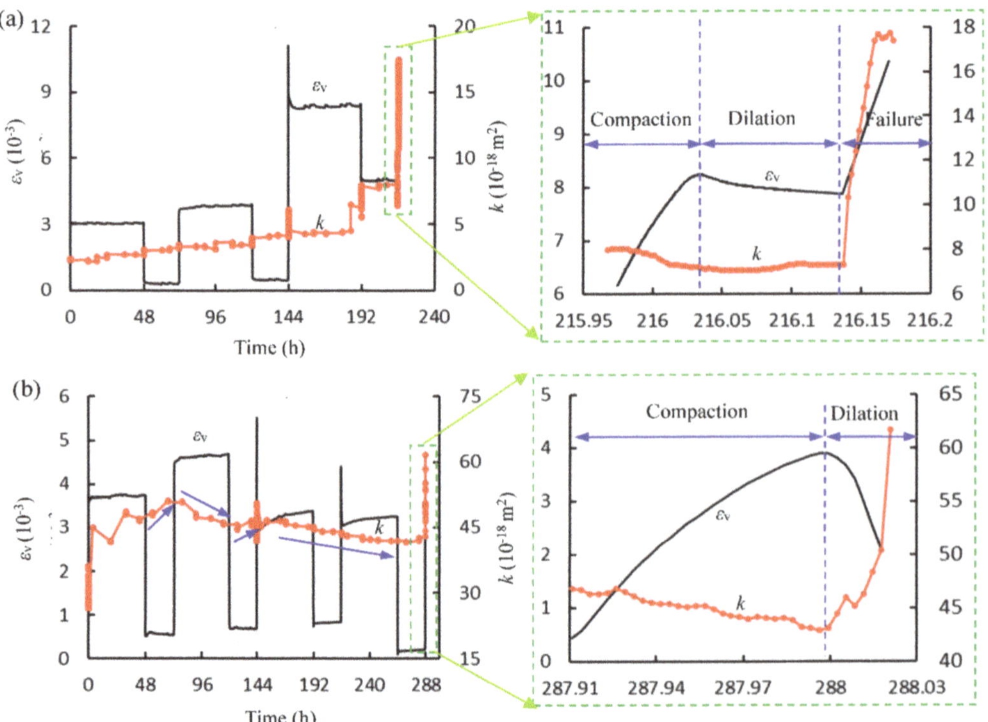

Figure 15. A sketch of crack evolution of pre-cracked sandstone in (**a**) loading and (**b**) unloading creep test.

4. Conclusions

We report tightly constrained experiments on heat-treated specimens containing pre-exiting cracks to explore the evolution of creep rupture and the evolution of permeability in the rock surrounding heated underground gasification chambers. Multi-step loading and unloading creep tests with concurrent measurement of gas permeability were completed under a confining pressure of 30 MPa. The following conclusions are drawn based on the experimental results.

(1) The SEM results indicate that the threshold temperature that is required to produce thermally induced cracks in this study is 500 °C—this is verified by measurement of the damage variables of the various specimens. Peak strength increases as pre-treatment temperature increases from 250 to 750 °C and then decreases as the pre-treatment temperature reaches 1000 °C. Hence, the critical temperature for strength reduction is 750 °C.

(2) Comparison of histories of creep, visco-elastic and visco-plastic strains highlight the existence of a strain jump at a prescribed deviatoric stress—where the intervening rock bridge between the twin starter-cracks fractures. However, the connecting of these pre-existing cracks does not result in complete failure of the specimen, but merely produces a larger irreversible visco-plastic strain and reduces ultimate strength of the specimen. As the deviatoric stress level further increases, the visco-plastic strain accounts for a significant proportion of the total creep strain, especially for the high pre-treatment temperature specimens, where the development of the visco-plastic strain ultimately leads to the time-dependent failure of the rock.

(3) The axial strain at each of the dual peak strengths increases with pre-treatment temperature as it increases from 250 to 1000 °C; the thermal pre-treatment produces thermal cracks with their closure resulting in increased instantaneous elastic strains and instantaneous plastic strains. The steady-state creep rates increase slowly with stress ratio significantly below the failure stress ratio but rises suddenly as the specimen approaches failure, however, pre-treatment temperature has no clear nor apparent influence on steady creep strain rates.

(4) Rock specimens subject to hotter pre-treatments exhibit higher permeabilities, since they contain more thermal cracks and provide a greater number of connected channels for gas transport. The average permeability of specimens pre-heated to 250 °C, before failure, is 5×10^{-18} m^2 which is almost an order of magnitude lower than that for the permeability at 500 °C (45×10^{-18} m^2).

(5) The pre-existing cracks close under compression with a coplanar shear crack linking these formerly separate cracks. In addition, it is clear that the specimens pre-treated at higher temperatures accommodate greater damage.

(6) UCG-CCS can support the implementation of a carbon neutral and energy production strategy, however, the caprock may contain many thermal cracks, which may cause CO_2 leakage, and it is recommended to wait a period of time after coal gasification for the thermal cracks to close under compressive crustal stress before utilizing the gasification chamber.

Author Contributions: Conceptualization, S.-Q.Y.; formal analysis, investigation and methodology, S.-Q.Y. and J.-Z.T.; writing—original draft preparation, S.-Q.Y. and J.-Z.T.; writing—review and editing, D.E.; visualization, D.E.; supervision, D.E. All authors have read and agreed to the published version of the manuscript.

Funding: This research was supported by the National Natural Science Foundation of China (42077231) and the Fundamental Research Funds for the Central Universities (2021ZDPYJQ002).

Institutional Review Board Statement: Not applicable.

Informed Consent Statement: Not applicable.

Data Availability Statement: Archive of State Key Laboratory for Geomechanics and Deep Underground Engineering, School of Mechanics and Civil Engineering, China University of Mining and Technology, Xuzhou, China.

Acknowledgments: The authors would also like to express their sincere gratitude to the editor and anonymous reviewers for their valuable comments, which have greatly improved this paper.

References

1. Perkins, G. Underground coal gasification-Part I: Field demonstrations and process performance. *Prog. Energy Combust. Sci.* **2018**, *67*, 158–187. [CrossRef]
2. Perkins, G. Underground coal gasification-Part II: Fundamental phenomena and modeling. *Prog. Energy Combust. Sci.* **2018**, *67*, 234–274. [CrossRef]
3. Friedmann, S.J.; Upadhye, R.; Kong, F.M. Prospects for underground coal gasification in carbon-constrained world. *Energy Procedia* **2009**, *1*, 4551–4557. [CrossRef]
4. Younger, P.L. Hydrogeological and Geomechanical Aspects of Underground Coal Gasification and its Direct Coupling to Carbon Capture and Storage. *Mine Water Environ.* **2011**, *30*, 127–140. [CrossRef]
5. Blinderman, M.S.; Friedman, S.J. Underground Coal Gasification with Carbon Capture and Storage: A Pathway to a Low-Cost, Low Carbon Gas for Power Generation and Chemical Syntheses. In Proceedings of the NETL 5th Annual Conference on Carbon Sequestration, Alexandria, VA, USA, 8–11 May 2006. Exchange Monitor Publications.
6. Chopra, P.N. High-temperature transient creep in olivine rocks. *Tectonophysics* **1997**, *279*, 93–111. [CrossRef]
7. Zhang, Z.X.; Yu, J.; Kou, S.Q.; Lindqvist, P.A. Effects of high temperature on dynamic rock fracture. *Int. J. Rock Mech Min. Sci* **2001**, *38*, 211–225. [CrossRef]
8. Xu, X.L.; Kang, Z.X.; Ji, M.; Ge, W.X.; Chen, J. Research of microcosmic mechanism of brittle-plastic transition for granite under high temperature. *Procedia Earth Planet. Sci* **2009**, *1*, 432–437.
9. Shao, S.S.; Ranjith, P.G.; Wasantha, P.L.P.; Chen, B.K. Experimental and numerical studies on the mechanical behavior of Australian Strathbogie granite at high temperatures: An application to geothermal energy. *Geothermics* **2015**, *54*, 96–108. [CrossRef]
10. Ding, Q.L.; Ju, F.; Song, S.B.; Yu, B.Y.; Ma, D. An experimental study of fractured sandstone permeability after high-temperature treatment under different confining pressures. *J. Nat. Gas. Sci. Eng.* **2016**, *34*, 55–63. [CrossRef]
11. Yang, S.Q.; Xu, P.; Li, Y.B.; Huang, Y.H. Experimental investigation on triaxial mechanical and permeability behavior of sandstone after exposure to different high temperature treatments. *Geothermics* **2017**, *69*, 93–109. [CrossRef]
12. Sun, Q.; Lü, C.; Cao, L.; Li, W.; Geng, J.; Zhang, W. Thermal properties of sandstone after treatment at high temperature. *Int. J. of Rock Mech. Min. Sci.* **2016**, *85*, 60–66. [CrossRef]
13. Heap, M.J.; Baud, P.; Meredith, P.G.; Bell, A.F.; Main, I.G. Time dependent brittle creep in Darley Dale sandstone. *J. Geophys. Res.* **2009**, *114*, 4288–4309. [CrossRef]
14. Heap, M.J.; Baud, P.; Meredith, P.G. Influence of temperature on brittle creep in sandstones. *Geophys. Res. Lett.* **2009**, *36*, L19305. [CrossRef]
15. Chen, L.; Wang, C.P.; Liu, J.F.; Li, Y.; Liu, J.; Wang, J. Effects of temperature and stress on the time-dependent behavior of Beishan granite. *Int. J. Rock Mech. Min. Sci.* **2017**, *93*, 316–323. [CrossRef]
16. Yang, S.Q.; Hu, B. Creep and long-term permeability of a red sandstone subjected to cyclic loading after thermal treatments. *Rock Mech Rock Eng* **2018**, *51*, 2981–3004. [CrossRef]
17. Ashby, M.F.; Hallam, S.D. The failure of brittle solids containing small cracks under compressive stress states. *Acta Met.* **1986**, *34*, 497–510. [CrossRef]
18. Hoek, E.; Martin, C.D. Fracture initiation and propagation in intact rock-a review. *J. Rock Mech. Geotech. Eng.* **2014**, *6*, 287–300. [CrossRef]
19. Bobet, A.; Einstein, H.H. Fracture coalescence in rock-type materials under uniaxial and biaxial compression. *Int. J. Rock Mech. Min. Sci.* **1998**, *35*, 863–889. [CrossRef]
20. Sagong, M.; Bobet, A. Coalescence of multiple flaws in a rock-model material in uniaxial compression. *Int. J. Rock Mech. Min. Sci.* **2002**, *39*, 229–241. [CrossRef]
21. Zhao, Y.; Zhang, L.; Wang, W.; Wan, W.; Li, S.; Ma, W.; Wang, Y. Creep Behavior of Intact and Cracked Limestone Under Multi-Level Loading and Unloading Cycles. *Rock Mech. Rock Eng.* **2017**, *50*, 1409–1424. [CrossRef]
22. Yang, S.Q.; Huang, Y.H.; Tian, W.L.; Zhu, J.B. An experimental investigation on strength, deformation and crack evolution behavior of sandstone containing two oval flaws under uniaxial compression. *Eng. Geol.* **2017**, *217*, 35–48. [CrossRef]
23. Lin, Q.X.; Liu, Y.M.; Tham, L.G.; Tang, C.A.; Lee, P.K.K.; Wang, J. Time-dependent strength degradation of granite. *J. Int. J. Rock Mech Min. Sci.* **2009**, *46*, 1103–1114. [CrossRef]
24. Chen, S.; Yang, C.; Wang, G. Evolution of thermal damage and permeability of Beishan granite. *Appl. Therm. Eng.* **2017**, *110*, 1533–1542. [CrossRef]
25. Fabre, G.; Pellet, F. Creep and time-dependent damage in argillaceous rocks. *Int. J. Rock Mech. Min. Sci.* **2006**, *43*, 950–960. [CrossRef]
26. Zhang, X.Z.; Wong, L.N.Y.; Wang, S.J.; Han, G.Y. Engineering properties of quartz mica schist. *Eng. Geol.* **2011**, *121*, 135–149. [CrossRef]
27. Zhang, Y.; Xu, W.Y.; Gu, J.J.; Wang, W. Triaxial creep tests of weak sandstone from fracture zone of high dam foundation. *J. Cent. South. Univ.* **2013**, *20*, 2528–2536. [CrossRef]

28. Yang, S.Q.; Hu, B. Creep and permeability evolution behavior of red sandstone containing a single fissure under a confining pressure of 30 MPa. *Sci. Rep.* **2020**, *10*, 1900. [CrossRef] [PubMed]

29. Otto, C.; Kempka, T. Thermo-mechanical simulations of rock behavior in underground coal gasification show negligible impact of temperature-dependent parameters on permeability changes. *Energies* **2015**, *8*, 5800–5827. [CrossRef]

30. Xu, P.; Yang, S.Q. Permeability evolution of sandstone under short-term and long-term triaxial compression. *Int. J. Rock Mech. Min. Sci.* **2016**, *85*, 152–164. [CrossRef]

31. Zhou, H.; Jia, Y.; Shao, J.F. A unified elastic–plastic and viscoplastic damage model for quasi-brittle rocks. *Int. J. Rock Mech. Min. Sci.* **2008**, *45*, 1237–1251. [CrossRef]

32. Ma, L.J.; Liu, X.Y.; Fang, Q.; Xu, H.F.; Xia, H.M.; Li, E.B.; Yang, S.G.; Li, W.P. A new elasto-viscoplastic damage model combined with the generalized Hoek–Brown failure criterion for bedded rock salt and its application. *Rock Mech. Rock Eng.* **2013**, *46*, 53–66. [CrossRef]

33. Zhang, L.; Liu, Y.R.; Yang, Q. A creep model with damage based on internal variable theory and its fundamental properties. *Mech. Mater.* **2014**, *78*, 44–55. [CrossRef]

34. Kachanov, L.M. Time of the rupture process under creep conditions. Izvestiia Akademii Nauk SSSR, Otdelenie Teckhnicheskikh Nauk. *Mater. Sci. Appl.* **1958**, *8*, 26–31.

35. Jin, J.F.; Li, X.B.; Yin, Z.Q.; Zou, Y. A method for defining rock damage variable by wave impedance under cyclic impact loadings. *Chin. J. Rock Soil Mech.* **2011**, *32*, 1385–1393. (In Chinese)

Article

A New Anelasticity Model for Wave Propagation in Partially Saturated Rocks

Chunfang Wu [1], Jing Ba [1,*], Xiaoqin Zhong [2], José M. Carcione [1,3], Lin Zhang [1] and Chuantong Ruan [1,4]

[1] School of Earth Sciences and Engineering, Hohai University, Nanjing 211100, China; chunfang@hhu.edu.cn (C.W.); jcarcione@libero.it (J.M.C.); zlin@hhu.edu.cn (L.Z.); rct@hhu.edu.cn (C.R.)
[2] Exploration and Development Research Institute of PetroChina Changqing Oilfield Company, Xi'an 710018, China; zhongxiaoqin_cq@petrochina.com.cn
[3] National Institute of Oceanography and Applied Geophysics—OGS, Geophysics, 34010 Trieste, Italy
[4] School of Mathematics and Statistics, Zhoukou Normal University, Zhoukou 466001, China
* Correspondence: jba@hhu.edu.cn

Abstract: Elastic wave propagation in partially saturated reservoir rocks induces fluid flow in multi-scale pore spaces, leading to wave anelasticity (velocity dispersion and attenuation). The propagation characteristics cannot be described by a single-scale flow-induced dissipation mechanism. To overcome this problem, we combine the White patchy-saturation theory and the squirt flow model to obtain a new anelasticity theory for wave propagation. We consider a tight sandstone Qingyang area, Ordos Basin, and perform ultrasonic measurements at partial saturation and different confining pressures, where the rock properties are obtained at full-gas saturation. The comparison between the experimental data and the theoretical results yields a fairly good agreement, indicating the efficacy of the new theory.

Keywords: partial saturation; patchy saturation; squirt flow; P-wave velocity dispersion and attenuation; anelasticity; ultrasonic measurements

Citation: Wu, C.; Ba, J.; Zhong, X.; Carcione, J.M.; Zhang, L.; Ruan, C. A New Anelasticity Model for Wave Propagation in Partially Saturated Rocks. *Energies* **2021**, *14*, 7619. https://doi.org/10.3390/en14227619

Academic Editor: Eugen Rusu

Received: 5 October 2021
Accepted: 6 November 2021
Published: 15 November 2021

Publisher's Note: MDPI stays neutral with regard to jurisdictional claims in published maps and institutional affiliations.

1. Introduction

Seismic waves induce fluid flow and anelasticity (the wave-velocity dispersion and dissipation factor) in rocks saturated with immiscible fluids [1–8]. The level of anelasticity depends on the in situ pressure, fluid content and type, and pore structure. This subject is highly relevant to petroleum exploration and production.

WIFF (wave-induced fluid flow) occurs at various spatial scales that can be categorized as macroscopic, mesoscopic, and microscopic [9]. The first is the wavelength-scale equilibration process occurring between the peaks and troughs of a P-wave, while the mesoscopic length is much larger than the typical pore size but smaller than the wavelength. The microscopic scale is of the same order of magnitude as the pore and grain sizes.

The macroscopic mechanism has been discussed by Biot [10–12] and is often referred to as the Biot relaxation peak (usually at kHz dominant frequencies). The basic assumptions are that the rock frame is homogeneous and isotropic, and the relative motion between the grains and the pore fluid is governed by Darcy's law. Local fluid flow on meso- and micro-scales are neglected, and consequently, the Biot peak cannot explain the observed wave anelasticity at all frequencies [13].

Partial saturation leads to fluid heterogeneity at the mesoscopic scale and the pressure difference between fluid phases causes wave dissipation at low frequencies [9,14–19]. White [20] proposed the first patchy-saturation model (the White model, spherical pockets). Dutta and Odé [21] reformulated this model by using the Biot theory, while Johnson [22] generalized it to patches of arbitrary geometry by using a branch function. Liu et al. [23] analyzed the effect of the fluid properties.

Moreover, dissimilar pores, with different shapes (micro-fractures and intergranular pores) and/or orientations, also cause mesoscopic pressure gradients and squirt flow, resulting in dissipation. At the pore level, dissipation can be described with squirt flow models [24,25]. Dvorkin and Nur [26] unified the Biot and squirt flows and proposed the BISQ model (Biot/squirt), which describes anelasticity at some frequency ranges. However, the low-frequency P-wave velocity prediction from the BISQ model is smaller than the Gassmann velocity [27], while it is consistent with the Biot one at high frequencies. Dvorkin et al. [28] extended the BISQ model to partially saturated rocks by incorporating the Wood equation [29] and proposed that the squirt flow length can be related to water saturation. Dvorkin et al. [30] reformulated the BISQ model to achieve consistency with the Gassmann velocity at the low-frequency limit. However, the P-wave velocity obtained with this model is higher than the theoretical high limit at high frequencies (when all the cracks are closed, and the P-wave velocity value is determined by the Biot model) [31]. Wu et al. [31] proposed a reformulated modified frame squirt flow model (MFS) to solve the problem.

Mavko and Jizba [32] introduced a modified frame to estimate the high-frequency unrelaxed dry rock shear and bulk moduli (M-J model), where cracks are saturated and the stiff pores are drained. To obtain the wet rock properties from the M-J model, Gurevich et al. [33] used the pressure relaxation method of Murphy et al. [34]. The model can be applied in a broad frequency range. Wu et al. [31] presented a reformulated modified frame squirt flow model, but although the prediction is acceptable at ultrasonic frequencies, Pride et al. [35] showed that the attenuation is significantly underestimated at seismic frequencies because the mesoscopic mechanism is not taken into account.

Wave anelasticity is mainly due to the effect of multi-scale fluid flow [36–40]. Rubino and Holliger [41] studied the problem at the micro and meso scales, analyzing the effects of the pore aspect ratio, while Li et al. [42] studied wave velocity in fractured poroelastic media saturated with immiscible fluids. Recently, Sun [43] proposed a model which considers the three loss mechanisms, i.e., the Biot, squirt flow, and mesoscopic relaxation peaks, in the framework of a double-porosity theory. This low-velocity limit does not honor Gassmann velocity.

We briefly review the propagation models at different scales, and propose a new one, based on the White theory and a reformulated modified frame squirt flow model (see Figure 1). Based on the numerical examples, the wave propagation characteristics of the new model and the effect of permeability and the outer diameter of the patch are analyzed. The P-wave velocity and attenuation with varying saturations are measured at different confining pressures. The crack properties and squirt flow length are obtained from the experimental data. The new model is applied to ultrasonic measurements performed on tight sandstone from the Qingyang area of the Ordos basin. The comparison between the experimental data and the theoretical results are made, so as to verify the capability of the new model in the description of those wave properties.

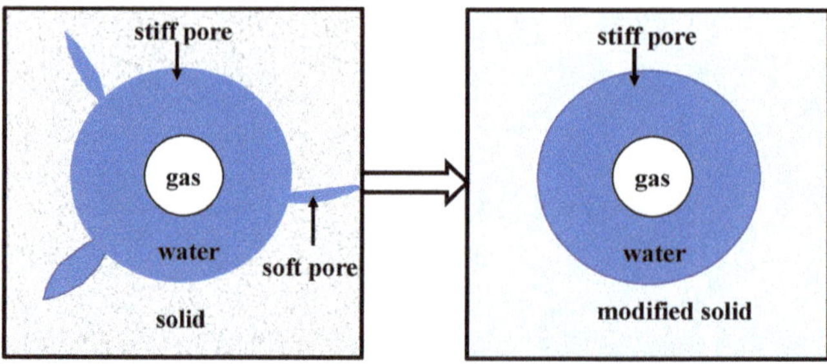

Figure 1. A new model based on a reformulated modified frame squirt flow (MFS) model with the White theory. The effects of squirt flow occurring between soft and stiff pores in the water-saturated host medium are incorporated by using an equivalent host medium of a modified solid (the MFS model). On the other hand, the White theory describes the anelasticity due to the patchy saturation of the immiscible fluid mixture.

2. Model

2.1. Patchy-Saturation (White) Model

White [20] proposed a patchy saturation model, by considering flow in a concentric spherical model where the inner sphere is saturated with one fluid type (gas), and the outer shell is saturated with a liquid (water), where the frame is assumed to be homogeneous. Let a and b be the inner and outer diameters, such as $(b > a)$, and the gas saturation is $S_g = a^3/b^3$. Dutta and Odé [21] modified the White model based on the Biot model, and obtained the following wet rock bulk and shear moduli:

$$K^*(\omega) = \frac{K_\infty}{1 - WK_\infty},\tag{1}$$

$$G^*(\omega) = G_{dry},\tag{2}$$

respectively, where K_∞ is the bulk modulus at the high-frequency limit, G_{dry} is the dry rock shear modulus, and W is a complex function of porosity, permeability, fluid viscosity, etc. (see Appendix A in Carcione et al. [44], and the Section 2.3).

2.2. Squirt Flow Model

Th flow between microcrack and grain contacts back and forth to stiff (equant) pores induces dissipation even for a single saturating fluid. The microcracks are incorporated into an effective rock skeleton, containing only stiff pores.

The reformulated modified frame squirt flow model considers both the squirt and Biot flows. According to Dvorkin et al. [30] and a boundary condition given by Gurevich et al. [33], the modified bulk modulus is (Wu et al. [31]):

$$K_{ms} = K_{msd} + \frac{\alpha_c{}^2 F_c}{\phi_c}\left[1 - \frac{2J_1(\lambda R)}{\lambda R J_0(\lambda R)}\right],\tag{3}$$

where $K_{msd} = \left(1/K_0 - 1/K_{hp} + 1/K_{dry}\right)^{-1}$, K_0 is the bulk modulus of the mineral mixture, K_{dry} is the dry rock bulk modulus, $F_c = \left(1/K_{fl} + 1/(\phi_c Q_c)\right)^{-1}$, ϕ_c is the microcrack porosity, $\alpha_c = 1 - K_{msd}/K_0$, $Q_c = K_0/(\alpha_c - \phi_c)$, R is characteristic squirt flow length, ω is the angular frequency, $\lambda^2 = i\omega\eta\phi_c/\kappa\left(1/K_{fl} + 1/(\phi_c Q_c)\right)$, η is the fluid viscosity, κ is the permeability, K_{fl} is the bulk modulus of fluid, and J_0 and J_1 are the zero- and first-order Bessel functions, respectively.

The modified dry-frame bulk and shear moduli are (Wu et al. [31]):

$$\frac{1}{K_{md}} = \frac{1}{K_{ms}} + \frac{1}{K_{hp}} - \frac{1}{K_0},\tag{4}$$

$$\frac{1}{G_{md}} = \frac{1}{G_{dry}} - \frac{4}{15}\left(\frac{1}{K_{dry}} - \frac{1}{K_{md}}\right),\tag{5}$$

respectively, where K_{hp} is the high-pressure modulus [33]. The P-wave phase velocity and attenuation can then be obtained according to Toksöz and Johnston [45] as

$$V_{phP1,2} = \frac{1}{\mathrm{Re}(X_{1,2})},\ a_{1,2} = \omega\mathrm{Im}(X_{1,2}),\tag{6}$$

where

$$X_{1,2} = \sqrt{Y_{1,2}},\ Y_{1,2} = -\frac{B}{2A} \pm \sqrt{\left(\frac{B}{2A}\right)^2 - \frac{C}{A}},\ A = \frac{\phi F M_{dry}}{\rho_{22}^2},$$

$$B = \frac{F\left(2\alpha_{md} - \phi - \phi\frac{\rho_{11}}{\rho_{22}}\right) - \left(M_{dry} + F\frac{\alpha_{md}^2}{\phi}\right)\left(1 + \frac{\rho_a}{\rho_{22}} + i\frac{\omega_c}{\omega}\right)}{\rho_{22}},$$

$$C = \frac{\rho_{11}}{\rho_{22}} + \left(1 + \frac{\rho_{11}}{\rho_{22}}\right)\left(\frac{\rho_a}{\rho_{22}} + i\frac{\omega_c}{\omega}\right), \quad \rho_{11} = (1-\phi)\rho_s, \quad \rho_{22} = \phi\rho_{fl},$$

$$F = \left(1/K_{fl} + (\alpha_{md} - \phi)/(\phi K_0)\right)^{-1},$$

where ρ_a is the additional coupling density, $\omega_c = \eta\phi/(\kappa\rho_{fl})$ is the characteristic frequency, $\alpha_{md} = 1 - K_{md}/K_0$, ϕ is porosity, M_{dry} is the uniaxial modulus of the rock skeleton under drained conditions, and ρ_s and ρ_{fl} are the mineral density and fluid density, respectively.

2.3. Patch-Saturation and Squirt Flow Models Combined

The White model assumes a uniform rock skeleton and that the area outside the inclusion is fully saturated with water. Therefore, the modified dry rock moduli (4) and (5) are used in the White model, thus combining the micro and meso descriptions of anelasticity. If subindices 1 and 2 refer to the gas-inclusion region and host medium (water), respectively, we have the wet rock moduli

$$K(\omega) = \frac{K_\infty}{1 - WK_\infty} \tag{7}$$

$$G(\omega) = G_{md}, \tag{8}$$

where

$$K_\infty = \frac{K_{G2}(3K_{G1} + 4G_{md}) + 4G_{md}(K_{G1} - K_{G2})S_g}{(3K_{G1} + 4G_{md}) - 3(K_{G1} - K_{G2})S_g} \tag{9}$$

$$W = \frac{3ia\kappa(R_1 - R_2)(F_1 - F_2)}{b^3\omega(\eta_1 Z_1 - \eta_2 Z_2)}. \tag{10}$$

Moreover,

$$K_{G1} = \frac{K_0 - K_{md} + \phi K_{md}\left(K_0/K_{fl1} - 1\right)}{1 - \phi - K_{md}/K_0 + \phi K_0/K_{fl1}} \tag{11}$$

$$K_{G2} = \frac{K_0 - K_{md} + \phi K_{md}\left(K_0/K_{fl2} - 1\right)}{1 - \phi - K_{md}/K_0 + \phi K_0/K_{fl2}} \tag{12}$$

are Gassmann moduli, where K_{fl1} and K_{fl2} are fluid moduli,

$$R_1 = \frac{(K_{G1} - K_{md})(3K_{G2} + 4G_{md})}{(1 - K_{md}/K_0)\left[K_{G2}(3K_{G1} + 4G_{md}) + 4G_{md}(K_{G1} - K_{G2})S_g\right]} \tag{13}$$

$$R_2 = \frac{(K_{G2} - K_{md})(3K_{G1} + 4G_{md})}{(1 - K_{md}/K_0)\left[K_{G2}(3K_{G1} + 4G_{md}) + 4G_{md}(K_{G1} - K_{G2})S_g\right]} \tag{14}$$

$$F_1 = \frac{(1 - K_{md}/K_0)K_{A1}}{K_{G1}} \tag{15}$$

$$F_2 = \frac{(1 - K_{md}/K_0)K_{A2}}{K_{G2}} \tag{16}$$

$$Z_1 = \frac{1 - \exp(-2\gamma_1 a)}{(\gamma_1 a - 1) + (\gamma_1 a + 1)\exp(-2\gamma_1 a)} \tag{17}$$

$$Z_2 = \frac{(\gamma_2 b + 1) + (\gamma_2 b - 1)\exp[-2\gamma_2(b - a)]}{(\gamma_2 b + 1)(\gamma_2 a - 1) - (\gamma_2 b - 1)(\gamma_2 a + 1) - \exp[-2\gamma_2(b - a)]} \tag{18}$$

$$\gamma_1 = \sqrt{i\omega\eta_1/\kappa K_{E1}} \tag{19}$$

$$\gamma_2 = \sqrt{i\omega\eta_2/\kappa K_{E2}},\tag{20}$$

where η_1 and η_2 are fluid viscosities, and

$$K_{E1} = \left[1 - \frac{K_{fl1}(1 - K_{G1}/K_0)(1 - K_{md}/K_0)}{\phi K_{G1}\left(1 - K_{fl1}/K_0\right)}\right]K_{A1}\tag{21}$$

$$K_{E2} = \left[1 - \frac{K_{fl2}(1 - K_{G2}/K_0)(1 - K_{md}/K_0)}{\phi K_{G2}\left(1 - K_{fl2}/K_0\right)}\right]K_{A2}\tag{22}$$

$$\frac{1}{K_{A1}} = \left(\frac{\phi}{K_{fl1}} + \frac{1-\phi}{K_0} - \frac{K_{md}}{K_0^2}\right)\tag{23}$$

$$\frac{1}{K_{A2}} = \left(\frac{\phi}{K_{fl2}} + \frac{1-\phi}{K_0} - \frac{K_{md}}{K_0^2}\right).\tag{24}$$

According to Wood [29], the effective bulk modulus of the gas-water mixture can be calculated from

$$\frac{1}{K_{fl}} = \frac{S_g}{K_{fl1}} + \frac{S_w}{K_{fl2}}\tag{25}$$

where S_w is the water saturation.

Finally, the P-wave phase velocity and attenuation are

$$V_p = \sqrt{\frac{\mathrm{Re}(K(\omega) + 4G(\omega)/3)}{\rho}},\tag{26}$$

$$Q_p^{-1} = \frac{\mathrm{Im}(K(\omega) + 4G(\omega)/3)}{\mathrm{Re}(K(\omega) + 4G(\omega)/3)},\tag{27}$$

respectively, where $\rho = (1 - \phi)\rho_s + \phi(S_g\rho_1 + S_w\rho_2)$ is bulk density, and ρ_1 and ρ_2 are the fluid densities.

2.4. Results

The MFS model is directly applied in partially saturated reservoir rocks, where the gas–water mixture is obtained with the Wood equation (there are no gas pockets), and the properties are listed in Table 1. The numerical examples of the characteristics of wave prorogation by the proposed model are shown in Figure 2, and the effects of permeability and the outer diameter of the patch on the wave velocity and attenuation are shown in Figures 3 and 4, respectively.

Table 1. Rock physical properties.

Mineral density (kg/m^3)	2650	Porosity (%)	10
Mineral mixture bulk modulus (GPa)	38	Water bulk modulus (GPa)	2.25
Dry rock bulk modulus (GPa)	17	Gas bulk modulus (GPa)	0.0022
Dry rock shear modulus (GPa)	12.6	Water density (kg/m^3)	1000
Permeability (mD)	1	Gas density (kg/m^3)	1.2
Squirt flow length (mm)	0.01	Water viscosity (Pa·s)	0.001
High-pressure modulus (GPa)	22	Gas viscosity (Pa·s)	0.00011
Crack porosity (%)	0.02	External diameter (m)	0.0005

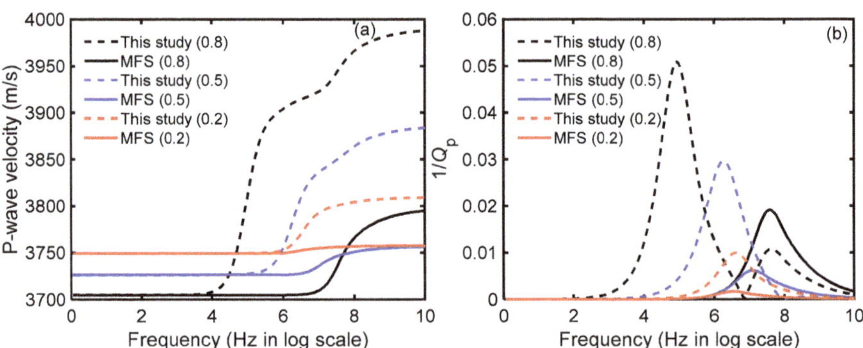

Figure 2. P-wave velocity (**a**) and attenuation (**b**) of the present and MFS models. The number between parentheses indicates water saturation.

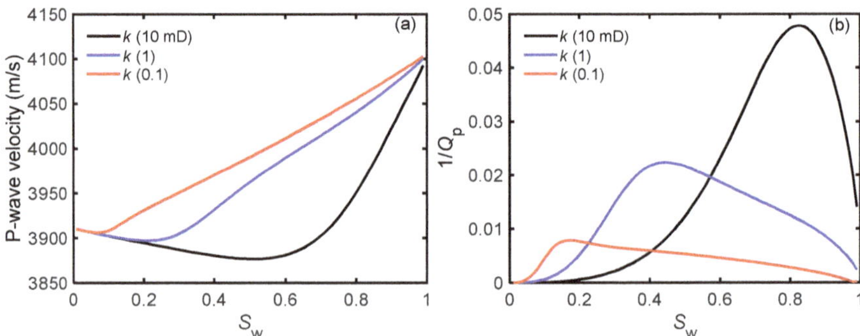

Figure 3. P-wave velocity (**a**) and attenuation (**b**) of the present model as a function of water saturation and various permeabilities.

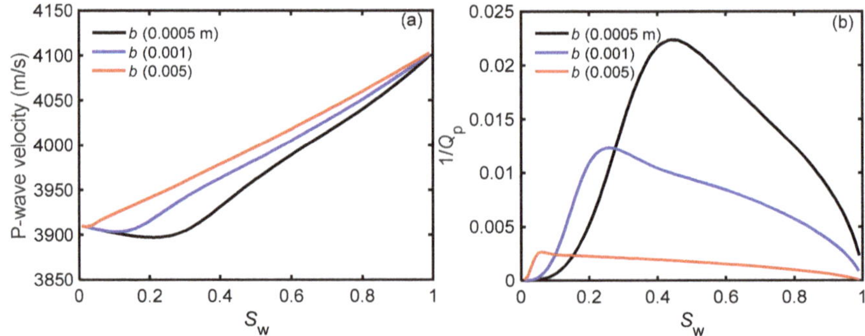

Figure 4. P-wave velocity (**a**) and attenuation (**b**) of the present model as a function of water saturation for different values of the outer diameter.

Figure 2 compares the P-wave velocity (a) and attenuation (b) of the present model with those of the MFS model, where the number between parentheses indicates water saturation. The velocities coincide at low frequencies and increase with saturation, with those of the present model higher at high frequencies. Two inflection points are clearly observed, corresponding to the mesoscopic and squirt flow attenuation peaks when the saturation is 80%, the first being the stronger point. The attenuation of the present model is higher than that of the MFS one.

Figure 3 shows the effect of permeability, where we can see that attenuation has a maximum at a given saturation which increases with permeability. Figure 4 displays the same quantities as a function of saturation for different outer diameters (*b*). The velocity increases with *b*, and the attenuation decreases.

3. Ultrasonic Data

3.1. Rock Specimen and Experiment

A tight sandstone sample (S2-9) from the Qingyang area, Ordos Basin, was tested. The sample was processed into a cylinder with a diameter of 25.2 mm and a length of 50 mm. An aluminum standard with the same shape and size was processed corresponding to the specimen. The sample was composed of quartz, feldspar, and interstitial materials (mainly carbonate minerals and clay), and its porosity was 8.85%. A thin section is shown in Figure 5. The experimental set-up consisted of a pulse generator, a temperature control unit, a confining pressure control unit, a pore pressure control unit, and an ultrasonic wave test unit [46,47].

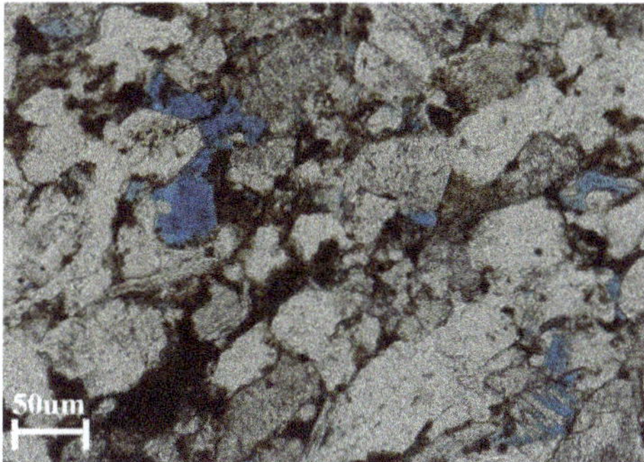

Figure 5. Thin section image of a tight sandstone.

The piezoelectric ultrasonic wave transducers were glued to the top and bottom of the sample, sealed with a rubber sleeve. An electrical pulse was applied to the source transducer to generate the ultrasonic P-waves. A digital oscilloscope was used to display and record the waveforms from the receiver. The temperature, pore, and confining pressures were controlled by the appropriate units [48]. The pore pressure was 15 MPa, the effective pressures were 5, 15, 25, 35 and 45 MPa, the temperature was 20 °C, and the waveforms were recorded after we maintained the experimental conditions for half an hour. For the partial gas–water saturation tests, the samples were first saturated with water by using the vacuum pressure saturation method and then placed in an oven to vary the saturation. The approach of Ba et al. [19] was adopted to quantify the fluid content. The sample was tested around six different water saturation conditions, 0%, 20%, 40%, 60%, 80%, and 100%. The wave velocities were obtained from the travel times and the spectral-ratio method was used to obtain the dissipation factor.

3.2. Experimental Results

Figure 6 shows the velocity as a function of water saturation and effective pressure. As expected, the P-wave velocity increases with water saturation and pressure, approaching a linear trend at high pressures [49,50], since microcracks close. In the partially saturated rock, the rock pore spaces contain air (with a lower bulk modulus and a lower P-wave velocity) and water (with a higher bulk modulus and a higher P-wave velocity). With the increase in water saturation, the volume ratio of water increases and that of air decreases while the rock skeleton stays unchanged. Generally, the P-wave velocity increases with the water saturation. The influence of effective pressure on the stiff pores is small and can be neglected [50–52]. The S-wave velocity also increases with effective pressure, but decreases as saturation increases, due to the density effect.

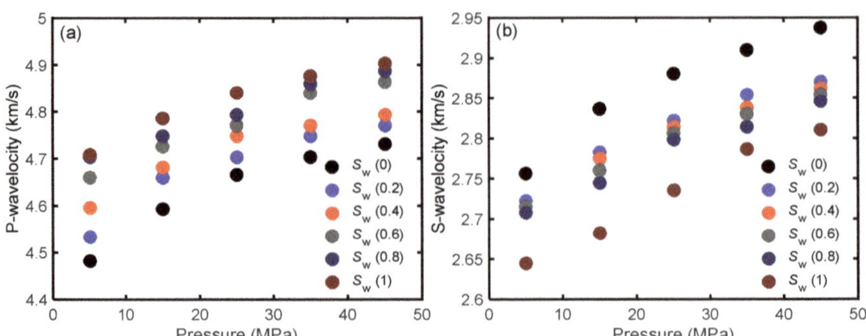

Figure 6. P-(**a**) and S-(**b**) wave velocities as a function of effective pressure at different water saturations.

The spectral ratio method is applied to calculate the dissipation factor [53,54]. We have

$$\ln\left(\frac{A_1(f)}{A_2(f)}\right) = -\frac{\pi x}{QV}f + \ln\frac{G_1(x)}{G_2(x)} \tag{28}$$

where f is the frequency, $A_1(f)$ and $A_2(f)$ are the amplitude spectra of the rock sample and standard, respectively, Q is the quality factor, x is the propagation distance, V is the wave velocity, and $G_1(x)$ and $G_2(x)$ are the sample and standard geometrical factors, respectively. As shown in Figure 7, attenuation decreases with effective pressure. Its behavior versus saturation is similar to that of Figure 4. The attenuation variations with respect to effective pressure and saturation are similar to those of the sandstone samples analyzed by Pang et al. [55] and Amalokwu et al. [56].

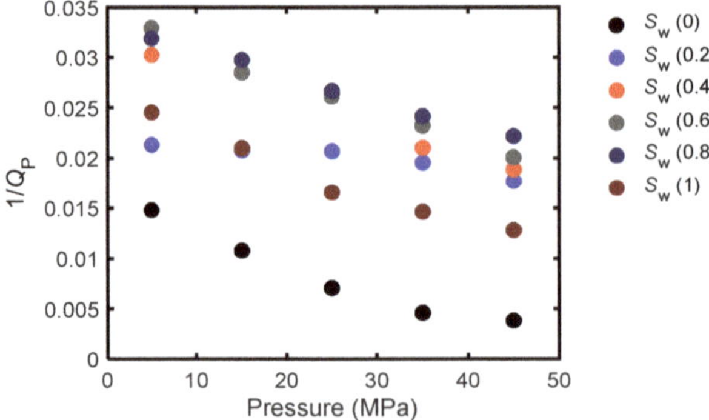

Figure 7. P-wave attenuation as a function of effective pressure and saturation.

3.3. Crack Parameters and Squirt Flow Length

The parameters of the present model can be obtained from the experimental data. They involve the skeleton bulk and shear moduli at different pressures, the dry rock bulk modulus with microcracks closed, the microcrack porosity, the squirt flow length, etc. The following steps are considered.

(1) The dry rock bulk and shear moduli are calculated from the velocities as

$$K_{dry} = \left(V_{pd}^2 - \frac{4}{3} V_{sd}^2 \right) \rho \tag{29}$$

$$G_{dry} = V_{sd}^2 \rho \tag{30}$$

where $\rho = \rho_s (1 - \phi)$, V_{pd} and V_{sd} are the P- and S-wave velocities of full gas saturation, respectively.

(2) The high-pressure dry rock bulk modulus, when the microcracks are closed, can be obtained from the linear trend of the dry rock velocities.

(3) The microcrack porosity and density is estimated at different pressures by using the DZ model [52], based on the experiment data (see Appendix A).

(4) The characteristic squirt flow length is estimated. This is an important parameter of the model and can be obtained with a least square method by matching the reformulated modified frame squirt flow model prediction with the experimental data at full water saturation.

The DZ model is applied to calculate the microcrack density (Figure 8a) and porosity (Figure 8b) based on the experimental data at different effective pressures. Their variations are more significant in the low-pressure range. As pressure increases, both quantities decrease. The microcrack density and porosity decreases, which can be attributed to the closure of microcracks [57,58].

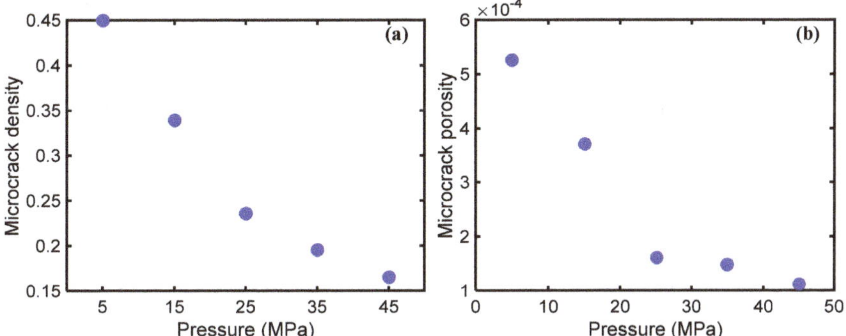

Figure 8. Microcrack density (**a**) and porosity (**b**) as a function of effective pressure.

The dry rock density of sample S2-9 is 2410 kg/m^3, and the bulk modulus of the mineral mixture is 39 GPa. The fluid properties are determined from the empirical equations of Batzle and Wang [59]. Figure 9 displays the P-wave velocity as a function of the effective pressure, where the squirt flow lengths are obtained by matching the theoretical results to the experimental data. It shows that the sample can be characterized by a constant squirt flow length at different pressures [31]. The characteristic length of sample S2-9 is 0.45 mm. This quantity is not so relevant to the pressure and it can be considered as an intrinsic rock property [26].

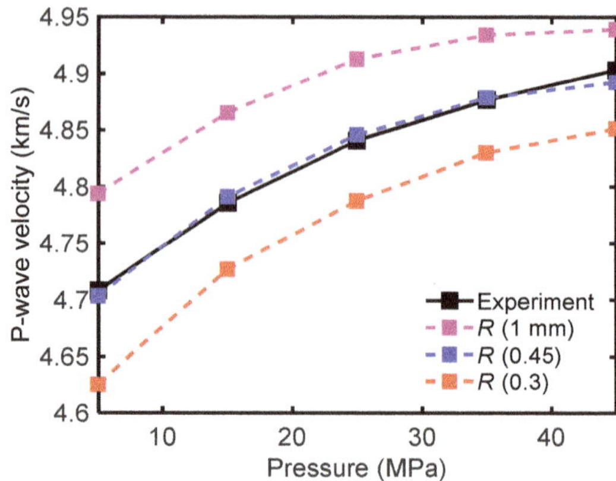

Figure 9. P-wave velocity as a function of the effective pressure compared to the experimental data. Results at different squirt flow lengths are shown.

4. Comparison between Theory and Experiment

4.1. Effect of Saturation

The present model is used to calculate the P-wave velocity and attenuation of sample S2-9 at 5 MPa effective pressure. The dry rock bulk modulus is 20.5 GPa, the Poisson ratio is 0.15, the permeability is 0.177 mD, the outer diameter is 0.12 mm, the high-pressure modulus is 23 GPa, and the fluid properties are listed in Table 1. Figure 10 displays the results for different models. The Gassmann–Hill curve does not consider the wave-induced fluid flow. The MFS curve coincides with the Gassmann–Wood curve at low saturations (Figure 10a), while the present model has a velocity of the order of the Gassmann–Hill curve. When saturation increases, the rock is stiffened by the microscopic fluid flow, resulting in a velocity increase.

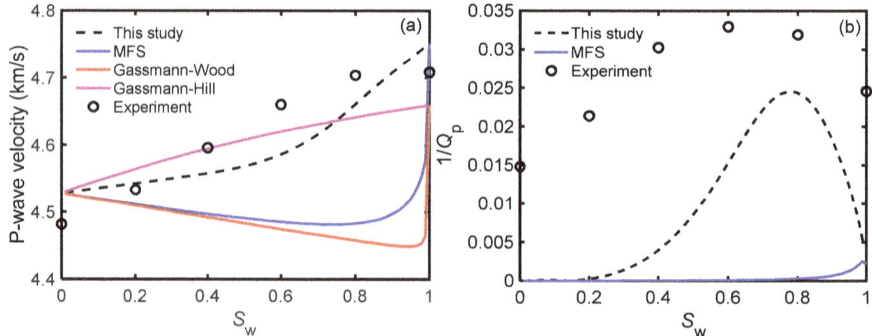

Figure 10. P-wave velocity (**a**) and attenuation (**b**) as a function of water saturation at 5 MPa. The open circles correspond to the experimental data.

The influence of fluid flow is determined by fluid pressure gradients at the interface between different fluid phases, where the fast P-wave converts to the slow (diffusive) Biot wave (mesoscopic loss) [15,60]. Attenuation has a maximum at a given saturation due to the mesoscopic loss mechanism, absent in the MFS model, whereas at full gas or water saturation, the results are similar. The attenuation curves are similar to those of the sandstone samples analyzed by Amalokwu et al. [56].

Compared with the simplified models, the present model provides a good match between the theory and the ultrasonic data for a tight sandstone, mainly the P-wave velocity as a function of saturation at an effective pressure of 5 MPa, showing the effectiveness of the squirt flow model combined with the White theory. However, attenuation is underestimated by the model due to the fact that the spatial variations in mineral grain and porosity are not considered.

4.2. Effect of Effective Pressure

In this example, the outer diameters at effective pressures of 15, 25, 35, and 45 MPa are 0.14, 0.16, 0.18 and 0.2 mm, respectively. Figures 11–14 display the P-wave velocity and attenuation as a function of water saturation at these pressures. The overall trend is similar to that at 5 MPa. As pressure increases, the MFS velocities approach the Gassmann–Wood velocities, and the P-wave velocity predictions from the new model increases; however, the attenuation decreases where most microcracks close, and the squirt flow effects are inhibited. Therefore, the characteristics of wave propagation from the new model are similar to those of the experimental data in Figures 11–14.

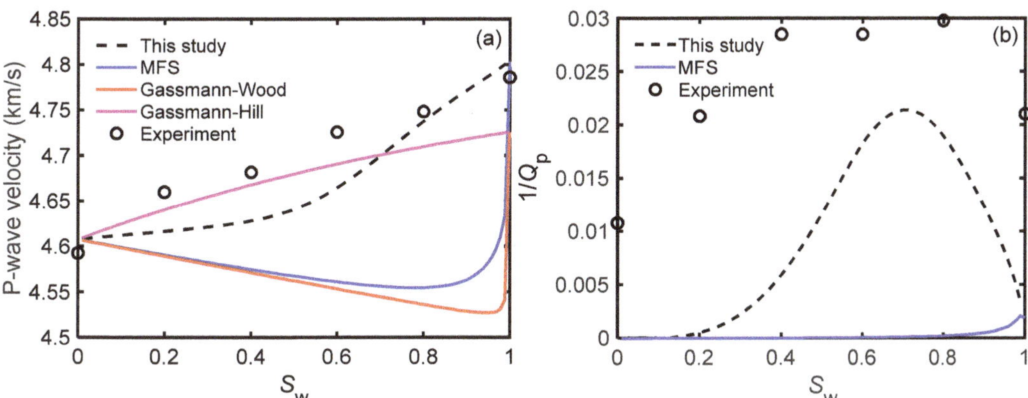

Figure 11. P-wave velocity (**a**) and attenuation (**b**) as a function of water saturation at 15 MPa. The open circles correspond to the experimental data.

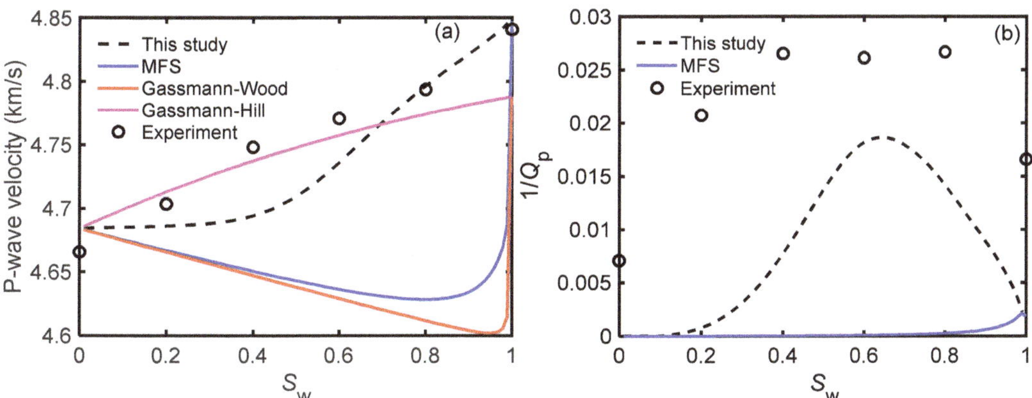

Figure 12. P-wave velocity (**a**) and attenuation (**b**) as a function of water saturation at 25 MPa. The open circles correspond to the experimental data.

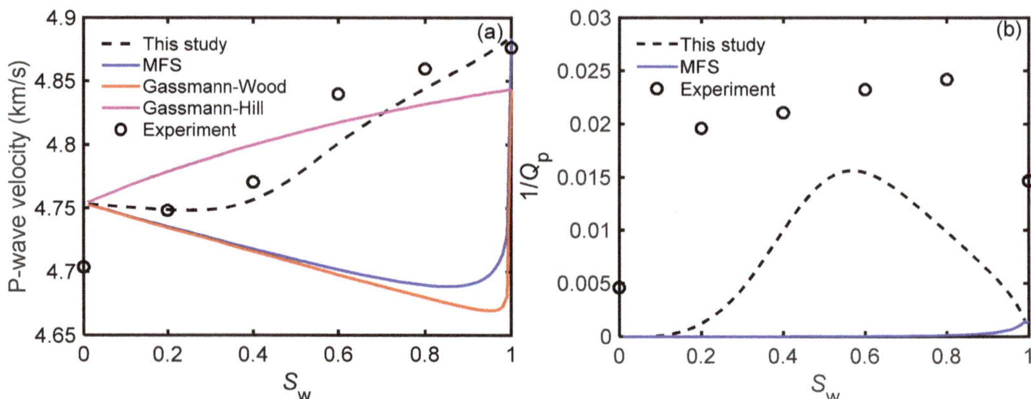

Figure 13. P-wave velocity (**a**) and attenuation (**b**) as a function of water saturation at 35 MPa. The open circles correspond to the experimental data.

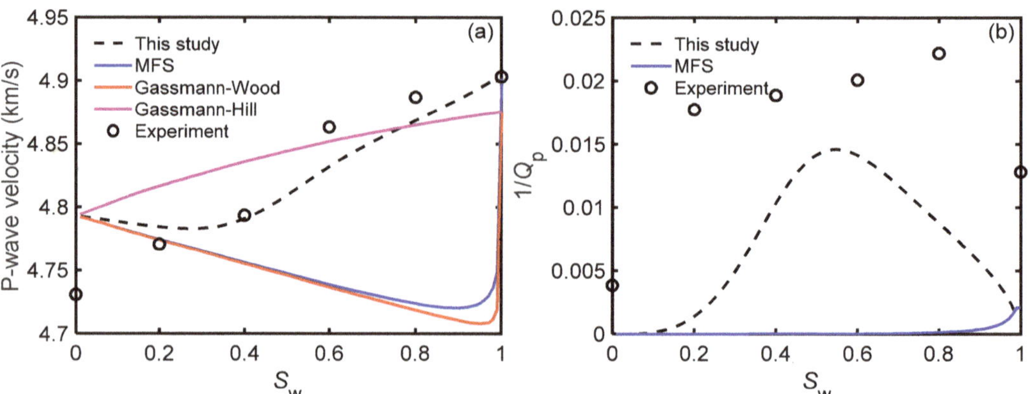

Figure 14. P-wave velocity (**a**) and attenuation (**b**) as a function of water saturation at 45 MPa. The open circles correspond to the experimental data.

4.3. Crossplots

Figure 15 shows crossplots of the measured and theoretical velocities, showing a good agreement.

Attenuation crossplots are displayed in Figure 16. The attenuation prediction from the present model is less than the experimental one, particularly at low effective pressures. This can be due to the fact that the model considers only the mesoscopic loss caused by partial saturation. Additional attenuation may be due to the presence of many minerals, and the microcrack content, shape, and distribution [15,51,61].

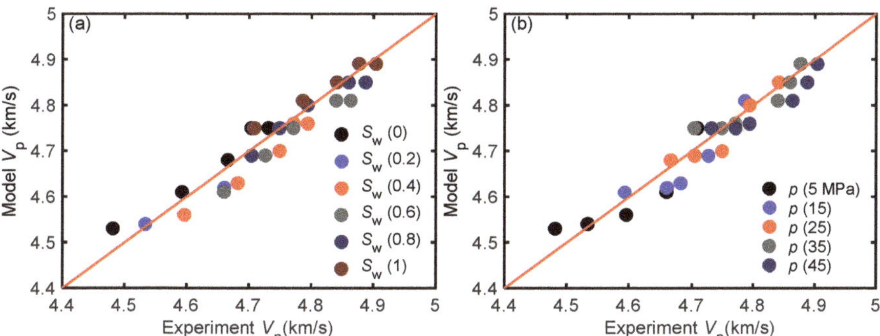

Figure 15. Crossplots of the measured and theoretical velocities at different water saturations (**a**) and effective pressures (**b**).

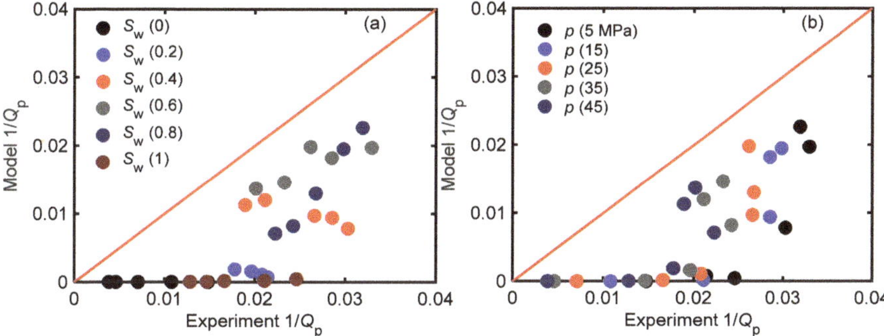

Figure 16. Crossplots of the measured and theoretical dissipation factors at different water saturations (**a**) and effective pressures (**b**).

5. Conclusions

We combine the reformulated modified frame squirt flow model and White's mesoscopic loss theory (spherical gas pockets) to develop a new model for describing wave anelasticity in partially saturated rocks. The microcrack properties and characteristic squirt flow lengths are obtained from experimental data at different effective pressures and water saturations. We compare the results with those of simplified models, showing that the present model provides a good match between the theory and the ultrasonic data for a tight sandstone, mainly the P-wave velocity as a function of saturation and pressure. Attenuation is underestimated by the model due to the fact that mesoscopic loss (fast P- to slow P-wave conversion) due to spatial variations in mineral grain and porosity are not considered. The new model can be used to predict the characteristics of wave propagation in partially saturated tight sandstones, mainly the P-wave velocity. Moreover, a better description at high frequencies (from tens of kHz) should consider the Biot attenuation peak. The generalization of the model will be the task of a future paper.

Due to the complex microstructures and fabric heterogeneity of tight sandstone, the proposed model cannot fully describe the experimental measurement data at low effective pressures. The theories cannot perfectly match real rocks, and there might be errors/defects in the experiment measurements. In the related engineering applications of hydrocarbon reservoir exploration, the methods of big data analytics and machine learning may be applied in combination with the theoretical models, so as to improve the applicability of the model and the accuracy of reservoir property prediction or interpretation.

Author Contributions: Conceptualization, J.B. and C.W.; Data curation, J.B. and X.Z.; Formal analysis, C.W., J.B., X.Z., J.M.C., L.Z. and C.R.; Funding acquisition, J.B.; Investigation, C.W., J.B., J.M.C. and L.Z.; Methodology, C.W., J.B. and J.M.C.; Project administration, J.B. and X.Z.; Resources, J.B. and X.Z.; Supervision, J.B.; Validation, C.W. and J.M.C.; Writing—original draft, C.W. and J.B.; Writing—review & editing, J.B., J.M.C. and L.Z. All authors have read and agreed to the published version of the manuscript.

Funding: This work was funded by the National Natural Science Foundation of China, grant number 41974123 and 42174161, the Jiangsu Province Outstanding Youth Fund Project, grant number BK20200021, the National Science and Technology Major Project of China, grant number 2017ZX05069-002, and Sinopec Key Laboratory of Geophysics.

Institutional Review Board Statement: Not applicable.

Informed Consent Statement: Not applicable.

Data Availability Statement: The data relevant with this study can be accessed by contacting the corresponding author.

Acknowledgments: This work is supported by the National Natural Science Foundation of China (Grant no. 41974123, 42174161), the Jiangsu Province Outstanding Youth Fund Project (Grant no. BK20200021), the National Science and Technology Major Project of China (Grant no. 2017ZX05069-002) and Sinopec Key Laboratory of Geophysics.

Conflicts of Interest: The authors declare no conflict of interest.

Appendix A. Symbols

Table A1. List of symbols.

S_g	gas saturation
b	outer diameter of the patch
$K^*(\omega)$	wet rock bulk modulus with the meso description of anelasticity
K_∞	bulk modulus at the high-frequency limit
G_{dry}	dry rock shear modulus
K_0	bulk modulus of the mineral mixture
ϕ_c	microcrack porosity
ω	angular frequency
κ	permeability
J_0	zero-order Bessel function
K_{md}	modified dry-frame bulk modulus
K_{hp}	high-pressure modulus
ω_c	characteristic frequency
ρ_s	mineral density
$K(\omega)$	wet rock bulk modulus with the micro and meso description of anelasticity
K_{fl1}	gas fluid modulus
η_1	gas viscosity
S_w	water saturation
ρ_2	water density
a	inner diameter of the patch
ϕ	porosity
$G^*(\omega)$	wet rock shear modulus with the meso description of anelasticity
W	a complex function of porosity, permeability and fluid viscosity, etc.
K_{ms}	modified bulk modulus
K_{dry}	dry rock bulk modulus
R	characteristic squirt flow length
η	fluid viscosity
K_{fl}	bulk modulus of the fluid
J_1	first-order Bessel function
G_{md}	modified dry-frame shear modulus
ρ_a	additional coupling density

Table A1. *Cont.*

M_{dry}	uniaxial modulus of the rock skeleton under drained conditions
ρ_{fl}	fluid density
$G(\omega)$	wet rock shear modulus with the micro and meso description of anelasticity
K_{fl2}	water fluid modulus
η_2	water viscosity
ρ_1	gas density
ρ	bulk density

Microcrack Porosity Estimation at Different Effective Pressures

-The aspect ratio of the stiff pores is estimated. Based on the MT model (Mori and Tanaka), the quantitative relation between elastic moduli and stiff porosity is established. The effective bulk and shear moduli of the host medium are

$$\frac{1}{K_{stiff}^{MT}} = \frac{1}{K_0}\left(1 + \frac{\phi_s}{1-\phi_s}P\right) \tag{A1}$$

$$\frac{1}{G_{stiff}^{MT}} = \frac{1}{G_0}\left(1 + \frac{\phi_s}{1-\phi_s}Q\right), \tag{A2}$$

respectively; G_0 is the shear modulus and ϕ_s is the stiff porosity.

$$P = \frac{(1-v)}{6(1-2v)} \times \frac{4(1+v)+2\gamma^2(7-2v)-[3(1+4v)+12\gamma^2(2-\gamma)]g}{2\gamma^2+(1-4\gamma^2)g+(\gamma^2-1)(1+\gamma)g^2} \tag{A3}$$

$$Q = \frac{4(\gamma^2-1)(1-\gamma)}{15\left\{8(\gamma-1)+2\gamma^2(3-4v)+[(7-8v)-4\gamma^2(1-2v)]g\right\}} \\ \times\left\{\frac{8(1-v)+2\gamma^2(3+4v)+[(8v-1)-4\gamma^2(5+2v)]g+6(\gamma^2-1)(1+v)g^2}{2\gamma^2+(1-4\gamma^2)g+(\gamma^2-1)(1+\gamma)g^2}\right. \\ \left.-3\left[\frac{8(v-1)+2\gamma^2(5-4v)+[3(1-2v)+6\gamma^2(v-1)]g}{-2\gamma^2+[(2-\gamma)+\gamma^2(1+v)]g}\right]\right\} \tag{A4}$$

where γ is the spheroidal aspect ratio, and v is the Poisson ratio of the grains, i.e.,

$$v = (3K_0 - 2G_0)/(6K_0 + 2G_0),$$

$$g = \begin{cases} \frac{\gamma}{(1-\gamma^2)^{3/2}}\left(\arccos\gamma - \gamma\sqrt{1-\gamma^2}\right)(\gamma<1) \\ \frac{\gamma}{(1-\gamma^2)^{3/2}}\left(\gamma\sqrt{1-\gamma^2}-\mathrm{arccosh}\gamma\right)(\gamma>1) \end{cases} \tag{A5}$$

Microcracks are included into the host material by neglecting the interactions between cracks and pores. The effective moduli (host with cracks) are

$$\frac{1}{K_{eff}^{MT}} = \frac{1}{K_{stiff}^{MT}}\left(1 + \frac{16\left(1-\left(v_{stiff}^{MT}\right)^2\right)\Gamma}{9\left(1-2v_{stiff}^{MT}\right)}\right) \tag{A6}$$

$$\frac{1}{G_{eff}^{MT}} = \frac{1}{G_{stiff}^{MT}}\left(1 + \frac{32\left(1-v_{stiff}^{MT}\right)\left(5-v_{stiff}^{MT}\right)\Gamma}{45\left(2-v_{stiff}^{MT}\right)}\right) \tag{A7}$$

where $v_{stiff}^{MT} = \left(3K_{stiff}^{MT}-2\mu_{stiff}^{MT}\right)/\left(6K_{stiff}^{MT}+2\mu_{stiff}^{MT}\right)$ and Γ is the microcrack density. When all the microcracks close at high pressures, a least square method is used to obtain the optimal aspect ratio of the stiff pores by using Equations (A1) and (A2).

-We obtain the cumulative microcrack density at different pressures by a least square method. Then the moduli can be obtained with Equations (A6) and (A7).

-The relation between effective pressure and microcrack density is established. The microcrack density obeys [62]

$$\Gamma = \Gamma^i e^{-p/\hat{p}} \tag{A8}$$

where Γ^i is the initial value when the effective pressure is zero, p is the effective pressure, and $\hat{p}$ is a constant.

-The microcrack aspect ratio distribution is computed. When effective pressure increases, microcracks gradually close. The minimum initial aspect ratio of the open microcracks is given by

$$\gamma_p^i = \frac{3}{4\pi} \int_{\Gamma^i}^{\Gamma} \frac{\left(1/K(\Gamma) - 1/K_{eff}^{hp}\right)}{\Gamma} \frac{dp}{d\Gamma} d\Gamma \tag{A9}$$

where $K(\Gamma)$ is the effective bulk modulus which can be obtained from Equation (A1).

Substituting Equation (A8) into (A9), we obtain

$$\gamma_p^i = \frac{3}{4\pi} \int_{\Gamma}^{\Gamma^i} \frac{\left(1/K(\Gamma) - 1/K_{eff}^{hp}\right)\hat{p}}{\Gamma^2} d\Gamma \tag{A10}$$

and by integrating Equation (A10) from Γ to Γ^i,

$$\gamma_p^i = \frac{4\hat{p}\left[1 - \left(v_{eff}^{hp}\right)^2 \ln\left(\frac{\Gamma^i}{\Gamma}\right)\right]}{3\pi K_{eff}^{hp}\left[1 - 2v_{eff}^{hp}\right]} \tag{A11}$$

where v_{eff}^{hp} is the effective Poisson ratio at high pressures, i.e., $v_{eff}^{hp} = \left(3K_{eff}^{hp} - 2G_{eff}^{hp}\right)/\left(6K_{eff}^{hp} + 2G_{eff}^{hp}\right)$.

Combining Equations (A8) and (A11), the relation between the minimum initial aspect ratio and the effective pressure can be obtained as

$$\gamma_p^i = \frac{4\left[1 - \left(v_{eff}^{hp}\right)^2\right]p}{\pi E_{eff}^{hp}} \tag{A12}$$

where $E_{eff}^{hp} = 3K_{eff}^{hp}\left[1 - 2v_{eff}^{hp}\right]$ is the effective Young modulus at high pressures. The cumulative microcrack density decreases with pressure. If pressure changes from zero to dp, the corresponding reduction of the cumulative microcrack density is $d\Gamma$. When the pressure increment is small enough, it can be considered that the decrease of microcrack density is mainly due to the closure of microcracks with an aspect ratio less than the minimum initial aspect ratio. David and Zimmerman [52] related the microcrack porosity and density as

$$\phi_c = \frac{4\pi\gamma}{3}\Gamma \tag{A13}$$

Therefore, the microcrack properties can be obtained from the acoustic wave velocities as a function of the effective pressure.

References

1. Winkler, K.W. Dispersion analysis of velocity and attenuation in Berea sandstone. *J. Geophys. Res. Solid Earth* **1985**, *90*, 6793–6800. [CrossRef]
2. Knight, R.; Nolen-Hoeksema, R. A laboratory study of the dependence of elastic wave velocities on pore scale fluid distribution. *Geophys. Res. Lett.* **1990**, *17*, 1529–1532. [CrossRef]
3. Gist, G.A. Interpreting laboratory velocity measurements inpartially gas-saturated rocks. *Geophysics* **1994**, *59*, 1100–1109. [CrossRef]
4. Mavko, G.; Nolen-Hoeksema, R. Estimating seismic velocities at ultrasonic frequencies in partially saturated rocks. *Geophysics* **1994**, *59*, 252–258. [CrossRef]

5. Zhao, H.B.; Wang, X.M.; Chen, S.M.; Li, L.L. Acoustic response characteristics of unsaturated porous media. *Sci. China Phys. Mech. Astron.* **2010**, *53*, 1388–1396. [CrossRef]
6. Ba, J.; Yan, X.F.; Chen, Z.Y.; Xu, G.C.; Bian, C.S.; Cao, H.; Yao, F.C.; Sun, W.T. Rock physics model and gas saturation inversion for heterogeneous gas reservoirs. *Chin. J. Geophys.* **2013**, *56*, 1696–1706. (In Chinese)
7. Sun, W.T.; Ba, J.; Müller, T.M.; Carcione, J.M.; Cao, H. Comparison of P-wave attenuation models of wave-induced flow. *Geophys. Prospect.* **2014**, *63*, 378–390. [CrossRef]
8. Cheng, W.; Ba, J.; Fu, L.Y.; Lebedev, M. Wave-velocity dispersion and rock microstructure. *J. Pet. Sci. Eng.* **2019**, *183*, 106466. [CrossRef]
9. Müller, T.M.; Gurevich, B.; Lebedev, M. Seismic wave attenuation and dispersion resulting from wave-induced flow in porous rocks-a review. *Geophysics* **2010**, *75*, 75A147–75A164. [CrossRef]
10. Biot, M.A. Theory of propagation of elastic waves in a fluid-saturated porous solid, I: Low frequency range. *J. Acoust. Soc. Am.* **1956**, *28*, 168–178. [CrossRef]
11. Biot, M.A. Theory of propagation of elastic waves in a fluid-saturated porous solid, II: Higher frequency range. *J. Acoust. Soc. Am.* **1956**, *28*, 179–191. [CrossRef]
12. Biot, M.A. Mechanics of deformation and acoustic propagation in porous media. *J. Appl. Phys.* **1962**, *33*, 1482–1498. [CrossRef]
13. Mavko, G.; Mukerji, T.; Dvorkin, J. *The Rock Physics Handbook: Tools for Seismic Analysis of Porous Media*, 2nd ed.; Cambridge University Press: Cambridge, UK, 2009.
14. Murphy, W.F. Effects of partial water saturation on attenuation in Massilon sandstone and Vycor porous glass. *J. Acoust. Soc. Am.* **1982**, *71*, 1458–1468. [CrossRef]
15. Carcione, J.M.; Picotti, S. P-wave seismic attenuation by slow-wave diffusion: Effects of inhomogeneous rock properties. *Geophysics* **2006**, *71*, 1–8. [CrossRef]
16. Deng, J.X.; Wang, S.X.; Du, W. A study of the influence of mesoscopic pore fluid flow on the propagation properties of compressional wave-A case of periodic layered porous media. *Chin. J. Geophys.* **2012**, *55*, 2716–2727. (In Chinese)
17. Wang, D.X. Study on the rock physics model of gas reservoirs in tight sandstone. *Chin. J. Geophys.* **2016**, *59*, 4603–4622. (In Chinese)
18. Ba, J.; Xu, W.H.; Fu, L.Y.; Carcione, J.M.; Zhang, L. Rock anelasticity due to patchy saturation and fabric heterogeneity, A double double-porosity model of wave propagation. *J. Geophys. Res. Solid Earth* **2017**, *122*, 1949–1976. [CrossRef]
19. Ba, J.; Ma, R.P.; Carcione, J.M.; Picotti, S. Ultrasonic wave attenuation dependence on saturation in tight oil siltstones. *J. Pet. Sci. Eng.* **2019**, *179*, 1114–1122. [CrossRef]
20. White, J.E. Computed seismic speeds and attenuation in rocks with partial gas saturation. *Geophysics* **1975**, *40*, 224–232. [CrossRef]
21. Dutta, N.C.; Odé, H. Attenuation and dispersion of compressional waves in fluid-filled porous rocks with partial gas saturation (White model)-Part I: Biot theory. *Geophysics* **1979**, *44*, 1777–1788. [CrossRef]
22. Johnson, D.L. Theory of frequency dependent acoustics in patchy-saturated porous media. *J. Acoust. Soc. Am.* **2001**, *110*, 682–694. [CrossRef]
23. Liu, J.; Ma, J.W.; Yang, H.Z. Research on P-wave's propagation in White's sphere model with patchy saturation. *Chin. J. Geophys.* **2010**, *53*, 954–962. (In Chinese)
24. Mavko, G.; Nur, A. Melt squirt in the asthenosphere. *J. Geophys. Res.* **1975**, *80*, 1444–1448. [CrossRef]
25. Carcione, J.M.; Gurevich, B. Differential form and numerical implementation of Biot's poroelasticity equations with squirt dissipation. *Geophysics* **2011**, *76*, N55–N64. [CrossRef]
26. Dvorkin, J.; Nur, A. Dynamic poroelasticity: A unified model with the squirt and the Biot mechanics. *Geophysics* **1993**, *58*, 524–533. [CrossRef]
27. Gassmann, F. Uber die Elasticität Poröser Medien (On the elasticity of porous media. *Vierteljahrsschr. Nat. Ges. Zürich* **1951**, *96*, 1–23.
28. Dvorkin, J.; Nolen-Hoeksema, R.; Nur, A. The squirt-flow mechanism: Macroscopic description. *Geophysics* **1994**, *59*, 428–438. [CrossRef]
29. Wood, A.B. *A Textbook of Sound*; Bell: London, UK, 1941.
30. Dvorkin, J.; Mavkon, G.; Nur, A. Squirt flow in fully saturated rocks. *Geophysics* **1995**, *60*, 97–107. [CrossRef]
31. Wu, C.F.; Ba, J.; Carcione, J.M.; Fu, L.Y.; Chesnokov, E.M.; Zhang, L. A squirt-flow theory to model wave anelasticity in rocks. *Phys. Earth Planet. Inter.* **2020**, *301*, 106450. [CrossRef]
32. Mavko, G.; Jizba, D. Estimating grain-scale fluid effects on velocity dispersion in rocks. *Geophysics* **1991**, *56*, 1940–1949. [CrossRef]
33. Gurevich, B.; Makarynska, D.; de Paula, O.B.; Pervukhina, M. A simple model for squirt-flow dispersion and attenuation in fluid-saturated granular rocks. *Geophysics* **2010**, *75*, N109–N120. [CrossRef]
34. Murphy, W.F.; Winkler, K.W.; Kleinberg, R.L. Acoustic relaxation in sedimentary rocks, dependence on grain contacts and fluid saturation. *Geophysics* **1986**, *51*, 757–766. [CrossRef]
35. Pride, S.R.; Berryman, J.G.; Harris, J.M. Seismic attenuation due to wave-induced flow. *J. Geophys. Res. Solid Earth* **2004**, *109*, B01201. [CrossRef]
36. Le, R.; Guéguen, M.Y.; Chelidze, T. Elastic wave velocities in partially saturated rocks: Saturation hysteresis. *J. Geophys. Res. Solid Earth* **1996**, *101*, 837–844.

37. Tang, X.M. A unified theory for elastic wave propagation through porous media containing cracks-An extension of Biot's poroelastic wave theory. *Sci. China Earth Sci.* **2011**, *41*, 784–795. [CrossRef]
38. Jin, Z.Y.; Chapman, M.; Papageorgiou, G. Frequency-dependent anisotropy in a partially saturated fractured rock. *Geophys. J. Int.* **2018**, *215*, 1985–1998. [CrossRef]
39. Zhang, L.; Ba, J.; Fu, L.; Carcione, J.M.; Cao, C. Estimation of pore microstructure by using the static and dynamic moduli. *Int. J. Rock Mech. Min. Sci.* **2019**, *113*, 24–30. [CrossRef]
40. Zhang, L.; Ba, J.; Carcione, J.M. Wave propagation in infinituple-porosity media. *J. Geophys. Res. Solid Earth* **2021**, *126*, e2020JB021266. [CrossRef]
41. Rubino, J.G.; Holliger, K. Research note: Seismic attenuation due to wave-induced fluid flow at microscopic and mesoscopic scales. *Geophys. Prospect.* **2013**, *51*, 369–379. [CrossRef]
42. Li, D.Q.; Wei, J.X.; Di, B.R.; Ding, P.B.; Huang, S.Q.; Shuai, D. Experimental study and theoretical interpretation of saturation effect on ultrasonic velocity in tight sandstones under different pressure conditions. *Geophys. J. Int.* **2018**, *212*, 2226–2237. [CrossRef]
43. Sun, W.T. On the theory of Biot-patchy-squirt mechanism for wave propagation in partially saturated double-porosity medium. *Phys. Fluids* **2021**, *33*, 076603. [CrossRef]
44. Carcione, J.M.; Helle, H.B.; Pham, N.H. White's model for wave propagation in partially saturated rocks: Comparison with poroelastic numerical experiments. *Geophysics* **2003**, *68*, 1389–1398. [CrossRef]
45. Toksöz, M.N.; Johnston, D.H. *Seismic Wave Attenuation*; Geophysics Reprint Series; Society of Exploration Geophysicists: Tulsa, OK, USA, 1981.
46. Guo, M.Q.; Fu, L.Y.; Ba, J. Comparison of stress-associated coda attenuation and intrinsic attenuation from ultrasonic measurements. *Geophys. J. Int.* **2009**, *178*, 447–456. [CrossRef]
47. Yan, X.F.; Yao, F.C.; Cao, H.; Ba, J.; Hu, L.L.; Yang, Z.F. Analyzing the mid-low porosity sandstone dry frame in central Sichuan based on effective medium theory. *Appl. Geophys.* **2011**, *8*, 163–170. [CrossRef]
48. Ma, R.P.; Ba, J. Coda and intrinsic attenuations from ultrasonic measurements in tight siltstones. *J. Geophys. Res. Solid Earth* **2020**, *125*, e2019JB018825. [CrossRef]
49. Deng, J.X.; Zhou, H.; Wang, H.; Zhao, J.G. The influence of pore structure in reservoir sandstone on dispersion properties of elastic waves. *Chin. J. Geophys.* **2015**, *58*, 3389–3400. (In Chinese)
50. Song, L.T.; Wang, Y.; Liu, Z.H.; Wang, Q. Elastic anisotropy characteristics of tight sands under different confining pressures and fluid saturation states. *Chin. J. Geophys.* **2015**, *58*, 3401–3411. (In Chinese)
51. Chen, Y.; Huang, T.F.; Liu, E.R. *Rock Physics*; China University of Science and Technology Press: Hefei, China, 2009. (In Chinese)
52. David, E.C.; Zimmerman, R.W. Pore structure model for elastic wave velocities in fluid-saturated sandstones. *J. Geophys. Res. Solid Earth* **2012**, *117*, B07210. [CrossRef]
53. Ba, J.; Zhang, L.; Wang, D.; Yuan, Z.; Cheng, W.; Ma, R.; Wu, C. Experimental analysis on P-wave attenuation in carbonate rocks and reservoir identification. *J. Seism. Explor.* **2018**, *27*, 371–402.
54. Lucet, N.; Zinszner, B. Effects of heterogeneities and anisotropy on sonic and ultrasonic attenuation in rocks. *Geophysics* **1992**, *57*, 1018–1026. [CrossRef]
55. Pang, M.Q.; Ba, J.; Ma, R.P.; Chen, T.S. Analysis of attenuation rock-physics template of tight sandstones: Reservoir microcrack prediction. *Chin. J. Geophys.* **2020**, *63*, 281–295. (In Chinese)
56. Amalokwu, K.; Papageorgiou, G.; Chapman, M.; Best, A.I. Modelling ultrasonic laboratory measurements of the saturation dependence of elastic modulus: New insights and implications for wave propagation mechanisms. *Int. J. Greenh. Gas Control* **2017**, *59*, 148–159. [CrossRef]
57. Wei, Y.J.; Ba, J.; Ma, R.P.; Zhang, L.; Carcione, J.M.; Guo, M.Q. Effect of effective pressure change on pore structure and elastic wave responses in tight sandstones. *Chin. J. Geophys.* **2020**, *63*, 2810–2822. (In Chinese)
58. Sun, Y.Y.; Carcione, J.M.; Gurevich, B. Squirt-flow seismic dispersion models: A comparison. *Geophys. J. Int.* **2020**, *222*, 2068–2082. [CrossRef]
59. Batzle, M.L.; Wang, Z.J. Seismic properties of pore fluids. *Geophysics* **1992**, *57*, 1396–1408. [CrossRef]
60. Carcione, J.M. Wave Fields in Real Media. In *Theory and Numerical Simulation of Wave Propagation in Anisotropic, Anelastic, Porous and Electromagnetic Media*, 3rd ed.; Elsevier: Amsterdam, The Netherlands, 2014.
61. Helle, H.B.; Pham, N.H.; Carcione, J.M. Velocity and attenuation in partially saturated rocks: Poroelastic numerical experiments. *Geophys. Prospect.* **2003**, *55*, 551–566. [CrossRef]
62. Shapiro, S.A. Elastic piezosensitivity of porous and crackd rocks. *Geophysics* **2003**, *68*, 482–486. [CrossRef]

Article

Prediction of Conformance Control Performance for Cyclic-Steam-Stimulated Horizontal Well Using the XGBoost: A Case Study in the Chunfeng Heavy Oil Reservoir

Zehao Xie [1,2], Qihong Feng [1,2], Jiyuan Zhang [1,2,*], Xiaoxuan Shao [3,4], Xianmin Zhang [1,2] and Zenglin Wang [5]

[1] Key Laboratory of Unconventional Oil & Gas Development, China University of Petroleum (East China), Ministry of Education, Qingdao 266580, China; xiezehao19@163.com (Z.X.); fengqihong.upc@gmail.com (Q.F.); spemin@126.com (X.Z.)

[2] School of Petroleum Engineering, China University of Petroleum (East China), Qingdao 266580, China

[3] Key Laboratory of Deep Oil and Gas, China University of Petroleum (East China), Qingdao 266555, China; 13780752708@163.com

[4] School of Geosciences, China University of Petroleum (East China), Qingdao 266555, China

[5] Shengli Oilfield Company, Sinopec, Dongying 257001, China; wangzenglin.slyt@sinopec.com

* Correspondence: zhjy221@126.com

Citation: Xie, Z.; Feng, Q.; Zhang, J.; Shao, X.; Zhang, X.; Wang, Z. Prediction of Conformance Control Performance for Cyclic-Steam-Stimulated Horizontal Well Using the XGBoost: A Case Study in the Chunfeng Heavy Oil Reservoir. *Energies* **2021**, *14*, 8161. https://doi.org/10.3390/en14238161

Academic Editor: Reza Rezaee

Received: 14 October 2021
Accepted: 12 November 2021
Published: 5 December 2021

Abstract: Conformance control is an effective method to enhance heavy oil recovery for cyclic-steam-stimulated horizontal wells. The numerical simulation technique is frequently used prior to field applications to evaluate the incremental oil production with conformance control in order to ensure cost-efficiency. However, conventional numerical simulations require the use of specific thermal numerical simulators that are usually expensive and computationally inefficient. This paper proposed the use of the extreme gradient boosting (XGBoost) trees to estimate the incremental oil production of conformance control with N_2-foam and gel for cyclic-steam-stimulated horizontal wells. A database consisting of 1000 data points was constructed using numerical simulations based on the geological and fluid properties of the heavy oil reservoir in the Chunfeng Oilfield, which was then used for training and validating the XGBoost model. Results show that the XGBoost model is capable of estimating the incremental oil production with relatively high accuracy. The mean absolute errors (MAEs), mean relative errors (MRE) and correlation coefficients are 12.37/80.89 t, 0.09%/0.059% and 0.99/0.98 for the training/validation sets, respectively. The validity of the prediction model was further confirmed by comparison with numerical simulations for six real production wells in the Chunfeng Oilfield. The permutation indices (PI) based on the XGBoost model indicate that net to gross ratio (NTG) and the cumulative injection of the plugging agent exerts the most significant effects on the enhanced oil production. The proposed method can be easily transferred to other heavy oil reservoirs, provided efficient training data are available.

Keywords: heavy oil reservoirs; cyclic steam stimulation; conformance control; numerical simulation; extreme gradient boost (XGBoost) trees; prediction model

1. Introduction

Heavy oil and bitumen resources are estimated to be 158.43 billion ton, which account for more than 2/3 of the worldwide oil reserves (236.73 billion ton), according to OGL's oil reserves summary [1]. The efficient development of heavy oil reserves is considered as a significant means to add to world energy supply [2].

To date, thermal recovery is the primary method to improve the production of heavy oils [3], among which cyclic steam stimulation (CSS) has proven a cost-efficient technique widely applied in field practices [4,5]. The CSS technology was first applied on vertical wells [6,7] for heavy oil reservoirs with thick layers; it is, however, usually uneconomic to develop thin-layer heavy oil reservoirs, due to severe heat losses [8,9]. For thin-layer reservoirs, the use of horizontal wells has proven to be more cost-effective than vertical wells [10], thus the CSS integrated with horizontal wells has been widely used worldwide [8,9]. A significant issue with the CSS is that steam channeling exacerbated greatly after multiple cycles of steam injections, due to reservoir heterogeneities [11,12]. The channeled steams along high permeability areas lead to reducing the sweep efficient, and hence the oil recovery factors are approximately 10~20% [13,14].

The conformance control is an efficient technology to increase steam sweep efficiency and oil recovery factors in heavy oil reservoirs [15,16]. To date, the high-temperature-resistant gel [17,18] and foam [19,20] system are the primary agents that have been used to realize conformance control [21,22]. The gel system is typically formulated using polymer and cross-linker [23,24]. The primary plugging mechanism of gel is that the injected gel flows through the high-permeability channels and remains therein, which plugs steam channels effectively [25]. When gel is injected by injection wells, it can change the direction of flow to a lower permeability zone and block the offended areas [26]. As a result, the steam sweep efficiency is improved during steam injection operations and then excess coproduction of injection fluids (i.e., steam and water) is reduced [25]. However, the conventional cross-linker decreased the performance when the reservoir temperature rises over 120 °C [27]. A number of gel systems that can be used in high temperature conditions were invented [28,29] and have been implemented successfully on many heavy oil reservoirs [30,31].

Foam system usually includes foam and other gas, such as nitrogen (N_2) and carbon dioxide (CO_2), in addition to hydrocarbon gas [32,33]. Compared with CO_2 and hydrocarbon gas, N_2 is capable of better stability and flooding in high temperature reservoirs [34]. Foams injected into formation can increase the steam viscosity to stabilize the displacement process and reduce the capillary by the presence of surfactant [35,36]. Foams can also restrain steam overlying to the top of reservoirs and prevent steam channeling in the high permeability regions [37]. Meanwhile, the foams help reduce the heat loss of steam injection and steam migration, due to low thermal conductivity [38]. Compared with the gel systems, the N_2-foam system exhibits better temperature-resistance capability and is beneficial to reducing underground heat loss [39,40]. However, the presence of oil has a significant effect on the stability of foam [32,33]. Many experiments confirmed that they could overcome this problem by optimizing the foaming agent system [41,42]. The injected foam is capable of blocking the water flow pathways without affecting the oil, which therefore is advantageous to reducing produced water volume.

It is important to evaluate the improved oil production of conformance control prior to field implementation, in order to ensure the cost-efficiency. To date, preliminary evaluations are commonly undertaken with numerical simulations, which require specific numerical simulators that are usually expensive. Besides, numerical simulations are usually quite time-consuming. Thus, it should be of practical significance to construct an accurate and robust model for the fast prediction of the improved oil production of conformance control after CSS. This paper proposes the use of the extreme gradient boost (XGBoost) [43,44] to estimate the EOR of CSS, with a focus on the heavy oil reservoirs of the Chunfeng Oilfield. The validity of the prediction model was tested using both synthetic and real field production data. Sensitivity of the influencing factors was quantified using the permutation information (PI) method.

2. Methods

2.1. Database

Valid and extensive data are mandatory for the construction of a reliable prediction model based on supervised learning methods [45–47]. In this study, the datasets were constructed by numerically simulating the CSS and the subsequent conformance control process, based on the geological feature and fluid property of the Southern P601 block of the Chunfeng Oilfield [48].

2.1.1. Construction of the Geological Models

The reservoir models used for constructing the datasets were extracted from a pre-built base geological model for the Southern P601 block in the Chunfeng Oilfield, which is an extreme high-viscosity reservoir that is characterized with low thickness, high permeability and high oil viscosity (Table 1). The base model for the Southern P601 block was initially constructed using the sequential Gaussian–Bayesian simulation [49], based on the logging data of 98 wells (Appendix A). A number of 200 scenarios of the sub-model were extracted from the base model (Figure 1). Each extracted model consists of 40, 21 and 10 grid blocks in the x-, y- and z-directions, respectively. The dimension of each grid block is 10 m, 10 m and 0.5 m in the x-, y- and z-directions, respectively. The uncertainty of top depth, net to gross ratio, permeability and porosity were considered in real reservoirs (Figure 2).

Table 1. The properties of the Southern P601 block in the Chunfeng Oilfield.

Property	Minimum Value	Max Value
Depth (m)	500	700
Thickness (m)	3	6
Porosity	0.18	0.41
Permeability (mD)	340	14,200
Formation temperature (°C)	28	36
Formation pressure (MPa)	5.7	6.1
Viscosity at 34 °C (mPa·s)	55,000	57,211

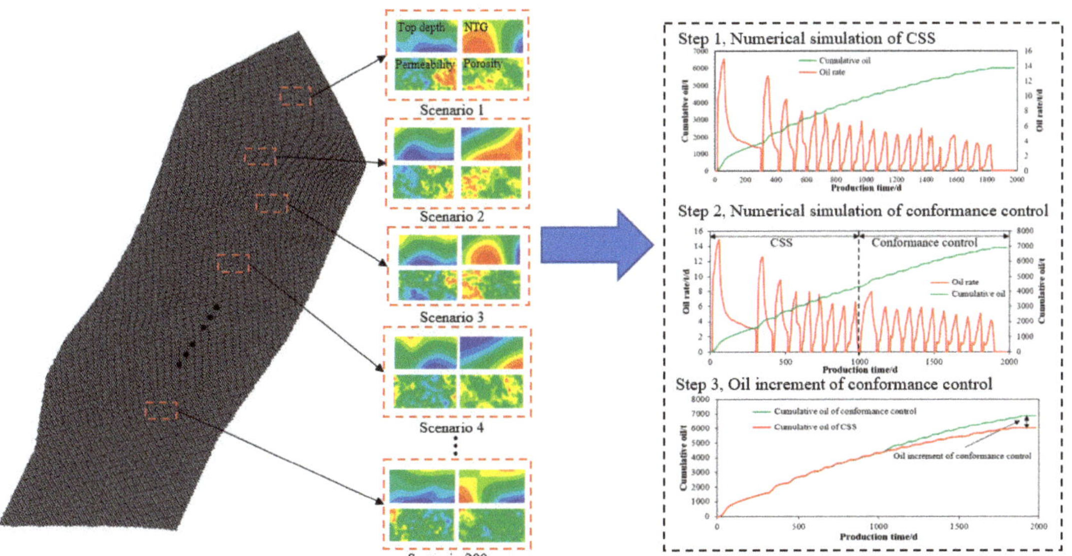

Figure 1. Workflow for predicting the performance of conformance control by numerical simulation, including extracted model, and calculating the oil increment of conformance control.

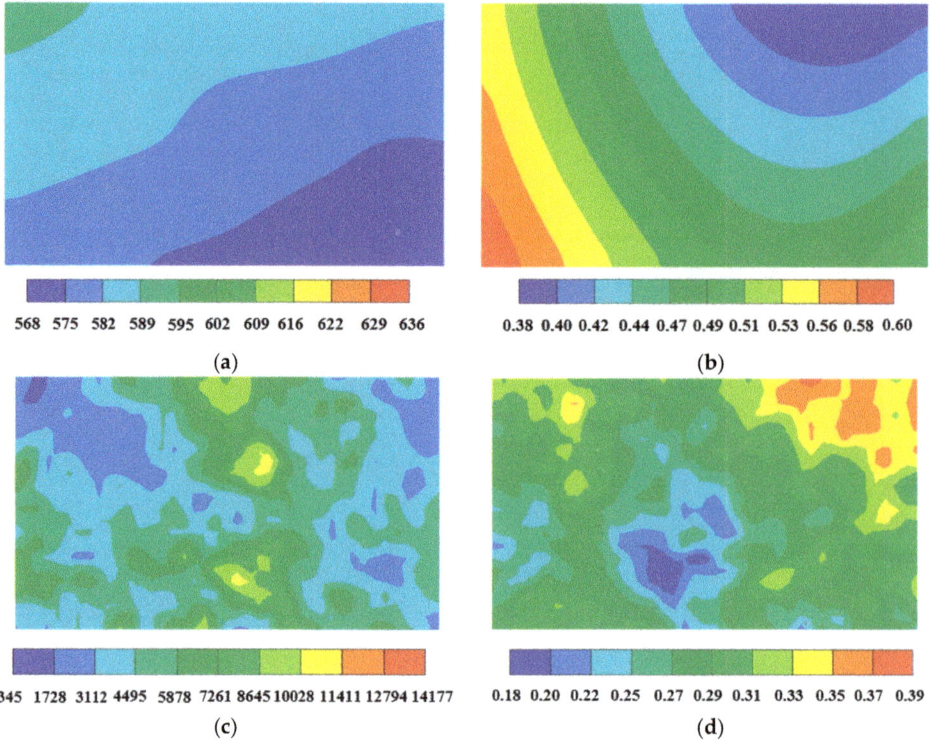

Figure 2. Distribution of top depth, NTG, permeability and porosity of the reservoir model. (**a**) Top depth distribution (579 m in average), (**b**) net to gross ratio distribution (0.47 in average), (**c**) permeability distribution (5405 mD in average), (**d**) porosity distribution (0.31 in average).

The heterogeneity has a great influence on production performance in heavy oil reservoirs. Therefore, it is very necessary to consider the heterogeneity in numerical simulation models of cyclic steam stimulations. To reflect the heterogeneity of heavy oil reservoirs, we calculate the average of top depth, NTG, permeability and porosity in numerical simulation models. Variation coefficient of permeability (v_k see Equation (1)) and permeability ratio (α_k see Equation (2)) are also important parameters for the heterogeneity of reservoirs. Therefore, we incorporate the average of NTG, permeability and porosity, variation coefficient of permeability and permeability ratio into the geological features and take them into account in XGBoost model training.

$$v_k = \frac{1}{\bar{\bar{k}}} \sqrt{\frac{1}{n} \sum_{j=1}^{n} \left(\overline{k_j} - \bar{k} \right)^2} \tag{1}$$

$$\alpha_k = \frac{k_{max}}{k_{min}} \tag{2}$$

where v_k is the variation coefficient of permeability; $\overline{k_j}$ is the average permeability of j layer; $\bar{k}$ is the average permeability of reservoir; k_{max} is the maximum permeability of reservoir; k_{min} is the minimum permeability of reservoir.

2.1.2. Numerical Simulation Setup

Numerical simulations were conducted using the CMG's STARS simulator in this study. The STARS is capable of simulating water/gas/oil flow, heat transfer, plugging agent chemical reactions and the associated plugging effects, which has been used widely for predicting the performance of thermal recovery in heavy oil reservoirs [50,51].

The injection of high-temperature steam into heavy oil reservoirs tends to reduce the oil viscosity and enlarge the two-phase flow span of relative permeability curves [52]. The dependence of oil viscosity on temperature is shown in Figure 3. For relative permeability, we defined saturation endpoints at different temperatures (Figure 4), and then relative permeability can be determined by endpoint scaling due to the similarity [53]. The heat loss through the roof and ceil layers was considered using the semi-analytical infinite-overburden heat loss model that was proposed by Vinsome, P.K.W. [54]. Key thermal parameters of the reservoir and formation fluids are given in Table 2.

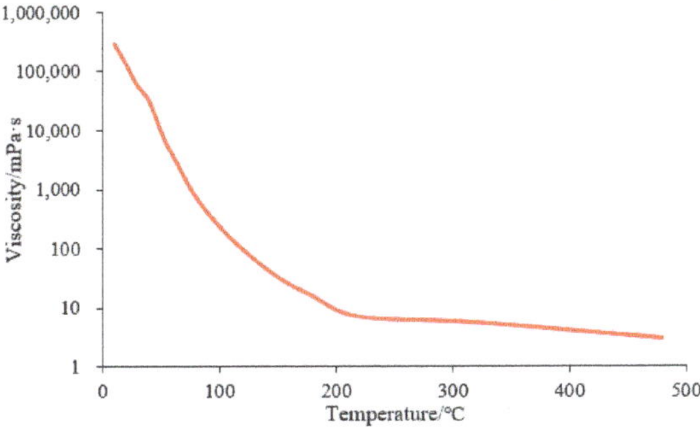

Figure 3. Oil viscosity versus temperature in the simulation model.

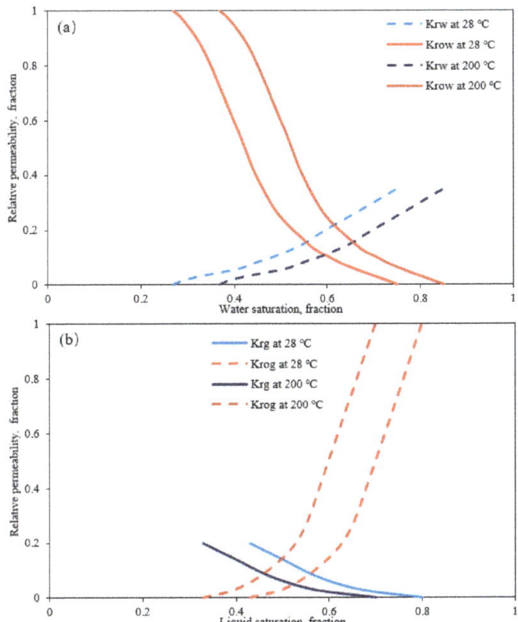

Figure 4. Relative permeability curves used in the simulation for (**a**) water–oil and (**b**) liquid–gas.

Table 2. The key thermal reservoir parameters.

Parameter	Value
Reservoir rock thermal conductivity (J/m·day·C)	1.634×10^5
Water phase thermal conductivity (J/m·day·C)	5.99×10^4
Oil phase thermal conductivity (J/m·day·C)	9.77×10^3
Gas phase thermal conductivity (J/m·day·C)	1.9×10^3
Volumetric heat capacity of over-/under-burden rock (J/m^3·C)	2.575×10^6
Thermal conductivity of over-/under-burden rock (J/m·day·C)	1.055×10^5
Thermal expansion coefficient	1×10^{-6}

In this study, we considered two popular plugging agents that have been widely used for conformance control, namely the gel and nitrogen foam. When cross-linker and polymer were injected underground, they formed a mixture in the high permeability region that is called in situ gel [55]. The blocking mechanisms of the gel are based on the adsorption of injection chemicals in the porous media and the residual resistance factor (RRF) that reduces the effective permeability [56,57]. We set that the value of RRF was 40 in our model. Meanwhile, a chemical reaction was set up to complete the underground gelation process. Firstly, three components of the gel are designed, including the xlinker, the xanthan and the gel generated by the reaction. The chemical reaction rate is set at 16, and different gel injections are simulated by controlling the injection amount of xlinker and xanthan. The concentration of xlinker and xanthan is 0.002% and 0.1%.

Two methods, namely the mechanism method and the empirical approach, are implemented in STARS to simulate nitrogen foam conformance control. In this study, the empirical approach was used, considering foam plugging water without plugging oil, through an interpolating relative permeability curve that decreased the fluidity of foam, which needed fewer parameters and conveniently used the field scale [53,58]. The relative permeability is interpolated based on a dimensionless "interpolation factor (FM)" that is shown on Equation (3) [53].

$$FM = \left\{ 1 + MRF \left(\frac{\omega s}{\omega s^{max}} \right)^{es} \left(\frac{S_0^{max} - S_o}{S_0^{max}} \right)^{eo} \left(\frac{N_c^{ref}}{N_c} \right)^{ev} \right\}^{-1} \tag{3}$$

where FM varying between 1 (no foam) and $(MRF)^{-1}$ (strongest foam) where MRF was the maximum mobility reduction factor obtained via maximum surfactant concentration (ωs^{max}) or capillary number (N_c^{ref}), and was valued at 100, 5×10^{-5} and 2×10^{-4}, respectively, in our model. The es, eo and ev were exponents, and were chosen as simply 1, 1 and 0.3, respectively. S_0^{max} was the maximum oil saturation above which no foam will form and the value of S_0^{max} was set at 0.6 in our model [59].

Three components (i.e., water, foaming agent and nitrogen) were used to generate a nitrogen foam system. We controlled the injection rate of foaming agent and nitrogen to simulate different injection rates of nitrogen foam. In the numerical simulation model, relative parameters with the nitrogen foam system were set as follows: an injection rate of nitrogen at 10,000 m^3/d, foaming agent concentration of 0.6%, injection rate of the foaming agent between 0.2 and 0.6 PV, injection temperature of 34 °C and injection mode of continuous injection.

To perform CSS in the numerical simulation model, we set up an injection well at the same location as the production well to simulate the injection of steam, and cycling group events were used to control the cycles switch. Cycling group events included three cycle parts (i.e., steam injection, soaking and production that were altered by setting the injection rate, soaking time and production rate). When the oil rate reached 3 m^3/d or the production time lasted 180 d, a new cycle was started. Different cycle parts parameters were set in different geological models, as operation features trained in XGBoost models. Steam quality and steam injection temperatures also have an effect on well production performance [60].

Hence, we also selected both steam quality and steam injection temperatures as operation features. With the production continuing, oil rate did not meet our requirements, and we took conformance control for production wells using different plugging agents in different development stages (i.e., water cut, oil rate and oil recovery). After conformance control in a cycle, the production well also performed ten cyclic steam stimulations in the same operation parameters. Another cyclic steam stimulation model that had the same production cycles was compared with the cumulative oil result of the model of conformance control to calculate the oil increment of measures (Figure 1). The parameters of development stages were also selected as training features to train the XGBoost model.

In conclusion, three types of feature parameters (i.e., geological parameters, operation and conformance control timing) will be input data. Considering a reasonable range of parameter variations, we generated 200 simulation cases for every plugging agent to randomly make up the input database. Ranges of the geological parameters, operation and conformance control timing are summarized in Table 3. According to the design above, we constructed 800 numerical simulation models to obtain oil increments of conformance control measure. Therefore, the datasets need to be collated after carrying out the numerical simulations. Consequently, a total of 400 completed samples, including a different plugging agent, composed of oil increment and corresponding parameters, constitute the production dynamic database required for XGBoost training.

Table 3. Summary of the database used for constructing the estimation model.

Parameter Types	Parameter Name	Minimum Value	Maximum Value
	Top depth (m)	509.91	592.81
	Porosity	0.29	0.36
Geologic parameters	Net to gross	0.36	0.80
	Permeability (mD)	4217.06	6747.66
	Variation coefficient of permeability	0.12	0.36
	Steam quality	0	1
Operation parameters for	Steam injection temperature (°C)	270	350
cyclic steam stimulation	Soak time (d)	2	7
	Cumulative injection of plugging agent (PV)	0.2	0.8
	Production rate (t/d)	25	40
Conformance control timing	Water cut (%)	60	95
	Oil rate (t/d)	3.49	18.83
	Oil recovery (%)	2.38	7.58

2.2. Principal of XGBoost Trees

The XGBoost is a supervised learning algorithm proposed by Chen [43] under the gradient boosting framework. The XGBoost integrates multiple classification and regression tree (CART) models to form a classifier with strong generalization abilities. Each CART consists of a root node, a set of internal nodes and a set of leaf nodes (Figure 5a). Given a dataset $\mathcal{D} = \{(\mathbf{X}_i, y_i)\}, (\mathbf{X}_i \in \mathbb{R}^m, y_i \in \mathbb{R}, i = 1, 2, \dots, n)$ that consists of n samples with m feature variables, the XGBoost output is computed as the sum of the predicted values of a number of K CARTs (Figure 5b), with the mathematical model expressed as

$$\hat{y}_i = \sum_{k=1}^{K} f_k(X_i), f_k \in \mathcal{F} \tag{4}$$

where f_k is the Kth independent tree; $\hat{y}_i$ is the output computed using XGBoost tree. The space of a CART tree ($\mathcal{F}$) is represented with

$$\mathcal{F} = \left\{ f(\mathbf{X}) = \omega_{q(\mathbf{X})} \right\}, \left(q : \mathbb{R}^m \to T, \omega \in \mathbb{R}^T \right) \tag{5}$$

where q is a decision rule that maps an example to a binary leaf index; $\omega_{q(\mathbf{x})}$ is the fractions of leaves that form a set; T is number of leaf nodes; ω is the weight of leaf.

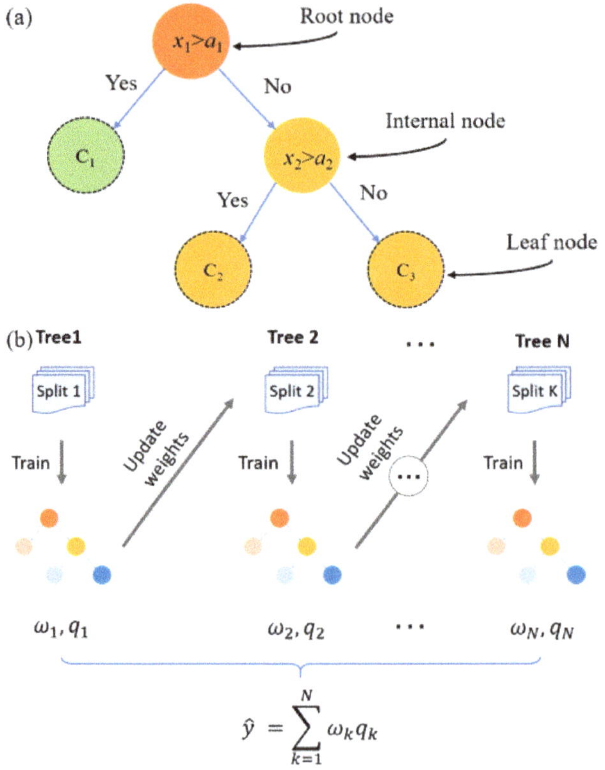

$$\hat{y} = \sum_{k=1}^{N} \omega_k q_k$$

Figure 5. Illustrations of the (**a**) CART model and (**b**) boosting ensemble trees [46,47].

In order to establish the prediction model f(x), the following objective function $\mathcal{L}(\phi)$ is to be minimized

$$\mathcal{L}(\phi) = \sum_{i}^{n} l(\hat{y}_i, y_i) + \sum_{k}^{K} \Omega(f_k) \tag{6}$$

$$\Omega(f) = \gamma T + \frac{1}{2}\lambda \| \omega \|^2 \tag{7}$$

where l is a differentiable convex loss function; Ω is regularized terms that limit the complexity of the model; γ is the coefficient of loss function; λ is the regularized term coefficient; ω is the weight of the leaf.

The first term on the right side of Equation (6) is the loss function term that is a differentiable convex function. For regression problems, the mean square error is common. By adding the loss function, we can obviously reduce the mean square error. The second term is the regularization term, which stands for the sum of the complexity of each CART. In the process of minimizing the objective, XGBoost applies a series of techniques to control the complexity of the model and prevent overfitting, e.g., regularization, optimize hyperparameters and set early stopping rounds [61–63]. For more details on the mathematical formulations of the XGBoost model, readers are referred to [43,46,47].

2.3. Construction of the Prediction Model

In this paper, the open source XGBoost package in Python [43] was implemented to construct the prediction model for the prediction of potential conformance control after multi-cycle steam stimulation on three types of input features, namely geological parameters, operation and conformance control timing. There are thirteen parameters in the database and the whole database was randomly divided into two parts, namely the training (80%) and testing (20%) sets. The 320 samples consisting the training set were used to train the XGBoost model and to determine the optimal hyperparameter values for the XGBoost trees; the remaining 180 samples constructing the testing set were used to examine the stability and robustness of the prediction model.

There is a type of parameter called hyperparameters in machine learning, which must be set manually before the process of learning. Empirically, the optimal hyperparameters can significantly improve the performance and effect of the XGBoost model. Learning rate (LR) can improve the generalization ability of the XGBoost model by reducing the feature weight. Min child weight (MCW) determines the sum of weight in a minimum child. A large MCW value makes the boost model avoid learning part of the special samples, while an exorbitant MCW value will lead to underfitting. Maximum tree depth (MTD) is connected with the complexity of the ensemble model, and increasing the MTD value can find more specific and more local samples [64–67]. The number of trees (n) is another important hyperparameter; eliminating potential overfitting requires one to add a larger n value and smaller LR value to the boosting model [68]. To get the optimal compound mode of these four key hyperparameters, we adopt the K-fold cross-validation integrated with the exhaustive grid search approach for the optimization [69–72].

In our XGBoost model, we first specify the range of hyperparameters that search the space with manual tuning. Five grid values in each of hyperparameters will be adjusted, and $5 \times 5 \times 5 \times 5 = 625$ searching scenarios were produced (Table 4). There were remaining hyperparameters that may exert minor effect on the performance, for which we adopted default values [53,68]. For each searching scenario, the K-fold (K = 5) cross-validation approach was applied to calculate the coefficient of determination (R^2) for each fold. The maximum averaged R^2 on the 5-fold subsets was the optimal compound mode of hyperparameters that will be set in our XGBoost model.

Table 4. Ranges of key hyperparameter values for the XGBoost used in cross-validation.

Hyperparameter	Description	Range
LR	Learning rate used for preventing over-fitting	0.01, 0.05, 0.10, 0.15, 0.2
n	The number of independent trees that form the ensemble	1000, 1500, 2000, 2500, 3000
MTD	The maximum number of edges from the leaf to the root node	2, 3, 4, 5, 6
MCW	The minimum sum of weights of all observations required in a child	1, 3, 5, 7, 9
γ	The minimum loss reduction term required to further split a leaf	0
Subsample	The ratio of observations for sampling to construct each tree	1
Column sample by tree	The ratio of columns for sampling to construct each tree	1
α	L1 regularization term on weights that is similar to lasso regression	0
λ	L2 regularization term on weights that is similar to ridge regression	1

2.4. Evaluation of the Prediction Model

To quantitatively evaluate the performance of the prediction model, three statistical matrices were used, namely the mean absolute error (MAE), mean relative error (MRE) and coefficient of determination (R^2), which are written as

$$\text{MAE} = \frac{1}{N}\sum_{i=1}^{N}|y_i - \hat{y}_i| \tag{8}$$

$$\text{MRE} = \frac{1}{N}\sum_{i=1}^{N}\left|\frac{y_i - \hat{y}_i}{y_i}\right| \tag{9}$$

$$R^2 = \sum_{i=1}^{N}(y_i - \hat{y}_i)^2 \Big/ \sum_{i=1}^{N}(y_i - \overline{y})^2 \tag{10}$$

where y_i and $\hat{y}_i$ are the true and prediction models of the oil increment of conformance control, respectively, $\bar{y}$ is the mean value of the calculated oil increment of conformance control, and n is the number of data sample points.

3. Results and Discussion

3.1. Training and Validation of the Prediction Model

The exhaustive grid search approach integrated with the five-fold cross-validation shows that the XGBoost predictions are obviously affected by the hyperparameter setting. Table 5 gives the top five scenarios with the highest averaged R^2. As can be observed, both the MTD (3) and the LR (<0.1) are relatively small among these top five scenarios, which is consistent with the previous studies [46,47,73] that recommended small MTD and LR values in order to attain the strong generalization capability. The MCM is consistent (7) among these five scenarios, which is larger than the default value given in [68]. A higher MCW value is beneficial to the improvement of generalization capability for the XGBoost model [74,75]. The number of trees (n) exhibits significant variations among these five scenarios, which indicates that this hyperparameter exhibits a minor effect on the prediction accuracy for the specific problem in this paper.

Table 5. Top five R^2 for MTD, LR, MCW and n, determined with cross-validation.

No.	MTD	LR	MCW	n	Average R^2
1	3	0.01	7	3000	0.971
2	3	0.01	7	2500	0.970
3	3	0.01	7	2000	0.970
4	3	0.01	7	1500	0.969
5	3	0.05	7	1000	0.968

Figure 6 depicts the prediction results of different plugging agents for training sets and testing sets using the constructed XGBoost prediction model (with the hyperparameter values producing the highest R^2). It is shown that the sample points in the training sets for both N_2-foam and gel are approximately located on the 45-degree line (Figure 6), representing relatively high training accuracies. The majority of the data points in the testing sets are grouped around the 45-degree line, although several outliers deviate obviously from the 45-degree line. Generally, the distribution of the data points in the testing sets exhibits a more scattered pattern than the training sets, which may be attributed to the uncertainties with the XGBoost modelling process [68]. Recall that the datasets were generated using numerical simulations based on stochastic geological models and that the information on the spatial heterogeneity was not included in the model input. As the spatial heterogeneity exerts non-negligible effect on the thermal recovery of heavy oil [76,77], the exclusion of information on spatial distribution of formation properties such as NTG, porosity and permeability inevitably result in predication inaccuracies. To possibly eliminate uncertainties with modeling process, we calculated average values of formation properties as input parameters for XGBoost model. However, the same formation parameters and input parameters inevitably include different spatial distributions. Nonetheless, the evaluation matrices, as shown in Table 6, demonstrate overall acceptable error ranges for the validation sets, indicating the constructed models have relatively strong robustness and generalization capability in predicting the unseen data. Besides, this paper is targeted at developing a prediction model for the preliminary screening of the conformance control performance, in order to quickly determine the most suitable well(s) for possible field applications of conformance control; thus, the modeling accuracies are generally acceptable from the perspective of engineering applications.

3.2. Verification of the Model with Real CSS Horizontal Wells

In this section, the constructed prediction model was further verified with real CSS horizontal wells in the P601 heavy oil reservoir of the Chunfeng Oilfield. Field production practices suggest that the CSS horizontal wells in the target area can be generally grouped into three categories according to their production characteristics. The first type includes wells that exhibit relatively initial high oil production rates (>20 t/d) and subsequent sharply decreasing trends after approximately five to seven steam stimulation cycles (Figure 7a). The second type of wells are characterized with a gradual climbing trend of oil rates in the initial 5–7 steam stimulation cycles and then a decreasing trend after 7–10 steam stimulation cycles (Figure 7b). The peak oil rates are generally less than 20 t/d for these type of wells. The third type of wells demonstrate relatively low oil rates (<10 t/d) throughout the production life-span (Figure 7c), which were generally shut-in after only 5–10 steam stimulation cycles, due to uneconomic production rates (<2 t/d).

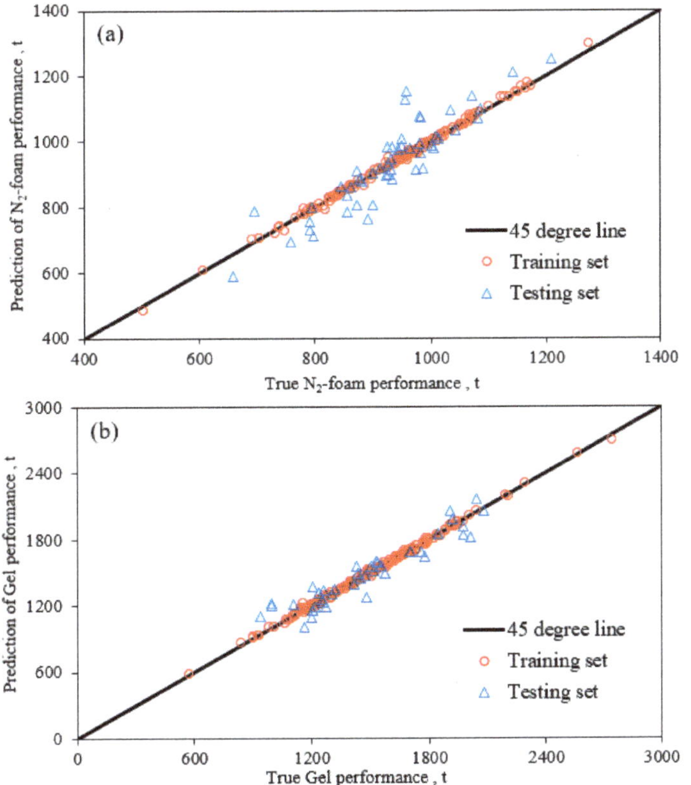

Figure 6. Cross plots of the true and prediction profile control oil increment using the XGBoost for the (**a**) N2-form and (**b**) gel.

Table 6. Summary of the evaluation matrices for N2-foam and gel.

	Matrices	N_2-Foam	Gel
	MAE, t	5.25	12.37
Training set	MRE, %	0.57	0.09
	R^2, fraction	0.995	0.999
	MAE, t	45.93	80.89
Testing set	MRE, %	5.01	0.059
	R^2, fraction	0.901	0.944
	MAE, t	21.41	26.07
Whole set	MRE, %	2.6	0.019
	R^2, fraction	0.967	0.988

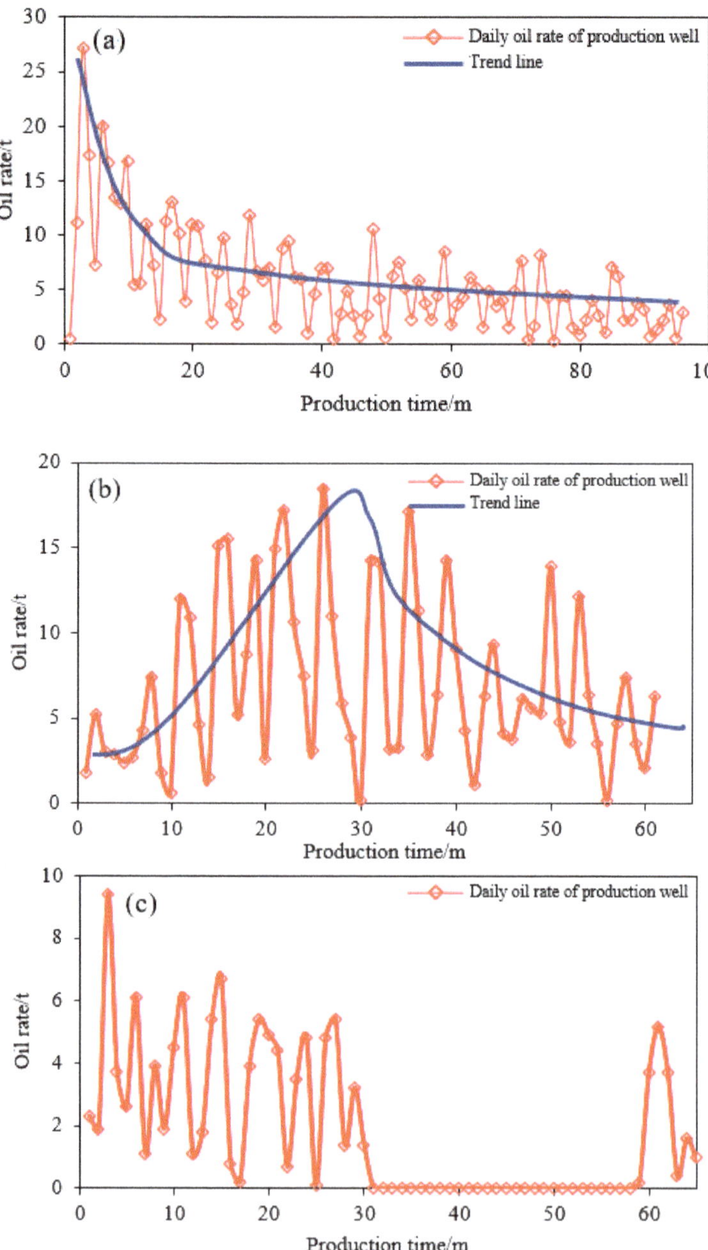

Figure 7. The production performance of three categories of wells. (**a**) The wells of relatively initial high oil production rates and subsequent sharply decreasing trends. (**b**) The wells of a gradual climbing trend of oil rates in the initial cycles and then a decreasing trend. (**c**) The wells of relatively low oil rates throughout the production life span.

A number of six horizontal wells with production characteristics that can be categorized into one of the above three types were picked out from the target block. History matching and subsequent conformance control simulations were conducted for these wells (Figure 8). For each well, the properties for the N_2-foam and gel were assigned with the same identical values as previously set. Operational parameters associated with N_2-foam were set with an injection rate and total injection volume of 10,000 m^3/d and 0.2 PV, respectively. Operational parameters associated with gel generation were the injection of polymer and xlinker, which were 0.2 PV and 0.02 PV, respectively.

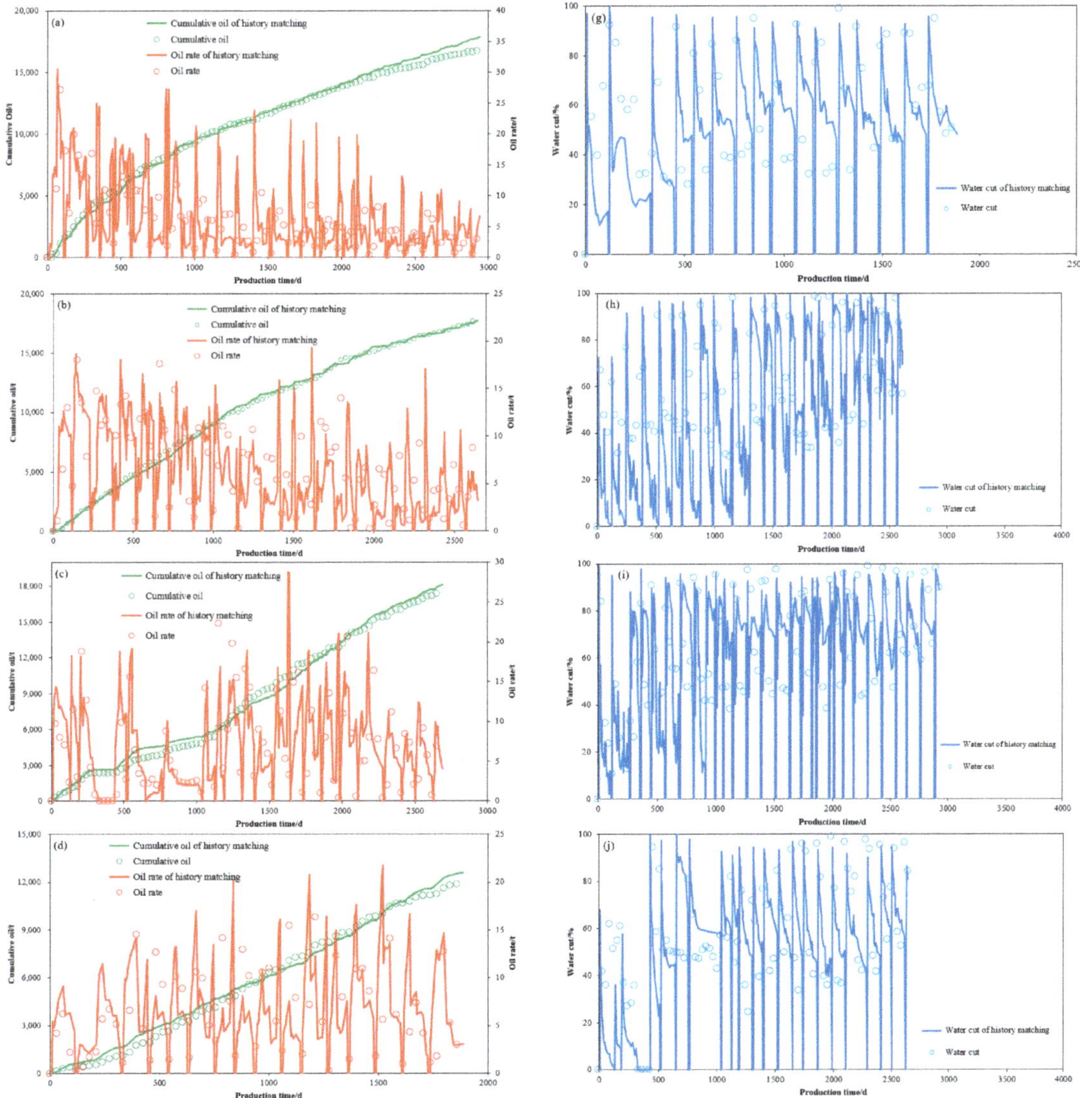

Figure 8. *Cont.*

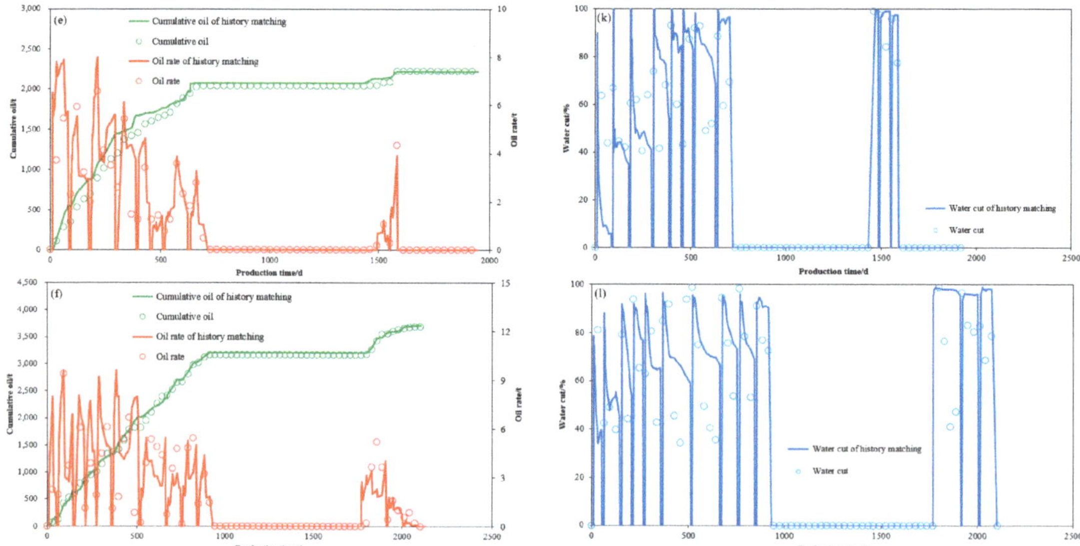

Figure 8. History matching of cumulative oil, oil rate, and water cut for three types of production wells. (**a**–**f**) History matching of cumulative oil and oil rates for P226, P235, P188, P185, X296 and X265, respectively. (**g**–**l**) History matching of water cut for P226, P235, P188, P185, X296 and X265, respectively. P226 well and P235 well belong to the first type, P188 well and P185 well belong to the second type and X296 and X265 belong to the third type.

Key reservoir parameters calibrated with history matching were used as inputs into the XGBoost model to estimate the conformance control performance. Figure 9 compares the predicted incremental oil productions using numerical simulations and using the XGBoost model. As can be observed, the incremental oil productions estimated with the XGBoost agree well with the simulated values for both the N_2-foam and gel agents. The MAE and MRE for the N_2-foam agent are 67.65 t and 7.99%, respectively. The MAE and MRE for gel agent are 132.68 t and 12.55%, respectively. These matrices suggest a relatively strong reliability of the constructed model for evaluating the conformance performance of real wells.

3.3. Quantitative Evaluation of the Input Feature Importance

In this section, the permutation importance (PI) [78,79] was used to quantify the effect of each input variable to the incremental oil production using different plugging agents. The PI is able to accurately evaluate the non-monotonicity of the input variable, which is superior to other commonly used measures, such as Pearson and Spearman correlation coefficients, which can only reflect linear correlations [78].

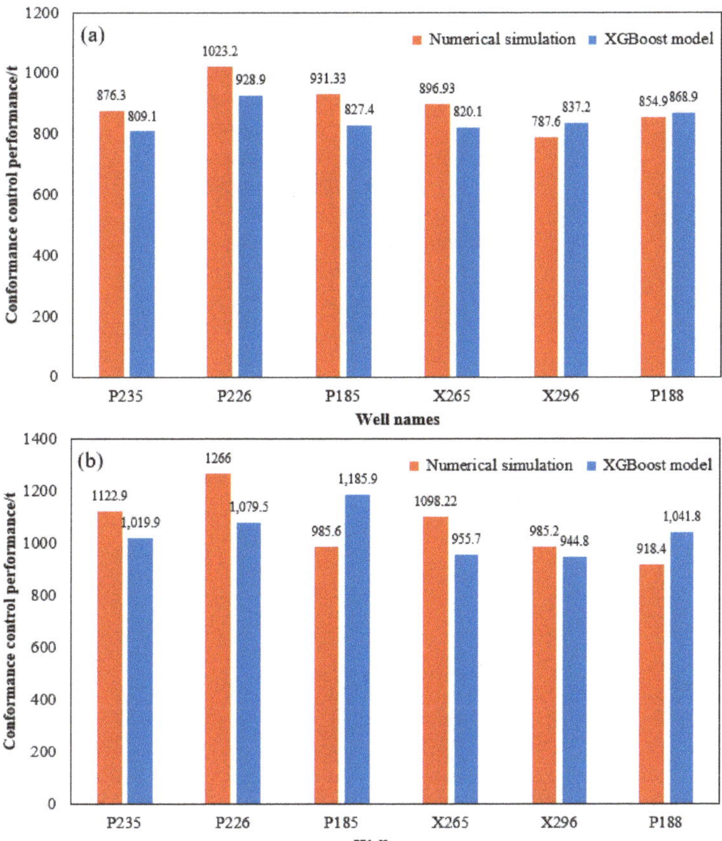

Figure 9. Comparisons of prediction results for profile control potential using the numerical simulation and XGBoost prediction model. (**a**) Potential of N_2-foam profile control. (**b**) Potential of gel profile control.

Figure 10a showed that the NTG has the highest PI value and exerted the most significant influence on the potential of N_2-foam conformance control among all the factors investigated. The net to gross can affect the value of geological reserves; therefore, the geological reserves can make a great impact on the potential of N_2-foam conformance control. The PI of the N_2-foam injection and variation coefficient of permeability are comparable, which are slightly higher than that of oil recovery and steam quality. The variation coefficient of permeability affected the degree of stratigraphic heterogeneity, and the heterogeneity, to some extent, impacted the potential of N_2-foam conformance control. The steam quality can influence oil recovery of thermal recovery in the heavy oil reservoir, so the PI of steam quality was similar to that of oil recovery. The PI of the oil rate was less than the PI of the parameters mentioned above, and the oil rate can also affect the potential of N_2-foam conformance control. The injector temperature, soaking time, porosity, water cut and production rate has the lower PI among all factors investigated and these parameters had little influence on the potential of N_2-foam conformance control. As a summary, the ranking of input variables in terms of decreasing importance to the potential of N_2-foam conformance control was net to gross>>N_2-foam injection> variation coefficient of permeability>oil recovery>steam quality>oil rate>injector temperature>soaking time>porosity>water cut>production rate.

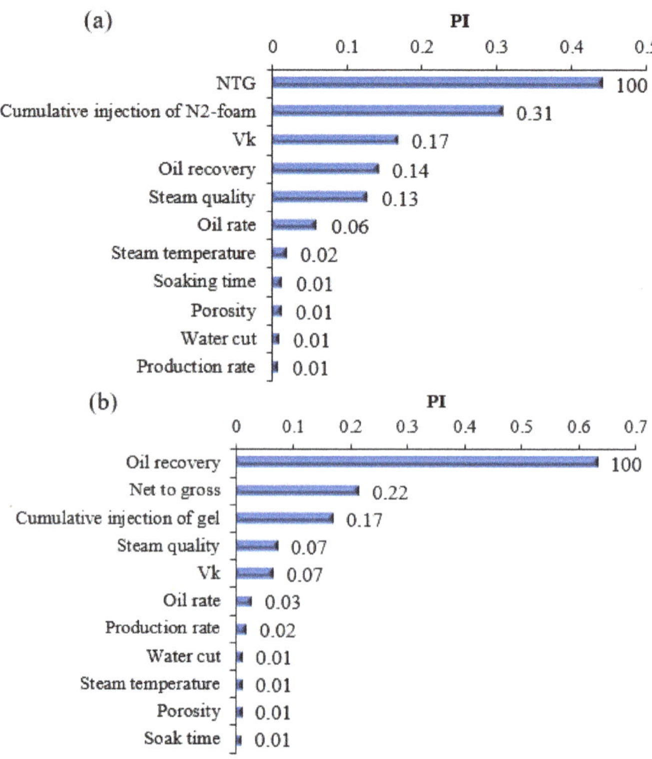

Figure 10. PI values for input variables in (**a**) N2-foam and (**b**) gel conformance control.

Figure 10b showed the results of permutation importance of gel conformance control. The process of calculation and sorting PI was similar to N_2-foam conformance control, which need not be specifically described again. As a short summary, the ranking of input variables in terms of decreasing importance to the potential of gel conformance control were: oil recovery >> Net to gross > gel injection > steam quality > variation coefficient of permeability > oil rate > production rate > water cut > injector temperature > porosity > soaking time. Compared with the PI of N_2-foam conformance control, the oil recovery before conformance control can exert more significant impact on the potential of gel conformance control. This is due to the different conformance control mechanisms of N_2-foam and gel. Gel injected into formation blocks the steam channel and achieves the conformance control, while N_2-foam implements conformance control through two processes, one is that N_2 is an inert gas which can reduce the heat loss and maintain high temperature in formation, another is that foam prevents the flow of water and does not affect the flow of oil. Gel conformance control cannot hold the process of thermal recovery but N_2-foam can keep this process. Therefore, the oil recovery of thermal recovery influenced by the potential of gel conformance control is more important than N_2-foam conformance control.

4. Conclusions

By coupling supervised learning and reservoir numerical simulation techniques, this paper proposes a fast and accurate method for predicting the potential of conformance control for heavy oil after multi-cycle steam stimulation. We used the K-fold cross-validation integrated with the exhaustive grid search approach to optimize the hyperparameters of XGBoost. After training the boosting trees using a database obtained from numerical simulations, the trained XGBoost model is capable of predicting the potential of conformance control for wells with better efficiency and accuracy. The performance of the new model was examined by statistical matrices, including mean absolute error (MAE), mean relative error (MRE) and coefficient of determination (R^2). In addition, we used PI to quantify the importance of each input variable for the potential of conformance control for N_2-foam and gel. Furthermore, this constructed model was implemented in real production wells of the Chunfeng oilfield and achieved excellent results. The key results are summarized as follows:

(1) The XGBoost model can predict the potential of conformance control by reproducing the underlying correlation between each input feature and oil increment measures. The statistical matrices (MAE, MRE and R^2) for N_2-foam are 5.25 t, 0.57% and 0.995 for the training set and 45.93 t, 5.01% and 0.901 for the testing set, respectively. The statistical matrices for the gel are 12.37 t, 0.09% and 0.999 for the training set and 80.89 t, 0.059% and 0.944147 for the testing set, respectively. For the two types of plugging agent, the absolute relative errors for most of the data samples are less than 10%, and the maximum relative error is less than 20%.

(2) The input variables in a sequence of decreasing importance to the potential of conformance control for N_2-foam, as quantified by the PI, are net to gross>>N_2-foam injection>variation coefficient of permeability>oil recovery>steam quality>oil rate>injector temperature>soaking time>porosity>water cut>production rate. While for gel the PI are oil recovery>>net to gross>gel injection>steam quality>variation coefficient of permeability>oil rate>production rate>water cut>injector temperature>porosity>soaking time. Different arrangements of PI are caused by the conformance control mechanism of N_2-foam and gel.

(3) The XGBoost model showed excellent performance that predicted the potential of conformance control for three types of real production wells in the Chunfeng oilfield. The maximum absolute error and relative error were 186.49 t and 17.28%, respectively. Comparing with the traditional numerical simulation method, our proposed model reduced the prediction time greatly, with similar prediction accuracy. This method can be applied in actual situations and provide a new view on the design of governance processes after multi-cycle steam stimulation.

Author Contributions: Conceptualization, Z.X. and J.Z.; methodology, Z.X. and J.Z.; software, Z.X. and X.S.; validation, Z.X. and X.S.; formal analysis, Z.X. and J.Z.; investigation, Z.X.; resources, Q.F. and X.Z.; data curation, Z.X.; writing—original draft preparation, Z.X.; writing—review and editing, J.Z. and X.S.; visualization, Z.X. and X.Z.; supervision, Q.F. and Z.W.; project administration, and Z.W.; funding acquisition, Q.F. and X.Z. All authors have read and agreed to the published version of the manuscript.

Funding: This study was completed under financial supports from the Fundamental Research Funds for the Central Universities (Grant No.21CX06021A).

Institutional Review Board Statement: Not applicable.

Informed Consent Statement: Not applicable.

Data Availability Statement: Data used in this study is available at request.

Conflicts of Interest: The authors declare that they have no known competing financial interests or personal relationships that could have appeared to influence the work reported in this paper.

Appendix A

Based on geological features of southern P601 block, we constructed the base geological model by using random geostatistical simulation (i.e., the sequential Gaussian–Bayesian simulation) [79]. The base geological model was imported to commercial numerical simulation software (i.e., CMG [38]) generating a base numerical simulation model. The grid top, net to gross ratio, permeability and porosity of numerical simulation model can be seen in Figures A1–A4, respectively.

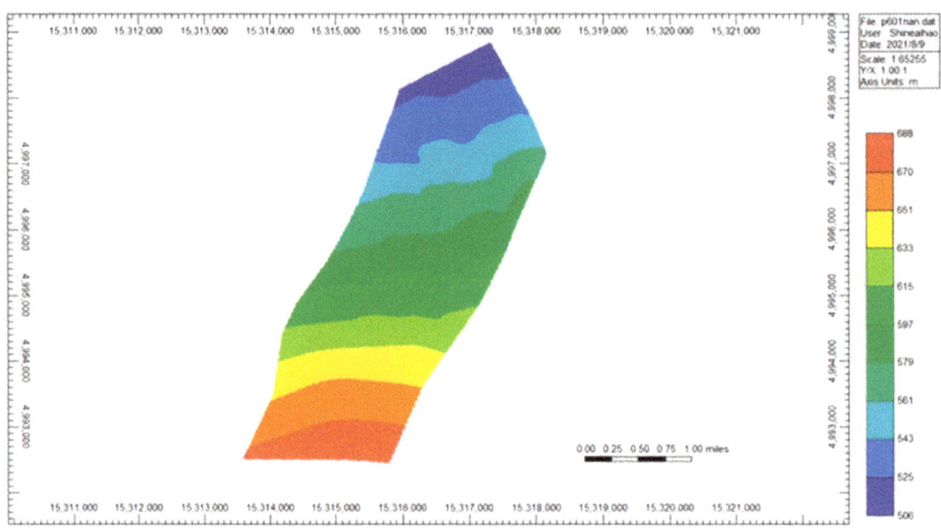

Figure A1. Grid top of numerical simulation model.

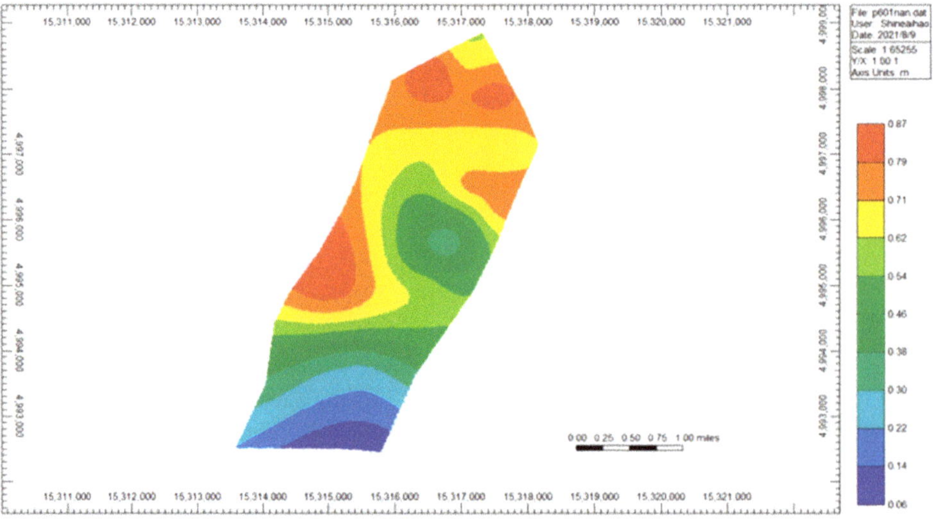

Figure A2. Net to gross ratio of numerical simulation model.

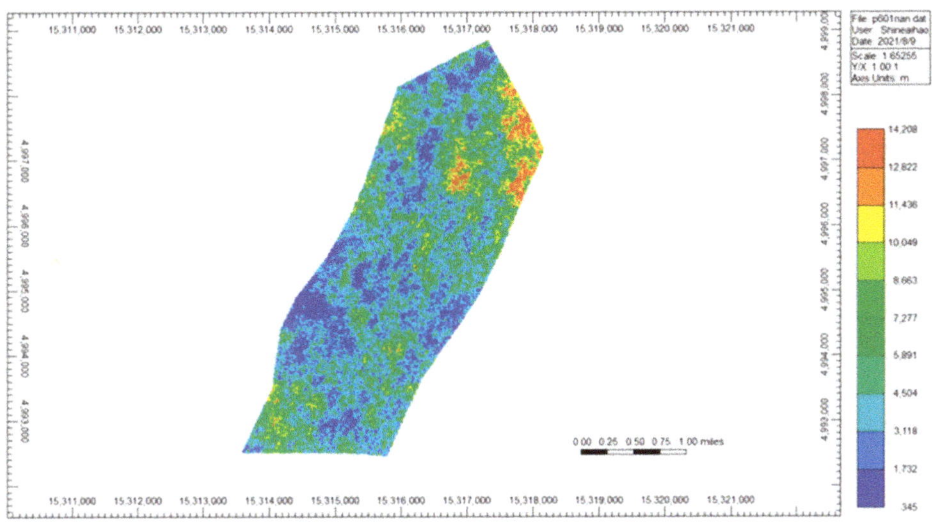

Figure A3. Permeability of numerical simulation model.

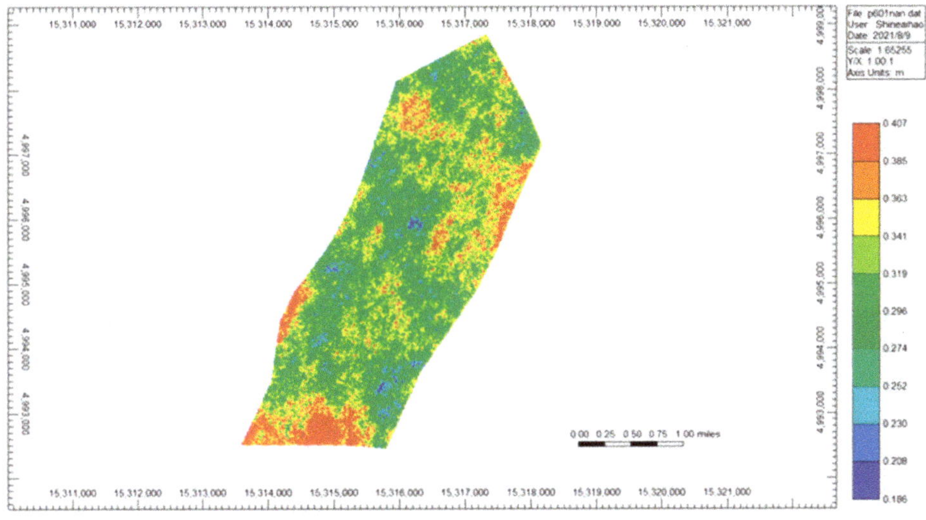

Figure A4. Porosity of numerical simulation model.

References

1. Xu, C.; Bell, L. Worldwide oil and gas reserves edge up, production down. *Oil Gas J.* **2020**, *12*, 14–18.
2. Dong, X.; Liu, H.; Chen, Z.; Wu, K.; Lu, N.; Zhang, Q. Enhanced oil recovery techniques for heavy oil and oilsands reservoirs after steam injection. *Appl. Energy* **2019**, *239*, 1190–1211. [CrossRef]
3. Farouq, S.A. Heavy oil—Evermore mobile. *J. Pet. Sci. Eng.* **2003**, *37*, 5–9.
4. Qian, S.; Ertekin, T. Structuring an artificial intelligence based decision making tool for cyclic steam stimulation processes. *J. Pet. Sci. Eng.* **2017**, *154*, 564–575.
5. Zhang, Q.; Liu, H.; Kang, X.; Liu, Y.; Dong, X.; Wang, Y.; Liu, S.; Li, G. An investigation of production performance by cyclic steam stimulation using horizontal well in heavy oil reservoirs. *Energy* **2021**, *218*, 119500. [CrossRef]
6. Catania, P. Predicted and actual productions of horizontal wells in heavy-oil fields. *Appl. Energy* **2000**, *65*, 29–43. [CrossRef]

7.   Escobar, E.; Valko, P.; Lee, W.J.; Rodriguez, M.G. *Optimization Methodology for Cyclic Steam Injection with Horizontal Wells*; Society of Petroleum Engineers SPE/CIM International Conference on Horizontal Well Technology: Calgary, AB, Canada, 2000.

8.   Chang, J. *Understanding HW-CSS for Thin Heavy Oil Reservoirs*; SPE Heavy Oil Conference-Canada: Calgary, AB, Canada, 2013.

9.   Pang, Z.; Jiang, Y.; Wang, B.; Cheng, G.; Yu, X. Experiments and analysis on development methods for horizontal well cyclic steam stimulation in heavy oil reservoir with edge water. *J. Pet. Sci. Eng.* **2020**, *188*, 106948. [CrossRef]

10.   Zhang, J.; Feng, Q.; Zhang, X.; Hu, Q.; Wen, S.; Chen, D.; Zhai, Y.; Yan, X. Multi-fractured horizontal well for improved coalbed methane production in eastern Ordos basin, China: Field observations and numerical simulations. *J. Pet. Sci. Eng.* **2020**, *194*, 107488. [CrossRef]

11.   Jha, R.K.; Kumar, M.; Benson, I.P.; Hanzlik, E.J. New Insights into Steam-Solvent Co-injection Process Mechanism. *SPE J.* **2012**, *18*, 867–877. [CrossRef]

12.   Li, S.; Li, Z.; Li, B. Experimental study and application of tannin foam for conformance modification in cyclic steam stimulated well. *J. Pet. Sci. Eng.* **2014**, *118*, 88–98. [CrossRef]

13.   Speight, J.G. *Enhanced Recovery Methods for Heavy Oil and Tar Sands*; Elsevier: Gulf Publishing Company: Houston, TX, USA, 2013.

14.   Butler, R.; Yee, C. Progress in the In Situ Recovery of Heavy Oils and Bitumen. *J. Can. Pet. Technol.* **2002**, *41*. [CrossRef]

15.   Zhao, G.; Dai, C.; Gu, C.; You, Q.; Sun, Y. Expandable graphite particles as a novel in-depth steam channeling control agent in heavy oil reservoirs. *Chem. Eng. J.* **2019**, *368*, 668–677. [CrossRef]

16.   Pang, Z.; Liu, H. The study on permeability reduction during steam injection in unconsolidated porous media. *J. Pet. Sci. Eng.* **2013**, *106*, 77–84. [CrossRef]

17.   Zhu, D.; Hou, J.; Chen, Y.; Zhao, S.; Bai, B. In Situ Surface Decorated Polymer Microsphere Technology for Enhanced Oil Recovery in High-Temperature Petroleum Reservoirs. *Energy Fuels* **2018**, *32*, 3312–3321. [CrossRef]

18.   Wang, Y.; Liu, H.; Pang, Z.; Gao, M. Visualization Study on Plugging Characteristics of Temperature-Resistant Gel during Steam Flooding. *Energy Fuels* **2016**, *30*, 6968–6976. [CrossRef]

19.   Cao, Y.; Liu, D.; Zhang, Z.; Wang, S.; Wang, Q.; Xia, D. Steam channeling control in the steam flooding of super heavy oil reservoirs, Shengli Oilfield. *Pet. Explor. Dev.* **2012**, *39*, 785–790. [CrossRef]

20.   Liang, S.; Hu, S.; Li, J.; Xu, G.; Zhang, B.; Zhao, Y.; Yan, H.; Li, J. Study on EOR method in offshore oilfield: Combination of polymer microspheres flooding and nitrogen foam flooding. *J. Pet. Sci. Eng.* **2019**, *178*, 629–639. [CrossRef]

21.   Zhang, X.; Zhang, Y.; Yue, Q.; Gao, Y.; Shen, D. Conformance Control of CSS and Steam Drive Process with a Carbamide Surfactant. *J. Can. Pet. Technol.* **2009**, *48*, 16–18. [CrossRef]

22.   Bai, B.; Liu, Y.; Coste, J.-P.; Li, L. Preformed Particle Gel for Conformance Control: Transport Mechanism through Porous Media. *SPE Reserv. Eval. Eng.* **2007**, *10*, 176–184. [CrossRef]

23.   El-Karsani, K.S.M.; Al-Muntasheri, G.A.; Hussein, I.A. Polymer Systems for Water Shutoff and Profile Modification: A Review over the Last Decade. *SPE J.* **2014**, *19*, 135–149. [CrossRef]

24.   Goudarzi, A.; Zhang, H.; Varavei, A.; Taksaudom, P.; Hu, Y.; Delshad, M.; Bai, B.; Sepehrnoori, K. A laboratory and simulation study of preformed particle gels for water conformance control. *Fuel* **2015**, *140*, 502–513. [CrossRef]

25.   Leng, J.; Wei, M.; Bai, B. Review of transport mechanisms and numerical simulation studies of preformed particle gel for conformance control. *J. Pet. Sci. Eng.* **2021**, *206*, 109051. [CrossRef]

26.   Abdulbaki, M.; Huh, C.; Sepehrnoori, K.; Delshad, M.; Varavei, A. A critical review on use of polymer microgels for conformance control purposes. *J. Pet. Sci. Eng.* **2014**, *122*, 741–753. [CrossRef]

27.   Zhu, D.; Bai, B.; Hou, J. Polymer Gel Systems for Water Management in High-Temperature Petroleum Reservoirs: A Chemical Review. *Energy Fuels* **2017**, *31*, 13063–13087. [CrossRef]

28.   Zhu, D.; Hou, J.; Wei, Q.; Wu, X.; Bai, B. Terpolymer Gel System Formed by Resorcinol–Hexamethylenetetramine for Water Management in Extremely High-Temperature Reservoirs. *Energy Fuels* **2017**, *31*, 1519–1528. [CrossRef]

29.   Ziegler, R. Technology Focus: High-Pressure/High-Temperature Challenges (April 2017). *J. Pet. Technol.* **2017**, *69*, 79. [CrossRef]

30.   Wang, C.; Liu, H.; Wang, J.; Hong, C.; Dong, X.; Meng, Q.; Liu, Y. A Novel High-temperature Gel to Control the Steam Channeling in Heavy Oil Reservoir. In Proceedings of the Society of Petroleum Engineers—SPE Heavy Oil Conference Canada, Calgary, AB, Canada, 10–12 June 2014.

31.   Liu, J.; Zhong, L.; Wang, C.; Li, S.; Wang, Q. Investigation of a high temperature gel system for application in saline oil and gas reservoirs for profile modification. *J. Pet. Sci. Eng.* **2020**, *195*, 107852. [CrossRef]

32.   Zhang, Y.; Wang, Y.; Xue, F.; Wang, Y.; Ren, B.; Zhang, L.; Ren, S. CO_2 foam flooding for improved oil recovery: Reservoir simulation models and influencing factors. *J. Pet. Sci. Eng.* **2015**, *133*, 838–850. [CrossRef]

33.   Abdelaal, A.; Gajbhiye, R.; Al-Shehri, D. Mixed CO_2/N_2 Foam for EOR as a Novel Solution for Supercritical CO_2 Foam Challenges in Sandstone Reservoirs. *ACS Omega* **2020**, *5*, 33140–33150. [CrossRef] [PubMed]

34.   Aarra, M.G.; Skauge, A.; Solbakken, J.; Ormehaug, P.A. Properties of N_2- and CO_2-foams as a function of pressure. *J. Pet. Sci. Eng.* **2014**, *116*, 72–80. [CrossRef]

35.   Farajzadeh, R.; Andrianov, A.; Krastev, R.; Hirasaki, G.; Rossen, W.R. Foam-Oil Interaction in Porous Media: Implications for Foam Assisted Enhanced Oil Recovery. *Adv. Colloid Interface Sci.* **2012**, *183–184*, 1–13. [CrossRef]

36.   Ding, L.; Maklad, M.; Guerillot, D. Revisit of Modeling Techniques for Foam Flow in Porous Media. In Proceedings of the 12th International Exergy, Energy and Environment Symposium (IEEES-12), Doha, Qatar, 20–24 December 2020.

37. Sander, P.; Clark, G.; Lau, E.C. Steam-Foam Diversion Process Developed to Overcome Steam Override in Athabasca. In Proceedings of the SPE Annual Technical Conference and Exhibition, Dallas, TX, USA,, 6–9 October 1991.
38. Pang, Z.; Liu, H.; Zhu, L. A laboratory study of enhancing heavy oil recovery with steam flooding by adding nitrogen foams. *J. Pet. Sci. Eng.* **2015**, *128*, 184–193. [CrossRef]
39. Sun, L.; Wei, P.; Pu, W.; Wang, B.; Wu, Y.; Tan, T. The oil recovery enhancement by nitrogen foam in high-temperature and high-salinity environments. *J. Pet. Sci. Eng.* **2016**, *147*, 485–494. [CrossRef]
40. De Haas, T.W.; Bao, B.; Ramirez, H.A.; Abedini, A.; Sinton, D. Screening High-Temperature Foams with Microfluidics for Thermal Recovery Processes. *Energy Fuels* **2021**, *35*, 7866–7873. [CrossRef]
41. Duan, X.; Hou, J.; Cheng, T.; Li, S.; Ma, Y. Evaluation of oil-tolerant foam for enhanced oil recovery: Laboratory study of a system of oil-tolerant foaming agents. *J. Pet. Sci. Eng.* **2014**, *122*, 428–438. [CrossRef]
42. Talebian, S.H.; Tan, I.M.; Sagir, M.; Muhammad, M. Static and dynamic foam/oil interactions: Potential of CO2-philic surfactants as mobility control agents. *J. Pet. Sci. Eng.* **2015**, *135*, 118–126. [CrossRef]
43. Chen, T.; Guestrin, C. XGBoost: A Scalable Tree Boosting System. In Proceedings of the 22nd ACM SIGKDD Conference on Knowledge Discovery and Data Mining, San Francisco, CA, USA, 13–17 August 2016.
44. Zhang, J.; Feng, Q.; Zhang, X.; Hu, Q.; Yang, J.; Wang, N. A Novel Data-Driven Method to Estimate Methane Adsorption Isotherm on Coals Using the Gradient Boosting Decision Tree: A Case Study in the Qinshui Basin, China. *Energies* **2020**, *13*, 5369. [CrossRef]
45. Niroomand-Toomaj, E.; Etemadi, A.; Shokrollahi, A. Radial basis function modeling approach to prognosticate the interfacial tension CO2/Aquifer Brine. *J. Mol. Liq.* **2017**, *238*, 540–544. [CrossRef]
46. Zhang, J.; Feng, Q.; Zhang, X.; Shu, C.; Wang, S.; Wu, K. A Supervised Learning Approach for Accurate Modeling of CO_2–Brine Interfacial Tension with Application in Identifying the Optimum Sequestration Depth in Saline Aquifers. *Energy Fuels* **2020**, *34*, 7353–7362. [CrossRef]
47. Zhang, J.; Sun, Y.; Shang, L.; Feng, Q.; Gong, L.; Wu, K. A unified intelligent model for estimating the (gas + n-alkane) interfacial tension based on the eXtreme gradient boosting (XGBoost) trees. *Fuel* **2020**, *282*, 118783. [CrossRef]
48. Wang, X.; Yang, Y.; Xi, W. Microbial enhanced oil recovery of oil-water transitional zone in thin-shallow extra heavy oil reservoirs: A case study of Chunfeng Oilfield in western margin of Junggar Basin, NW China. *Pet. Explor. Dev.* **2016**, *43*, 689–694. [CrossRef]
49. Rafnuss. Sequential Gaussian Simulation (SGS), GitHub. Available online: https://github.com/Rafnuss-PhD/SGS (accessed on 1 July 2021).
50. Wang, Y.; Ren, S.; Zhang, L.; Peng, X.; Pei, S.; Cui, G.; Liu, Y. Numerical study of air assisted cyclic steam stimulation process for heavy oil reservoirs: Recovery performance and energy efficiency analysis. *Fuel* **2018**, *211*, 471–483. [CrossRef]
51. Liu, P.; Zhang, Y.; Liu, P.; Zhou, Y.; Qi, Z.; Shi, L.; Xi, C.; Zhang, Z.; Wang, C.; Hua, D. Experimental and numerical investigation on extra-heavy oil recovery by steam injection using vertical injector -horizontal producer. *J. Pet. Sci. Eng.* **2021**, *205*, 108945. [CrossRef]
52. Sharoh, G.; Marquez, S.; Mohamed, O.; Almarshed, A. Delineation of most efficient recovery technique for typical heavy oil reservoir in the middle east region through compositional simulation of temperature-dependent relative permeabilities. *J. Pet. Sci. Eng.* **2020**, *186*, 106725.
53. CMG. *STARS User's Guide*; Computer Modeling Group Ltd.: Calgary, AB, Canada, 2015.
54. Vinsome, P.; Westerveld, J. A Simple Method for Predicting Cap and Base Rock Heat Losses In' Thermal Reservoir Simulators. *J. Can. Pet. Technol.* **1980**, *19*, PETSOC-80-03-04. [CrossRef]
55. Dheiaa, A.; Alameedy, U. Factors affecting gel strength design for conformance control: An integrated investigation. *J. Pet. Sci. Eng.* **2021**, *204*, 108711.
56. Herbas, J.; Moreno, R.; Romero, M.F.; Coombe, D.; Serna, A. Gel Performance Simulations and Laboratory/Field Studies to Design Water Conformance Treatments in Eastern Venezuelan HPHT Reservoirs. In Proceedings of the SPE/DOE Symposium on Improved Oil Recovery, Tulsa, OK, USA, 17–21 April 2004.
57. Scott, T.; Roberts, L.; Sharp, S.; Clifford, P.; Sorbie, K. In-Situ Gel Calculations in Complex Reservoir Systems Using a New Chemical Flood Simulator. *SPE Reserv. Eng.* **1987**, *2*, 634–646. [CrossRef]
58. Strebelle, S.; Journel, A. Reservoir Modeling Using Multiple-Point Statistics. In Proceedings of the SPE Annual Technical Conference and Exhibition, New Orleans, LA, USA, 30 September–2 October 2001.
59. Kular, G.; Lowe, K.; Coombe, D. Foam Application in an Oil Sands Steamflood Process. In Proceedings of the SPE Annual Technical Conference and Exhibition, San Antonio, TX, USA, 8–11 October 1989.
60. Kirmani, F.U.; Raza, A.; Gholami, R.; Haidar, M.Z.; Fareed, C.S. Analyzing the effect of steam quality and injection temperature on the performance of steam flooding. Energy Geoscience. *Energy Geosci.* **2021**, *2*, 83–86. [CrossRef]
61. Priscilla, C.V.; Prabha, P. Influence of Optimizing XGBoost to handle Class Imbalance in Credit Card Fraud Detection. In Proceedings of the IEEE International Conference on Smart Systems and Inventive Technology (ICSSIT), Tirunelveli, India, 20–22 August 2020.
62. Qin, C.; Zhang, Y.; Bao, F.; Zhang, C.; Liu, P.; Liu, P. XGBoost Optimized by Adaptive Particle Swarm Optimization for Credit Scoring. *Math. Probl. Eng.* **2021**, *2021*, 6655510. [CrossRef]
63. Wang, M.-X.; Huang, D.; Wang, G.; Li, D.-Q. SS-XGBoost: A Machine Learning Framework for Predicting Newmark Sliding Displacements of Slopes. *J. Geotech. Geoenviron. Eng.* **2020**, *146*, 04020074. [CrossRef]

64. Mo, H.; Sun, H.; Liu, J.; Wei, S. Developing window behavior models for residential buildings using XGBoost algorithm. *Energy Build.* **2019**, *205*, 109564. [CrossRef]
65. Liu, W.; Gu, J. Predictive model for water absorption in sublayers using a Joint Distribution Adaption based XGBoost transfer learning method. *J. Pet. Sci. Eng.* **2020**, *188*, 106937. [CrossRef]
66. Chung, Y.-S. Factor complexity of crash occurrence: An empirical demonstration using boosted regression trees. *Accid. Anal. Prev.* **2013**, *61*, 107–118. [CrossRef]
67. Pedregosa, F.; Varoquaux, G.; Gramfort, A.; Michel, V.; Thirion, B.; Grisel, O.; Blondel, M.; Prettenhofer, P.; Weiss, R.; Dubourg, V.; et al. Scikit-learn: Machine learning in python. *J. Mach. Learn. Res.* **2011**, *12*, 2825–2830.
68. Ding, C.; Cao, X.J.; Næss, P. Applying gradient boosting decision trees to examine non-linear effects of the built environment on driving distance in Oslo. *Transp. Res. Part. A Policy Pract.* **2018**, *110*, 107–117. [CrossRef]
69. Lim, S.; Chi, S. Xgboost application on bridge management systems for proactive damage prediction. *Adv. Eng. Inform.* **2019**, *41*, 100922. [CrossRef]
70. Lee, Y.; Ragguett, R.M.; Mansur, R.B.; Boutilier, J.J.; Rosenblat, J.D.; Trevizol, A.; Brietzke, E.; Lin, K.; Pan, Z.; Subramaniapillai, M.; et al. Applications of machine learning algorithms to predict therapeutic outcomes in depression: A meta-analysis and systematic review. *J. Affect. Disord.* **2018**, *241*, 519–532. [CrossRef] [PubMed]
71. Gao, J.; Liao, W.; Nuyttens, D.; Lootens, P.; Vangeyte, J.; Pižurica, A.; He, Y.; Pieters, J.G. Fusion of pixel and object-based features for weed mapping using unmanned aerial vehicle imagery. *Int. J. Appl. Earth Obs. Geoinf.* **2018**, *67*, 43–53. [CrossRef]
72. Xia, Y.; Liu, C.; Li, Y.; Liu, N. A boosted decision tree approach using Bayesian hyper-parameter optimization for credit scoring. *Expert Syst. Appl.* **2017**, *78*, 225–241. [CrossRef]
73. Tao, H.; Habib, M.; Aljarah, I.; Faris, H.; Afan, H.A.; Yaseen, Z.M. An intelligent evolutionary extreme gradient boosting algorithm development for modeling scour depths under submerged weir. *Inf. Sci.* **2021**, *570*, 172–184. [CrossRef]
74. Ye, Z.J.; Schuller, B.W. Capturing dynamics of post-earnings-announcement drift using a genetic algorithm-optimized XGBoost. *Expert Syst. Appl.* **2021**, *177*, 114892. [CrossRef]
75. Feng, G.; Li, Y.; Yang, Z. Performance evaluation of nitrogen-assisted steam flooding process in heavy oil reservoir via numerical simulation. *J. Pet. Sci. Eng.* **2020**, *189*, 106954. [CrossRef]
76. Lyu, X.; Liu, H.; Pang, Z.; Sun, Z. Visualized study of thermochemistry assisted steam flooding to improve oil recovery in heavy oil reservoir with glass micromodels. *Fuel* **2018**, *218*, 118–126. [CrossRef]
77. Breiman, L. Random forests. *Mach. Learn.* **2001**, *45*, 5–32. [CrossRef]
78. Altmann, A.; Tološi, L.; Sander, O.; Lengauer, T. Permutation importance: A corrected feature importance measure. *Bioinformatics* **2010**, *26*, 1340–1347. [CrossRef]
79. Mariethoz, G. A general parallelization strategy for random path based geostatistical simulation methods. *Comput. Geosci.* **2010**, *36*, 953–958. [CrossRef]

Article

Velocity Structure Revealing a Likely Mud Volcano off the Dongsha Island, the Northern South China Sea

Yuning Yan [1], Jianping Liao [1,2,*], Junhui Yu [3,4], Changliang Chen [3], Guangjian Zhong [5], Yanlin Wang [3,4] and Lixin Wang [2]

[1] Hunan Provincial Key Laboratory of Shale Gas Resource Utilization, Hunan University of Science and Technology, Xiangtan 411201, China; lordarthas1995@outlook.com
[2] SINOPEC Key Laboratory of Geophysics, Sinopec Geophysical Research Institute, Nanjing 211103, China; slwlx@126.com
[3] Southern Marine Science and Engineering Guangdong Laboratory (Guangzhou), Guangzhou 511458, China; jhyu@scsio.ac.cn (J.Y.); chenchangliang@scsio.ac.cn (C.C.); yanlinw@scsio.ac.cn (Y.W.)
[4] Key Laboratory of Ocean and Marginal Sea Geology, South China Sea Institute of Oceanology, Innovation Academy of South China Sea Ecology and Environmental Engineering, Chinese Academy of Sciences, Guangzhou 510301, China
[5] Guangzhou Marine Geological Survey, Guangzhou 510760, China; guangjianz@21cn.com
* Correspondence: 1020116@hnust.edu.cn

Abstract: The Dongsha Island (DS) is located in the mid-northern South China Sea continental margin. The waters around it are underlain by the Chaoshan Depression, a relict Mesozoic sedimentary basin, blanketed by thin Cenozoic sediments but populated with numerous submarine hills with yet less-known nature. A large hill, H110, 300 m high, 10 km wide, appearing in the southeast to the Dongsha Island, is crossed by an ocean bottom seismic and multiple channel seismic surveying lines. The first arrival tomography, using ocean bottom seismic data, showed two obvious phenomena below it: (1) a low-velocity (3.3 to 4 km/s) zone, with size of 20×3 km^2, centering at ~4.5 km depth and (2) an underlying high-velocity (5.5 to 6.3 km/s) zone of comparable size at ~7 km depth. MCS profiles show much-fragmented Cenozoic sequences, covering a wide chaotic reflection zone within the Mesozoic strata below hill H110. The low-velocity zone corresponds to the chaotic reflection zone and can be interpreted as of highly-fractured and fluid-rich Mesozoic layers. Samples dredged from H110 comprised of illite-bearing authigenic carbonate nodules and rich, deep-water organisms are indicative of hydrocarbon seepage from deep source. Therefore, H110 can be inferred as a mud volcano. The high-velocity zone is interpreted as of magma intrusion, considering that young magmatism was found enhanced over the southern CSD. Furthermore, the origin of H110 can be speculated as thermodynamically driven, i.e., magma from the depths intrudes into the thick Mesozoic strata and promotes petroleum generation, thus, driving mud volcanism. Mud volcanism at H110 and the occurrence of a low-velocity zone below it likely indicates the existence of Mesozoic hydrocarbon reservoir, which is in favor of the petroleum exploration.

Keywords: Dongsha Waters in the northern South China Sea margin; velocity inversion; mud volcano; magma intrusion; Mesozoic hydrocarbon

Citation: Yan, Y.; Liao, J.; Yu, J.; Chen, C.; Zhong, G.; Wang, Y.; Wang, L. Velocity Structure Revealing a Likely Mud Volcano off the Dongsha Island, the Northern South China Sea. *Energies* **2022**, *15*, 195. https://doi.org/10.3390/en15010195

Academic Editors: Alexei Milkov and Alireza Nouri

Received: 27 October 2021
Accepted: 21 December 2021
Published: 28 December 2021

Publisher's Note: MDPI stays neutral with regard to jurisdictional claims in published maps and institutional affiliations.

1. Introduction

Generally, a mud volcano is formed by eruption of mud, gas, and fluid, often containing hydrocarbon, water, and so on; thus, it is an indicator of petroleum leakage from deep [1]. Because of rich fluid filling within porous sediments, seismic wave velocity in a mud volcano is low [2], and reflection from it becomes chaotic or blank [3]. The erupted mud may contain petroleum-associated mineral assemblage from the source layer in the depths [4]. The leaked methane can feed methanotrophic and deep-water organisms, producing authigenic carbonates on the seabed. Mud volcanism can be triggered by plate compression [5], gravitational instability [6], and thermodynamic drive [7]. Submarine mud volcanoes have been widely found over the world, e.g., in the Gulf of Mexico, the Caspian Sea [8], and the onshore and offshore Southwest Taiwan Island [5], where petroleum or gas hydrate are rich.

The Dongsha Island (DS) is located in the mid northern South China Sea (SCS) continental margin (Figure 1), where the Chaoshan Depression, a Mesozoic basin, relicts [9]. It is considered an exploration target area for Mesozoic hydrocarbon [10,11]. The waters are rich with young submarine hills. Some hills, e.g., MV3, MV8, southwest of the DS (Figure 2), were found in recent studies as mud volcanoes featuring chaotic or blank reflection zones on multichannel seismic (MCS) profiles, gas plume on CHIRP (Compressed High-Intensity Radiated Pulse) sub-bottom profiles, and abundant authigenic carbonate and deep-water organisms in dredged samples from the seabed [12]. However, their dynamic origins remain unknown, due to lack of deeper imaging.

In the waters southeast of the DS, there are also several submarine hills (Figure 2). Whether they are volcano or mud volcano is unknown, due to less investigation, but important for exploration of Mesozoic petroleum. A topographically large one, H110, ~300 m high and ~10 km wide, is crossed by an OBS/MCS (Ocean Bottom Seismic/Multiple-Channel Seismic) coincident survey line L2016-2, which can be helpful for revealing the deep geologic structures and, therefore, understanding its nature and origin.

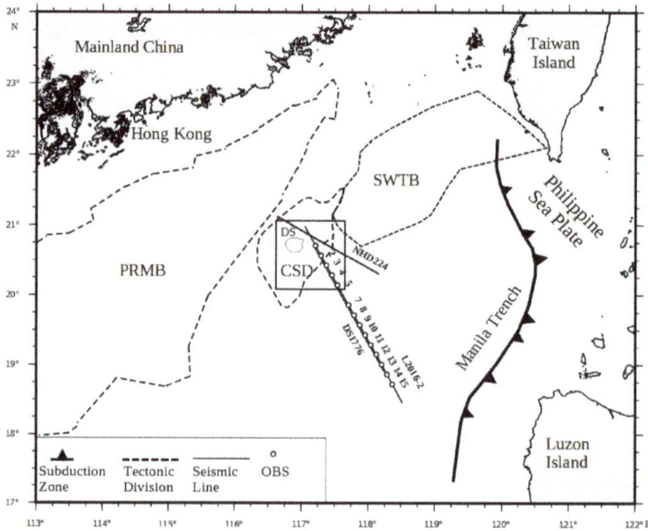

Figure 1. Locations of the Dongsha Island (DS), tectonic division [9], and seismic lines in the northeastern South China Sea. PRMB = Pearl River Mouth Basin; CSD = Chaoshan Depression; SWTB = Southwest Taiwan Basin; the box is the study area (also shown in Figure 2), and the black circle denotes the OBS station.

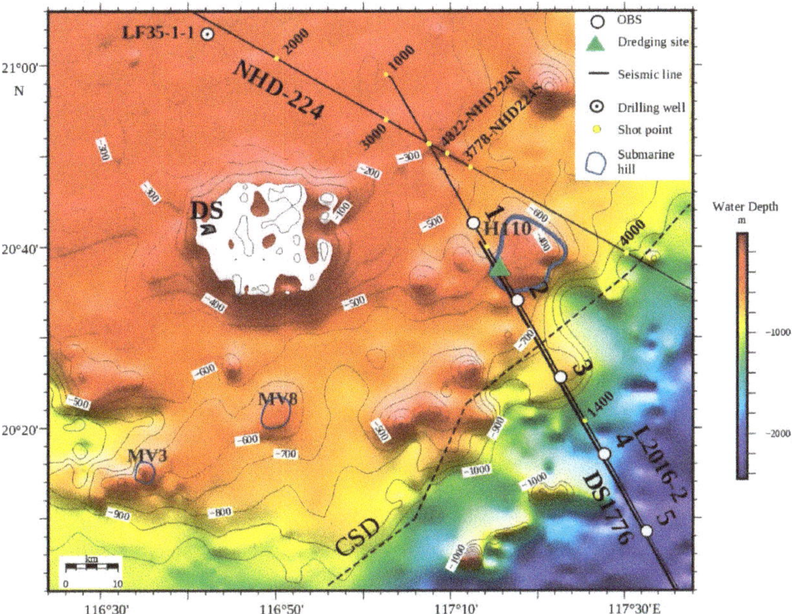

Figure 2. Bathymetry map of the DS waters (topographic data in 15″ × 15″ were downloaded from ftp://topex.ucsd.edu/pub/srtm15_plus/topo15.grd (accessed on 1 October 2014)). Hill MV3 and Hill MV8 were previously recognized mud volcanoes [12].

2. Geological Setting

The northern continental margin of the South China Sea has developed via magma-poor rifting [13] of the Mesozoic margin of South China, since the latest Mesozoic Era with a couple of major sedimentary basins formed [14] (Figure 1). The Chaoshan Depression (CSD) is neighbored by two major Cenozoic rift basins, the Pearl River Mouth Basin (PRMB) to the west and Southwest Taiwan Basin (SWTB) to the east, respectively. The PRMB is filled with thick Cenozoic sediments [15,16] and has become a major petroleum and gas production region. In the SWTB, many gas-seeping mud volcanoes are found along the northernmost accretion-subduction zone of the Manila Trench, due to convergence of the Philippine Sea Plate and the South China Sea [5].

Different from the PRMB and SWTB, where the crust has been highly faulted, detached, and depressed [15,16], the CSD remains, for the most part, less rifted and subsided [10]. It has developed with thin (~1 km) Cenozoic and thick (~5 km or more) Mesozoic Erathem [9–11]. In the northwestern CSD, a drilling well, LF35-1-1, penetrated the Early Cretaceous terrestrial and into Jurassic marine sequences, under thin Cenozoic cover [10]. Although the Early Cretaceous terrestrial contains a dense red bed, which features a seismic wave velocity of 4.0–5.5 km/s [9], a couple of interlayers in the underlying Jurassic marine strata were found with rich organic carbon [10]. In the southeast CSD, bottom simulating reflectors (BSR) were found and interpreted as originated from petroleum leakage from thick Mesozoic in the deep [11]. The Mesozoic basin seems to be an ample source and awaits the discovery of petroleum [12].

On the whole, the northern margin of South China Sea is magma-poor, but there are enhanced young magmatism locally, particularly around the southern CSD, in expressions of volcanoes, shallow intrusive, and lower crust underplating, featuring high-velocity (7.0–7.5 km/s) [13,17,18]. However, the impact of magmatism on the petroleum system in the CSD is less known.

The seismic survey line L2016-2 was initially designed to explore the magmatism over the DS water. Wan, et al. [18] has used the OBS data to construct forward and inversion velocity models and found two lower crust high-velocity (7.0–7.5 km/s) zones beneath the middle and lower slope, in the southeast, close to the CSD. Both zones were interpreted as formed by magma underplating. Their study focused on the crust but ignored the submarine hills, as well the effect of magmatic intrusion on the sedimentary sequences in the CSD.

3. Seismic Data

A seismic survey was conducted, onboard the RV Shiyan II of the South China Sea Institute of Oceanology (SCSIO), Chinese Academy of Sciences, along the Line L2016-2, in 2016. It crosses the northeast waters of the Dongsha Island and the lower continental slope to its southeast, totaling 320 km long. The source of three BOLT airguns, with a total volume of 4500 in^3, were shot 1572 times, spacing 200 m, in 10 m water depth. Data were simultaneously recorded using 14 ocean-bottom-seismometers (OBS) and a MCS streamer.

3.1. OBS Data
3.1.1. Collection, Pre-Processing, Phase Picks, and Analysis

The OBSs were developed by the Institute of Geology and Geophysics, China Academy of Sciences. Each OBS contains one hydrophone, as well as one vertical and two horizontal geophones. Fifteen OBSs were deployed, while fourteen of them were recovered, with OBS six lost. Location and clock time, during the deployment and recovery of the OBSs, were recorded in the log files.

In pre-processing, the raw OBS data were converted to SEGY format with UTIG's software OBSTOOL [19]. Bathymetry is constrained by the coincident MCS profile. Clock drift is calculated by the readouts from the log files and used to correct the time of the OBS data. The direct waves, in a 4 km window, centering on OBS deployment position, were used to relocate the OBS falls on the sea bottom. As results, the space drifts were calculated as 84 m (OBS 1), 194 m (OBS 2), 235 m (OBS 3), 1377 m (OBS 4), 757 m (OBS 5), 262 m (OBS 7), 307 m (OBS 8), 228 m (OBS 9), 210 m (OBS 10), 127 m (OBS 11), 173 m (OBS 12), 176 m (OBS 13), 248 m (OBS 14), and 102 m (OBS 15), respectively. Offsets were recalculated with the relocated OBS positions. A 5–15 Hz band-pass filter was tested and well applied to suppress high frequency noise.

Each OBS record was displayed with a reduction velocity of 8 km/s. Depending on the signal-to-noise ratio, either hydrophone or vertical components were selected to pick the first arrivals. The first arrivals were picked for all OBS records. Records of OBS 1–OBS 5, which are close to H110, were selected to be illustrated in Figure 3. According to phase velocity, the picks include direct waves, labelled as Pw (1.5 km/s), and refracted waves, in primitive recognition, from sedimentary layers (1.6–5.5 km/s), crystalline basement (6–7 km/s), and Mantle (>7 km/s), i.e., Psed, Pg, and Pn, respectively (Figure 3). OBS 1 and 2 sat on the two sides of H110 (Figure 3), their records show that the first arrival waves were clear in near offset (<30 km), with high signal-to-noise ratio (S/N), and visible in middle and far range (offset >30 km), with low S/N. Three sedimentary layer-refracted phase drops are found, with time jump from 2.25 s to 2.5 s at 12 km offset of OBS 1, 1.5 s to 1.75 s at −5 km, and 2.25 s to 2.75 s at 10 km of OBS 2. These drops complicate the first arrival phase curves, implicating complex velocity structure. OBS 3–5 also recorded rich phases through H110. OBS 3 records show regular linear noises, likely due to a machine problem, but the first arrivals are still legible. For the OBS 7–15 records, the first arrivals are generally well recognizable to large offset.

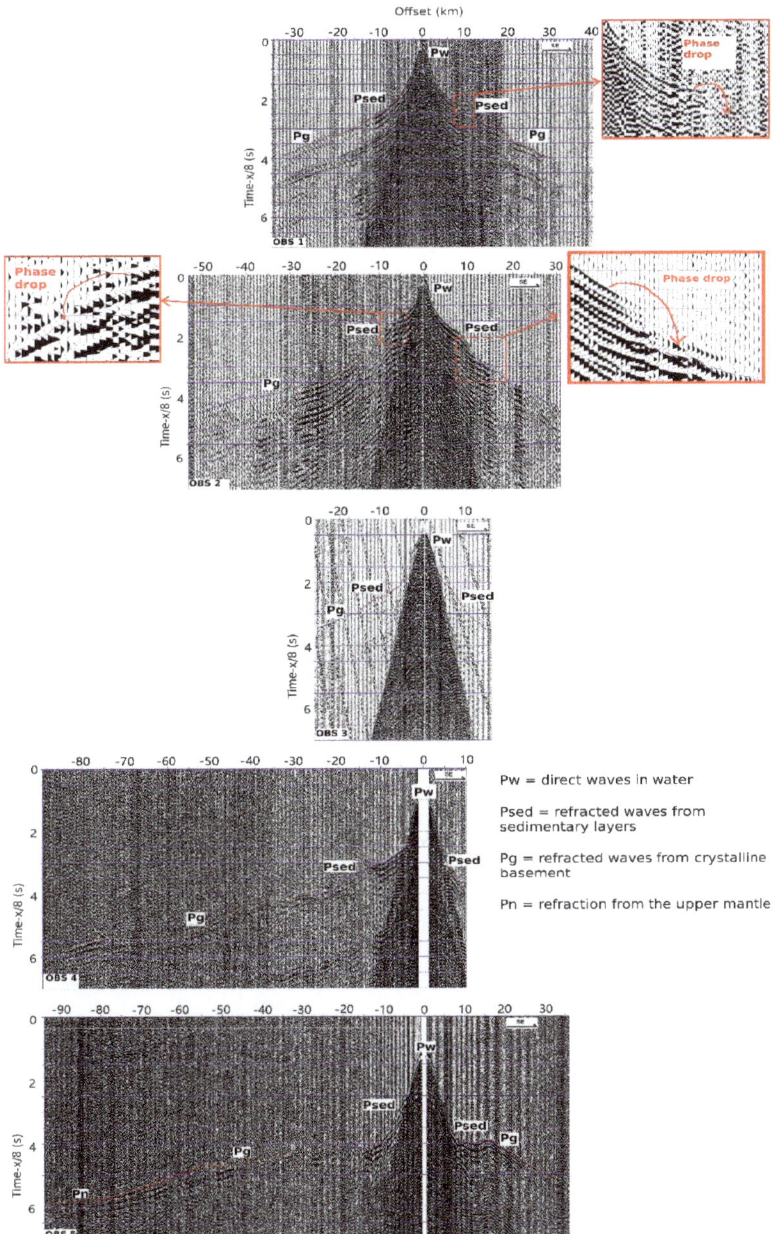

Figure 3. OBS 1–5 records. Dotted colour lines denote picks of the first arrivals.

3.1.2. Velocity Inversion

Velocity inversion requires an initial model. It is established in the following steps. Firstly, a 320 km × 30 km blank model is constructed using cell space 0.25 × 0.25 km. Secondary, it is endowed with bathymetry, converted from MCS time data, with a sea water velocity of 1.5 km/s. Thirdly, the geometry of Moho discontinuity is assumed, with a curve of three line sections. Fourthly, velocity is assigned by linear increase from 1.5 km/s to 8 km/s, with depth from seabed to Moho, and kept at 1.5 km/s in water and 8 km/s beneath Moho (Figure 4a). OBS locations are projected onto the profile of L2016-2.

According to the S/N ratio, uncertainty of the first arrival time picks is assigned as 0.02 s for offset <30 km and 0.1 s for offset >30 km, respectively. To run velocity inversion, the initial velocity model (Figure 4a), time picks of the first arrivals, location of OBS, and bathymetry were input together into FAST program [20]. The number of iterations and trade-off parameters (regularization factor) were tested to 20 and 5, respectively. To the 20th run, the root-mean-square misfit between the calculated and observed arrivals was 46.79 ms and normalized chi-square value (misfit of traveltimes to uncertainty of picks) was 1.546 ms, both of which are usually regarded as satisfactory [21]. The observed (black dot) and calculated (red dot) first arrivals for the initial (Figure 4b) and finial (Figure 4e) velocity model were plotted to show the inversion process. The initial and final ray paths are shown in Figure 4c,f.

The final velocity model (Figures 4d and 5) displays 2 remarkable features relating to H110. One is that the velocity generally increases from 1.5 km/s to 5.5 km/s to 7 km deep with an enclosed low-velocity zone 3 km thick, 20 km wide in depth of 3–5.5 km. Velocity in the zone decreased from 4 km/s (periphery) to 3.3 km/s at core. Another is two high-velocity (>5.5 km/s) zones. The west high-velocity zone (HVZ1) appears at 7–9 km and underlies slightly to east of the low-velocity zone below H110. The east one (HVZ2) centers at a model distance of 130 km, 3 km below the seabed.

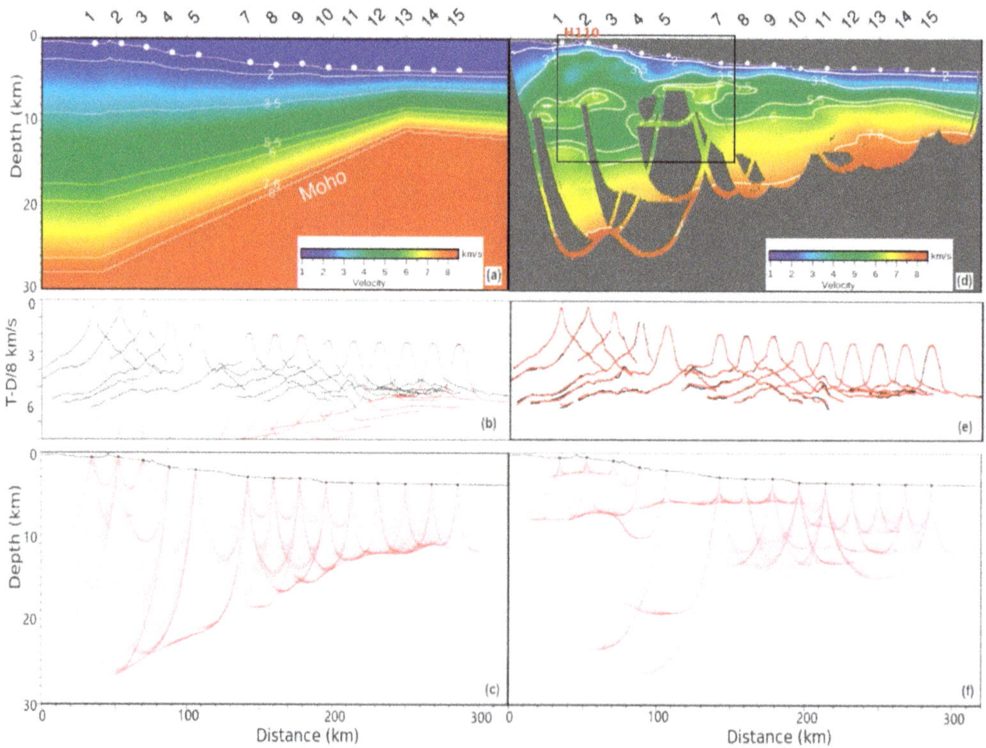

Figure 4. Velocity by iterative inversion. (**a**) an initial velocity model; (**b**) the observed (black dot) and calculated (red line) first arrivals for (**a**); (**c**) ray paths for (**a**); (**d**) inversed model in the 20th iteration; (**e**) observed and calculated first arrivals for (**d**); (**f**) ray paths for (**d**). The box around H110 in (**d**) is shown in Figure 5.

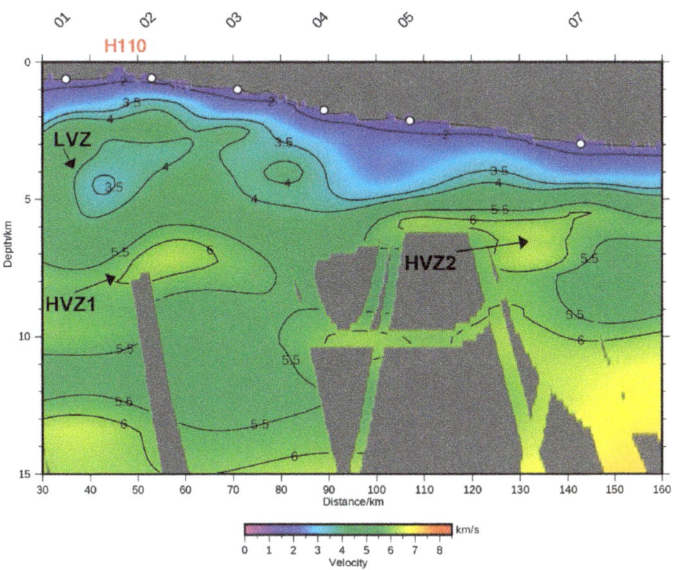

Figure 5. Zoomed velocity profile around the hill H110, showing a large low-velocity (LVZ, 3.3–4 km/s) zone, followed by a high-velocity (HVZ1, 5.5–6.3 km/s) zone at different levels below H110. Another high-velocity one (HVZ2, 5.5–6.3 km/s) appears to its east.

3.1.3. Checkerboard and Resolvability

To assess the resolution of the velocity inversion, the checkerboard method [21] was used. Synthetic perturbations of the inversed velocity were set in the range of −5% to 5%. Four checkerboard models, using sizes of 10 × 2.5 km, 20 × 5 km, 40 × 10 km, and 80 × 20 km, were tested (Figure 6). The resolvability was calculated with the Zelt (1998) [21] formulae of starting and recovered velocities. For the low-velocity zone (depth of 1–7 km) and the underlying high-velocity zone (depth of 7–9 km) beneath H110, the velocity resolvabilities were partly greater than 0.7 for 10 km cell size and universally higher for 20, 40, and 80 km cell sizes (Figure 7). The deeper part is less resolvable, due to fewer arrival picks and sparser ray paths.

It is noteworthy that below H110, the 20 × 3 km low-velocity (3.3–4 km/s) zone, appearing at 4.5 km depth and 30 × 3 km high-velocity (5.5–6.3 km/s) zone at 7 km depth (Figure 5), are basically well-recovered, with a confident resolvability of 0.7 in disturbance cell sizes of 10 × 2.5 km and 20 × 5 km, respectively (Figures 6 and 7).

3.2. Multichannel Seismic Data Processing and Reflection Characteristics

A multichannel seismic streamer, model Seal 428 (Sercel Corp, Nantes, France), 1.5 km long, with 120 channels, spacing at 12.5 m, was deployed 12 m deep in the water, with a minimum offset of 200 m, concurring with the OBS survey. The record time length was 14 s, and the sample interval is 2 ms.

To reveal the reflection structure of H110, MCS data of the northern section of L2016-2 were processed using the CGG-GeoVation (CGG, West Perthm, Australia). The routine procedures include amplitude compensation, bandpass filtering, CDP sorting, velocity analysis, NMO correction, stacking, and migrating. The migrated profile fairly clearly shows imaged structures before multiple waves, which smear the later portion, due to poor suppression with low coverage (only 4 folds).

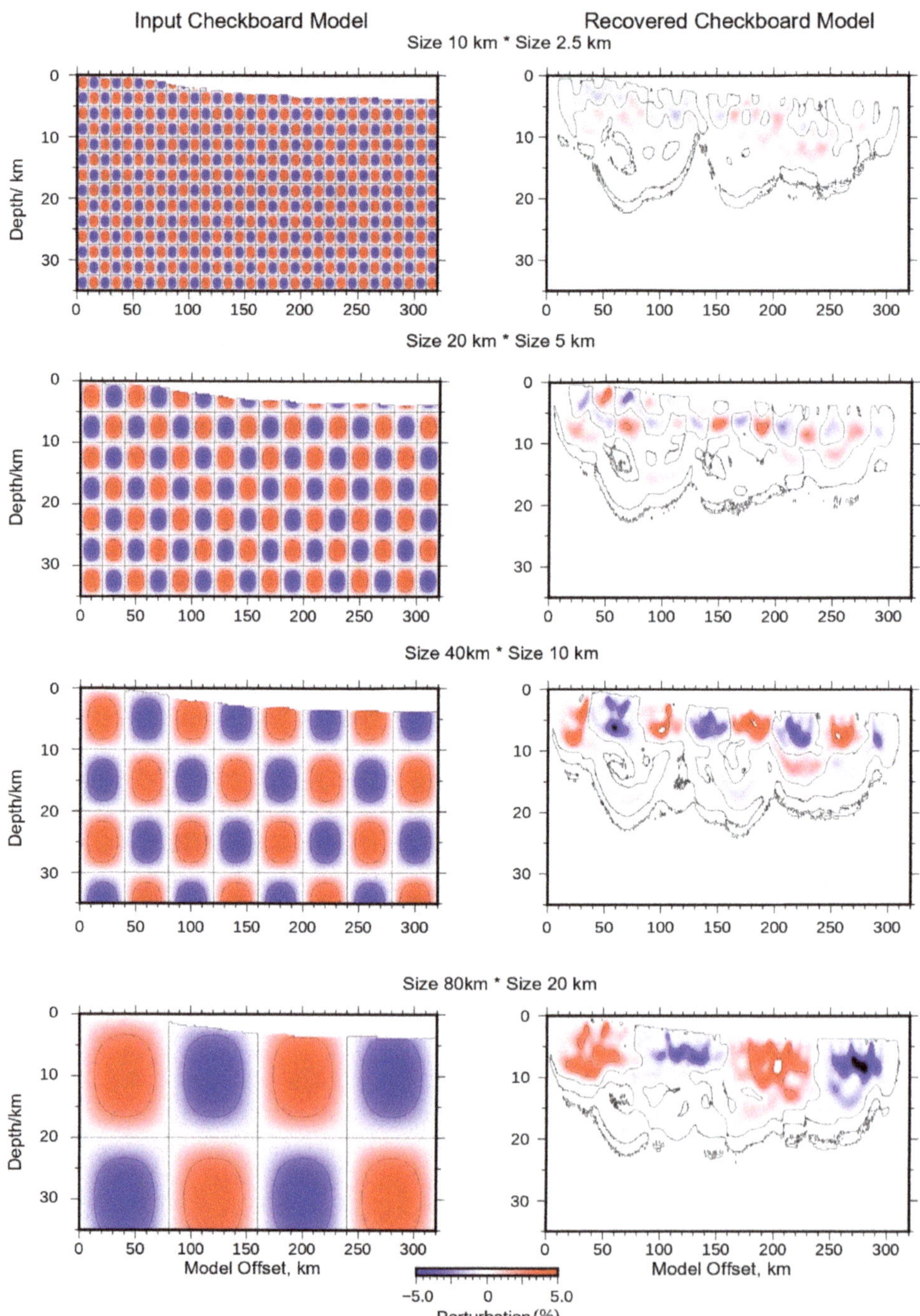

Figure 6. Four checkerboard models for velocity perturbation, using 10×2.5, 20×5, 40×10, and 80×20 km cell sizes. The left suites display starting models and the right ones show recovered models.

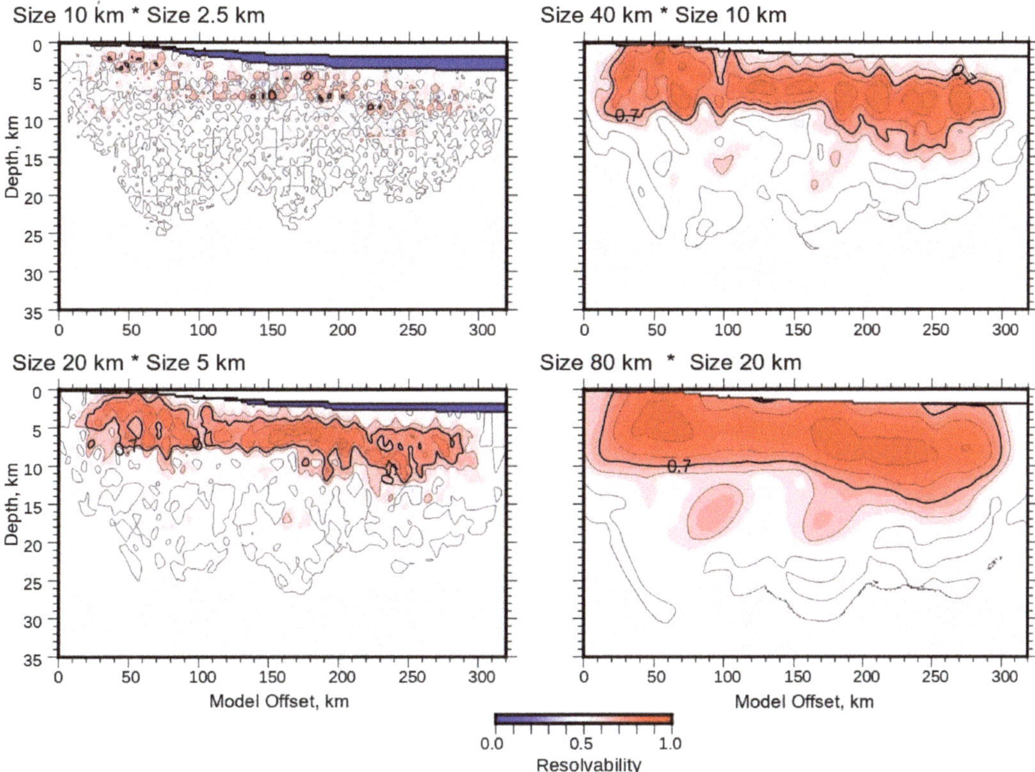

Figure 7. Model resolvability at four different disturbance cell sizes. When resolvability is greater or equal to 0.7, the velocity model is deemed recoverable. H110 is basically resolvable for perturbation, using cell size 20 km, and completely resolvable to ~9 km deep for coarser cell size 20, 40, and 80 km.

Two vintage MCS lines (NHD224, DS1776) crossing our research area were resorted to, in order supplement the study of H110 (Figures 1 and 2). The MCS acquisition parameters are listed in Table 1. Line NHD224 (Figure 8) runs north close to H110 and intersects with Line L2016-2, and Line DS1776 (Figure 9) coincides with it for the most part. Both them feature a clearer image with higher coverage.

Table 1. Acquisition parameter of MCS.

Line Name	L2016-2	NHD224-S	NHD224-N	DS1776
Channels	120	480	480	564
Channel spacing (m)	12.5	12.5	12.5	12.5
Coverage	4	60	120	141
Shot spacing (m)	200	50	25	25
Minimum offset (m)	200	250	250	200
Sample interval (ms)	2	2	2	1
Record time length (s)	14	12	9	8
Source	3 BOLT airguns	32 BOLT airguns	32 BOLT airguns	40 BOLT airguns
Airguns Volume (in^3)	4500	5080	5080	4100

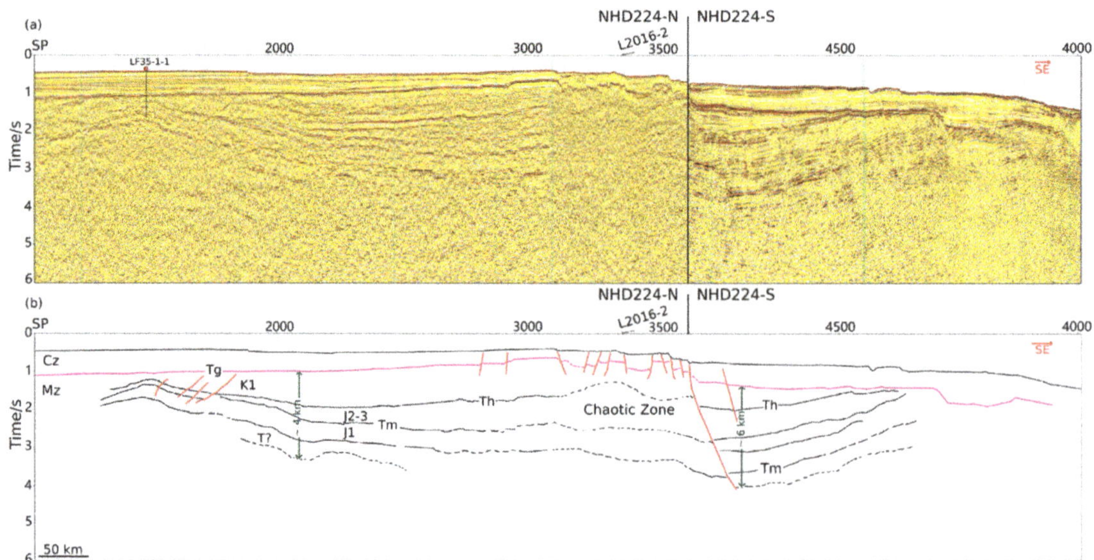

Figure 8. A MCS profile of NHD224 (**a**) that ran by a drilling well, LF35-1-1, and interpretation (**b**). Tg is the unconformity separating the Cenozoic (Cz) from the Mesozoic (Mz) Erathem; Th separates the Cretaceous (K1) from the Jurassic system (J); Tm separates the Middle and Higher (J2-J3) from the Lower Jurassic (J1) series; T? is likely Triassic; red lines denote faults. Interpretation is referred to by [9,11].

Line NHD224 starts from the vicinity of well LF35-1-1, where 1.6 km thick Mesozoic strata were drilled. The profile clearly shows several major reflectors, Tg, Th, and Tm, to depth (Figure 8). According to [11] Tg, featuring the strongest amplitude and highest continuity, is recognized as the base of the Cenozoic Erathem. Th and Tm, featuring high amplitude reflections, are the base of the Cretaceous and top of the early Jurassic (J1), respectively. The Mesozoic strata become thicker toward the southeast. In the range of shot point 3200–3778, the Cenozoic layers are cut by frequent faults, while the underlying Mesozoic Erathem appears too chaotic to specify any interface or major fault. In terms of the dips of the deep reflectors, this range sits a paleo-depocenter. On both sides of the chaotic zone, reflectors are visible deep to ~3 s and ~4 s, respectively, implying 4 and 6 km thick Mesozoic strata, using a velocity of 4 km/s. Hence, the thickness of the pristine Mesozoic sediment within the chaotic zone is maybe at least 5 km.

Figure 9a is a composite MCS profile, comprising of the northwestern section of Line L2016-2 (4 folds) and DS1776 (120 folds). It intersects with NHD224, in the mid-east of the CSD (Figure 2), where the Mesozoic strata are ~5 km or thicker, while the reflections are chaotic and weak. It reveals that there are several submarine hills, the highest one is H110, which is a well conic edifice in 2D view, featuring eruptive structure. Over H110, the overlying Cenozoic reflections are segmented by steep faults, while the deep reflections become chaotic, excepting only few localities, where highly folded layers are visible southeast of it. Given the thick Mesozoic layer around, the chaotic reflection zone below H110 is interpreted as comprising of highly fractured Mesozoic strata (Figure 9b).

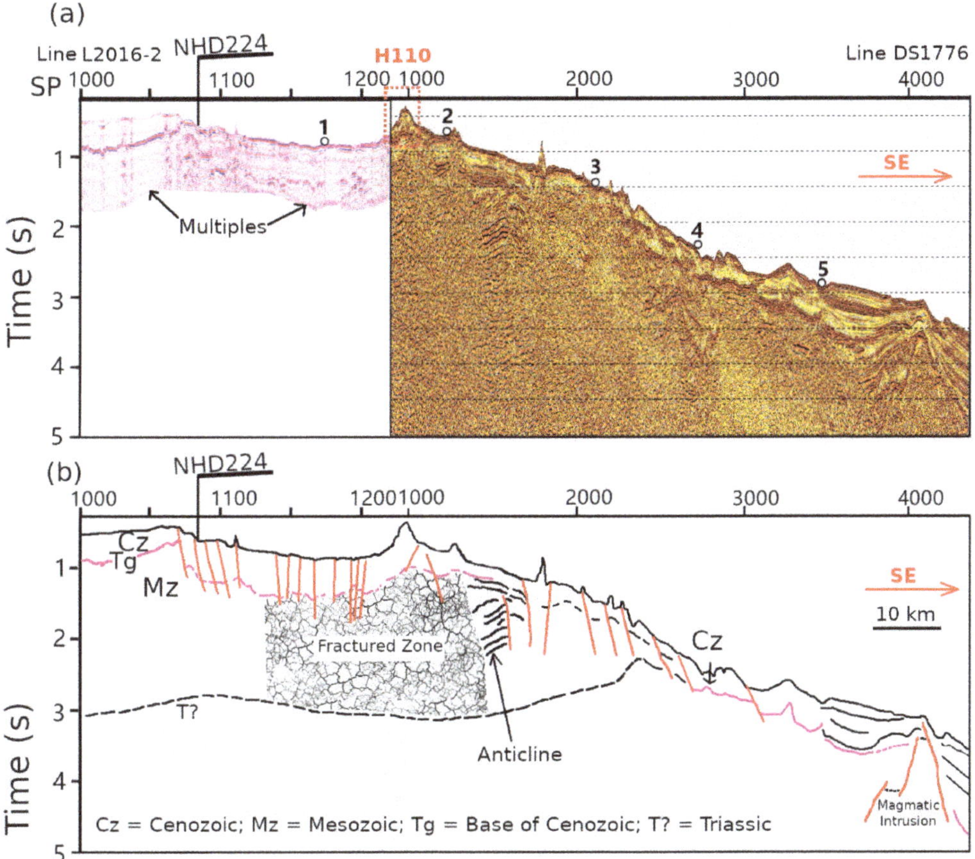

Figure 9. A composite MCS profile of L2016-2 and DS1776 (**a**) and interpretation (**b**). There shows several outcrop hills in the upper slope and buried ones in the lower slope. Because of multiples, the section below 1.8 s in the L2016-2 profile is not shown.

4. Discussion

4.1. Phase Drop and Low-Velocity Zone

Three sediment-related phase drops are visible in the records of the OBSs near H110 (Figure 3). Generally, phase drops can be caused by a major fault, steep interface, or low-velocity zone. The MCS profile (Figure 9a) shows neither major faults nor steep interface underlying H110 that would result in the phase drops, so they should have arisen from certain low-velocity variations, being consistent with the inversed low-velocity structure (Figure 5).

4.2. Velocity Structure and Sedimentary Layer Division

In CSD, the Mesozoic and Cenozoic strata feature much different velocity [9]. The Cenozoic strata comprise of the Neogene and Quaternary which are thin (totally <1 km) and shortly lithified, thus giving very low-velocity, while the Mesozoic layers remain great thickness (>5 km), despite high uplift and erosion. The Mesozoic strata includes Cretaceous red beds, which are of high-velocity (up to 5.5 km/s). A case profile shows velocity jumps to 4.5–5.0 km/s, for the Mesozoic from 1.8–2.4 km/s, for the mid-Miocene [9]. Over the CSD, the sedimentary column can be simplistically subdivided by seismic velocity as Cenozoic

(1.5 km/s to 3.5 km/s) and Mesozoic (3.5 km/s to 5.5 km/s). Applying such a rule of thumb to the velocity profile L2016-2 (Figure 6), the Cenozoic and Mesozoic strata beneath H110 can be roughly estimated as ~1.5 and 5.5 km thick, respectively, which is roughly comparable to the MCS interpretation (Figure 9).

4.3. Analysis of the Hill H110 and the Low-Velocity Zone

The submarine hill H110 features typical eruptive structure. An eruptive hill can be formed by magma volcanism, in most cases, or by mud volcanism, in other cases. Whether the hill is a magmatic volcano or mud volcano is discussed here.

The low-velocity zone below H110 centers at 4.5 km deep, a depth of the Mesozoic layers. It can be caused by geofluid or melt. Suppose the low-velocity (3.3–4 km/s) zone to be a melt-rich magma chamber, its appearance in shallow level (3–6 km, Figure 5) per magma should have given rise to intense magmatism upward with a great energy, most likely eruption as hill H110. If so, it would have cooled rapidly, making it hard to retain low-velocity in the long term. Yet, the cooled magma would have generated a high-velocity zone which, however, has not been resolved in our case nor in other reports [18]. Though not improbably, it is less likely an active magma chamber.

Additionally, the samples [22] dredged from top of H110 contain no igneous rocks from young volcanism. In fact, the samples (Figure 10) are rich of deep-water organisms and authigenic carbonate nodules, which are always generated by biological methane oxidation [23]. In terms of the X-ray diffraction (XRD) composition analysis (Table 2), beside calcite and high-magnesian calcite, the carbonate nodules contain abundant illite, kaolinite, and chlorite, which are usually sourced from buried and warmed deep strata [24]. The coeval mineral assemblage [25] and rich organisms can be deemed proxies of oil and gas leakage.

Figure 10. (**a**) A slice of authigenic carbonate nodule, sampled from the surface of H110 (processed and analyzed in SCISO). Along with calcite, the constituent minerals of the nodule contain illite, kaolinite, and chlorite, reflecting sources from deep substrates; (**b**) a microscopic thin section view of (**a**), showing rich biological fossils, which are unidentified.

The low-velocity zone in Figure 5 corresponds spatially to the fractured zone in Figure 9. The velocity value, 3.3–4 km/s, is slightly less than that of the surrounding Mesozoic strata (3.5–5.5 km/s) and can be correlated to high saturation porosity, 10–30% (Figure 11), according to the Mavko, et al. [26] review on reservoir rocks. Thus, it is most likely for rich geofluid to remain in the low-velocity zone below H110. According to the aforementioned interpretation of the MCS profile (Figure 9), the fractured zone lies within thick Mesozoic strata, mainly Cretaceous and Jurassic. As Jurassic interlayers were found with a high content of organic carbon (~1.5%) in Well LF35-1-1, and tend to be thicker southeastward [11]; it is reasonable that the low-velocity zone be rich of hydrocarbon (Figure 5). In terms of these features and the hydrocarbon leakage, hinted by the surface samples, H110 can be judged as a hydrocarbon-related mud volcano, rather than an igneous volcano.

Table 2. Minerals of representative samples (Figure 10a) from H110, analyzed using X-ray diffraction.

Mineral	Ratio
illite	14.50%
high-magnesian calcite	36.00%
calcite	31.80%
kaolinite	3.10%
chlorite	5.10%
quartz	6.40%
iron-bearing dolomite	3.00%

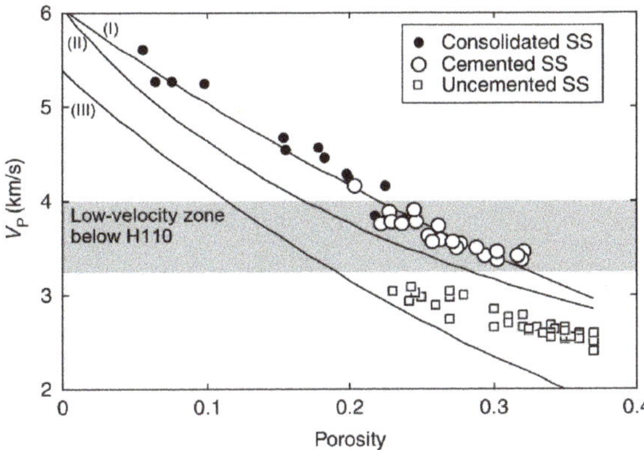

Figure 11. Projection of the low-velocity range below H110 onto the classical statistics [26] of velocity versus porosity in water-saturated sandstones. Curves I and II are based on equations for unconsolidated [27] and consolidated sandstones [28], while III is for quartz-calcite rocks [29].

4.4. Analysis of the High-Velocity Zone HVZ1

The west high-velocity (5.5–6.3 km/s) zone HVZ1 appears at 7–9 km deep, following the low-velocity zone below H110 (Figure 5). Its depth corresponds to the base of the Mesozoic strata, as well as the top upper crust. As the CSD is a relict Mesozoic Basin, suffering less crustal extension than those Cenozoic Basins, basement relief seems smooth within it, though usually ambiguous [9,17]. Thus, HVZ1 is less likely a protruding crustal block. Generally, the seismic velocity of igneous and metamorphic rocks is higher (>5.5 km/s) [29]. Ruled out as sedimentary rocks, HVZ1 can be interpreted as a thick batholith, caused by magma intrusion from deep crust (Figure 12), with considerations of enhanced, though limited, young magmatism around the southern CSD. For example, the east one (HVZ2, Figure 5) coincides spatially with the magmatic intrusive body at around SP 4000 of the MCS profile DS1776 (Figure 9). South of the CSD, magmatic underplating has been suggested in the lower crust by some workers [17,18], with discoveries of a high-velocity (>7 km/s) zone. Furthermore, a volcano-patched band, with high gravity anomalies, were found along the continental–ocean transition south of the CSD [13], within which IODP 368/369 penetrated basalts on a couple of sea mounts [30]. Additionally, a few post-spreading sills within the shallow sedimentary sequences were also recognized [31] in a few localities in the southern PRMB, which are close to the CSD. Its intrusion may follow the paleo-depocenter, where weak zone was inherited.

Considering that HVZ1 succeeds the low-velocity zone below H110, they must have a close three-party relationship, in terms of their origins, from which a scenario can be figured out as follows. As the post-spreading magma intruded the thick Mesozoic strata and solidified, the high-velocity zone HVZ1 formed at the base. When the thick host strata absorbed heat lastingly, the organic matter was promoted to transform into rich hydrocarbon. With accumulation of hydrothermal flow and petroleum, the geo-pressure was increased to produce a hydro-fracture in the host strata. As a result, the low-velocity zone was formed via a fractured zone filled with geofluid. As a comparison, the HVZ2 in the southeast is covered with considerably thinner sediments, where no heat-absorbing, low-velocity zone or mud volcanism appeared.

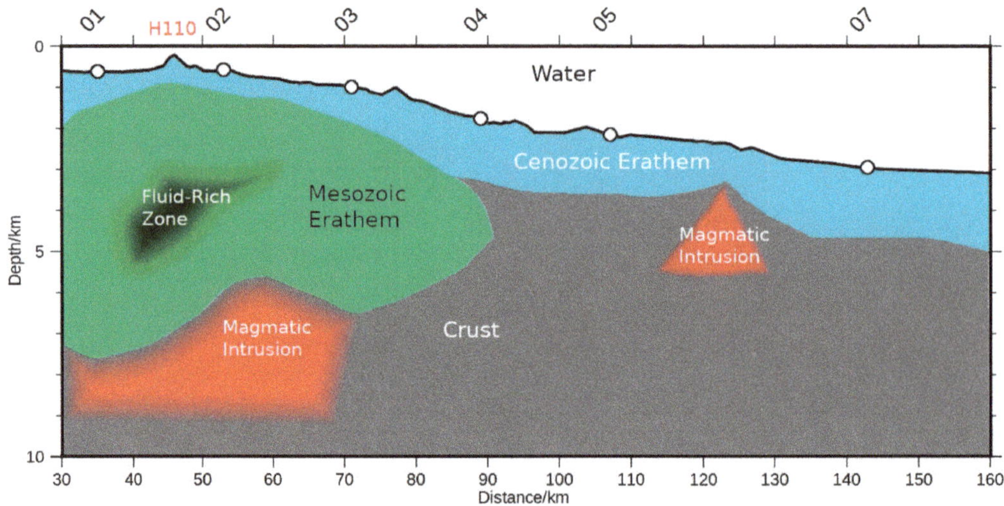

Figure 12. The geological interpretation alone Line L2016-2.

4.5. The Mechanism of Mud Volcano H110

With these arguments, a thermodynamic origin model (Figure 13) can be inferred from mud volcanism at H110, i.e., magma derived from deep crust promoted petroleum production before overpressure was achieved, and gas-bearing fluid were pressed upward along fractures and erupted to, finally, form a mud volcano (Figure 13). Similarly, a small magma intrusion in the depths has also been suggested to have stirred the catastrophic Lusi mud volcanism in Indonesia [2]. H110 represents a thermodynamic driving mud volcanism, different from those mud volcanoes in the Southwest Taiwan Basin, which were caused by plate collision [5]. It may apply to that of mud volcanos in the southwest of DS waters [12], in terms of their same geologic setting.

4.6. Mud Volcanism and Petroleum Exploration in the CSD

A mud volcano is always indicative of hydrocarbon leakage [8], often with methane hydrate forming, when in deep sea. In the southeastern CSD, the Cenozoic layers appear fragmented, often faulted to the seabed, reflecting recent activity. The existence of the large, low-velocity zone below H110 may imply that there is still abundant oil and gas remaining here, hopefully even as reservoirs in some anticlines (Figure 9). In a previous study, numerous mud volcanos were found, likely active, in the southwestern CSD [12,22]. These discoveries should be a favorable clue, in regard to Mesozoic hydrocarbon exploration there. Because the depth of H110 ranges from 300 to 600 m below sea level, it is also possible that gas hydrate accumulates in the deeper part.

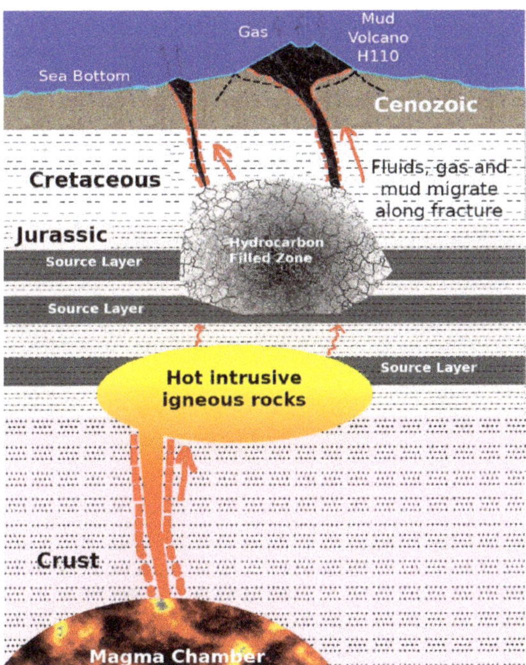

Figure 13. A schematic mud volcanism, driven by magma intrusion at the hill H110.

5. Conclusions

Both OBS and MCS data, acquired along seismic survey line L2016-2, were processed, with a focus on the submarine hill H110. The velocity tomography using the first arrivals of OBS data images a few large velocity structures under it with fairly high resolvability. Collaborating coincident OBS/MCS results and geochemical analysis of samples dredged from the seabed, three conclusions can be drawn, as follows.

(1) A large, low-velocity (3.3–4 km/s) zone, found below the conic hill H110, corresponds spatially to the fractured zone of Mesozoic strata in the MCS profile, likely a fluid-rich fractured zone. Dredged samples from the H110 surface, containing rich organisms, as well clay minerals, such as illite, kaolinite, and chlorite, are indicative hydrocarbon seepage. Therefore, the conic hill H110, featuring an eruptive structure, can be inferred as a mud volcano.

(2) A high-velocity (5.5–6.3 km/s) zone, which appears 7 km deep, under the low-velocity zone, is ascribed to magma intrusion. Thus, the mud volcanism at H110 may be thermodynamically driven by magma intrusion into the Mesozoic sedimentary layers, rich in organic matter.

(3) The mud volcanism and remains of the large low-velocity zone, associated with the Mesozoic basin, implies an abundant hydrocarbon source and potential reservoir, thus a likely favorable clue for Mesozoic petroleum exploration.

Author Contributions: Conceptualization, Y.Y.; data curation, Y.Y., J.Y. and C.C.; methodology, Y.Y. and J.Y.; software, Y.Y., J.Y. and C.C.; resources, J.Y., Y.W. and G.Z.; formal analysis, Y.Y., J.Y. and C.C.; Project administration, Y.W. and J.Y.; Supervision, L.W. and J.L.; Validation, J.Y.; visualization, Y.Y.; writing—original draft, Y.Y.; writing—reviewing & editing, J.L.; funding acquisition, J.L. and Y.W. All authors have read and agreed to the published version of the manuscript.

Funding: This work was financially supported by National Natural Science Foundation of China (41874156, 42074167, U1901217, 91855101); open fund of SINOPEC Key Laboratory of Geophysics (33550006-20-ZC0599-0012); Key Special Project for Introduced Talents Team of Southern Marine Science and Engineering Guangdong Laboratory (Guangzhou) (GML2019ZD0104); Special Support Program for Cultivating High-level Talents in Guangdong Province (2019BT02H594). The data and samples were collected with support from NSFC for sharing cruises (NORC2016-05, 06; NORC2017-05, 06, 07).

Data Availability Statement: The seismic data and samples are proprietary by SCISO and GMGS.

Acknowledgments: We appreciate 3 anonymous reviewers and associated editor for their crucial and constructive comments. Generic Mapping Tools (GMT) software is used to produce some of the figures in this study.

Conflicts of Interest: The authors declare no conflict of interest.

References

1. Mazzini, A.; Etiope, G. Mud volcanism: An updated review. *Earth-Sci. Rev.* **2017**, *128*, 81–112. [CrossRef]
2. Mohammad, J.F.; Anne, O.; Matteo, L.; Karyono, K.; Adriano, M. The Plumbing System Feeding the Lusi Eruption Revealed by Ambient Noise Tomography. *J. Geophys. Res. Solid Earth* **2017**, *122*, 8200–8213.
3. Somoza, L.; Díaz-del-Río, V.; León, R.; Ivanov, M.; Fernández-Puga, M.C.; Gardner, J.M.; Hernández-Molina, F.J.; Pinheiro, L.M.; Rodero, J.; Lobato, A.; et al. Seabed morphology and hydrocarbon seepage in the Gulf of Cádiz mud volcano area: Acoustic imagery, multibeam and ultra-high resolution seismic data. *Mar. Geol.* **2003**, *195*, 153–176. [CrossRef]
4. Dill, H.G.; Kaufhold, S. The Totumo mud volcano and its near-shore marine sedimentological setting (North Colombia)—From sedimentary volcanism to epithermal mineralization. *Sediment. Geol.* **2018**, *366*, 14–31. [CrossRef]
5. Chen, C.S.; Hsu, S.K.; Wang, Y.; Chung, S.H.; Chen, P.C.; Tsai, C.H.; Liu, C.S.; Lin, H.S.; Lee, Y.W. Distribution and characters of the mud diapirs and mud volcanoes off southwest Taiwan. *J. Asian Earth Sci.* **2014**, *92*, 201–214. [CrossRef]
6. Vogt, P.R.; Cherkashev, G.; Ginsvurg, G.; Ivanov, G.; Milkov, A.; Crane, K.; Sundvor, A.; Pimenov, N.; Egorov, A. Haakon Mosby Mud Volcano provides unusual example of venting. *EOS Trans. AGU* **2011**, *78*, 549–557. [CrossRef]
7. Mazzini, A.; Etiope, G.; Svensen, H. A new hydrothermal scenario for the 2006 Lusi eruption, Indonesia. Insights from gas geochemistry. *Earth Planet. Sci. Lett.* **2012**, *317–318*, 305–318. [CrossRef]
8. Milkov, A.V. Worldwide distribution of submarine mud volcanoes and associated gas hydrates. *Mar. Geol.* **2000**, *167*, 29–42. [CrossRef]
9. Yan, P.; Wang, L.L.; Wang, Y.L. Late Mesozoic compressional folds in Dongsha Waters, the northern margin of the South China Sea. *Tectonophysics* **2014**, *615–616*, 213–223. [CrossRef]
10. Hao, H.; Zhang, X.; You, W.; Wang, R. Characteristics and hydrocarbon potential of Mesozoic strata in eastern Pearl River Mouth Basin, Northern South China Sea. *J. Earth Sci.* **2009**, *20*, 117–123. [CrossRef]
11. Zhong, G.J.; Yan, P.; Sun, M.; Yu, J.H.; Feng, C.M.; Zhao, J.; Yi, H.; Zhao, Z.Q. Accumulation model of gas hydrate in the Chaoshan depression of South China Sea, China. *IOP Conf. Ser. Earth Environ. Sci.* **2020**, *569*, 1–13. [CrossRef]
12. Yan, P.; Wang, Y.L.; Liu, J.; Zhong, G.Z.; Liu, X.J. Discovery of the southwest Dongsha Island mud volcanoes amid the northern margin of the South China Sea. *Mar. Pet. Geol.* **2017**, *88*, 858–870. [CrossRef]
13. Yan, P.; Deng, H.; Zhang, Z.; Jiang, Y. The temporal and spatial distribution of volcanism in the South China Sea region. *J. Asian Earth Sci.* **2006**, *27*, 647–659. [CrossRef]
14. Taylor, B.; Hayes, D.E. Origin and History of the South China Sea Basin. In *The Tectonic and Geologic Evolution of Southeast Asian Seas and Islands: Part 2*; Lamont-Doherty Geological Observatory of Columbia University: New York, NY, USA, 1983.
15. Lei, C.; Ren, J.; Pang, X.; Chao, P.; Han, X. Continental rifting and sediment infill in the distal part of the northern South China Sea in the Western Pacific region: Challenge on the present-day models for the passive margins. *Mar. Pet. Geol.* **2018**, *93*, 166–181. [CrossRef]
16. Li, Y.; Abbas, A.; Li, C.-F.; Sun, T.; Zlotnik, S.; Song, T.; Zhang, L.; Yao, Z.; Yao, Y. Numerical modeling of failed rifts in the northern South China Sea margin: Implications for continental rifting and breakup. *J. Asian Earth Sci.* **2020**, *199*, 104402. [CrossRef]
17. Wei, X.D.; Ruan, A.G.; Zhao, M.H.; Qiu, X.L.; Li, J.B.; Zhu, J.J.; Wu, Z.; Ding, W. A wide-angle OBS profile across Dongsha Uplift and Chaoshan Depression in the mid-northern South China Sea. *Chin. J. Geophys.* **2011**, *54*, 1149–1160.
18. Wan, X.L.; Li, C.F.; Zhao, M.H.; He, E.Y.; Liu, S.Q.; Qiu, X.L.; Lu, Y.; Chen, N. Seismic velocity structure of the magnetic quiet zone and continent-ocean boundary in the northeastern South China Sea. *J. Geophys. Res. Solid Earth* **2019**, *124*, 11866–11899. [CrossRef]
19. Christeson, G. *OBSTOOL: Software for Processing UTIG OBS Data*; University of Texas Institute for Geophysics: Austin, TX, USA, 1995.
20. Zelt, C.A. Lateral velocity resolution from three-dimensional seismic refraction data. *Geophys. J. Int.* **1998**, *135*, 1101–1112. [CrossRef]
21. Zelt, C.A.; Barton, P.J. Three-dimensional seismic refraction tomography: A comparison of two methods applied to data from the Faeroe Basin. *J. Geophys. Res.* **1998**, *103*, 7187–7210. [CrossRef]
22. Yan, P.; Wang, Y.L.; Yu, J.H.; Luo, W.; Liu, X.J.; Jin, Y.B.; Li, P.C.; Liu, J.; Zhong, G.J.; Yi, H. Surveying mud volcanoes over the Dongsha Waters in the South China Sea. *J. Trop. Oceanogr.* **2021**, *40*, 34–43. (In Chinese)
23. Lein, A.Y. Authigenic Carbonate Formation in the Ocean. *Lithol. Miner. Resour.* **2004**, *39*, 1–30. [CrossRef]
24. Pevear, D.R. Illite and hydrocarbon exploration. *Proc. Natl. Acad. Sci. USA* **1999**, *96*, 3440–3446. [CrossRef] [PubMed]

25. Milkov, A.V.; Vogt, P.R.; Crane, K.; Lein, A.Y.; Sassen, R.; Cherkashev, G.A. Geological, geochemical, and microbial processes at the hydrate-bearing Håkon Mosby mud volcano: A review. *Chem. Geol.* **2004**, *205*, 347–366. [CrossRef]
26. Mavko, G.; Mukerji, T.; Dvorkin, J. Empirical relations. In *The Rock Physics Handbook: Tools for Seismic Analysis of Porous Media*; Cambridge University Press: Cambridge, UK, 2009.
27. Raymer, L.L.; Hunt, E.R.; Gardner, J. *An Improved Sonic Transit Time-To-Porosity Transform*; Silverchair: Lafayette, LA, USA, 1980.
28. Wyllie, M.; Gregory, A.; Gardner, L. Elastic wave velocities in heterogeneous and porous media. *Soc. Explor. Geophys.* **1956**, *21*, 1–209. [CrossRef]
29. Gardner, G.H.F.; Gardner, L.W.; Gregory, A.R. Formation velocity and density-the diagnostic basics for stratigraphic traps. *Geophysics* **1974**, *39*, 770–780. [CrossRef]
30. Larsen, H.C.; Mohn, G.; Nirrengarten, M.; Sun, Z.; Stock, J.; Jian, Z.; Klaus, A.; Alvarez-Zarikian, C.A.; Boaga, J.; Bowden, S.A.; et al. Rapid transition from continental breakup to igneous oceanic crust in the South China Sea. *Nat. Geosci.* **2018**, *11*, 782–789. [CrossRef]
31. Sun, Q.; Wu, S.; Cartwright, J.; Wang, S.; Lu, Y.; Chen, D.; Dong, D. Neogene igneous intrusions in the northern South China Sea: Evidence from high-resolution three dimensional seismic data. *Mar. Pet. Geol.* **2014**, *54*, 83–95. [CrossRef]

Article

Modal Analysis of Tubing Considering the Effect of Fluid–Structure Interaction

Jiehao Duan [1,2], Changjun Li [1,2,*] and Jin Jin [3]

[1] College of Petroleum Engineering, Southwest Petroleum University, Chengdu 610500, China; 201611000075@stu.swpu.edu.cn
[2] CNPC Key Laboratory of Oil & Gas Storage and Transportation, Southwest Petroleum University, Chengdu 610500, China
[3] CCDC Downhole Service Company, CNPC, Chengdu 610500, China; tt190710@126.com
* Correspondence: lichangjunemail@sina.com; Tel.: +86-187-2849-5286

Abstract: When tubing is in a high-temperature and high-pressure environment, it will be affected by the impact of non-constant fluid and other dynamic loads, which will easily cause the tubing to vibrate or even resonate, affecting the integrity of the wellbore and safe production. In the structural modal analysis of the tubing, the coupling effect of the fluid and the tubing needs to be considered at the same time. In this paper, a single tubing is taken as an example to simulate and analyze the modal changes of the tubing under dry mode and wet mode respectively, and the effects of fluid solid coupling effect, inlet pressure, and ambient temperature on the modal of the tubing are discussed. After considering the fluid–structure interaction effect, the natural frequency of tubing decreases, but the displacement is slightly larger. The greater the pressure in the tubing, the greater the equivalent stress on the tubing body, so the natural frequency is lower. Furthermore, after considering the fluid–solid coupling effect, the pressure in the tubing is the true pulsating pressure of the fluid. The prestress applied to the tubing wall changes with time, and the pressures at different parts are different. At this time, the tubing is changed at different frequencies. Vibration is prone to occur, that is, the natural frequency is smaller than the dry mode. The higher the temperature, the lower the rigidity of the tubing and the faster the strength attenuation, so the natural frequency is lower, and tubing is more prone to vibration. Both the stress intensity and the elastic strain increase with the increase of temperature, so the displacement of the tubing also increases.

Keywords: tubing; modal analysis; fluid-structure interaction; inlet pressure; temperature

Citation: Duan, J.; Li, C.; Jin, J. Modal Analysis of Tubing Considering the Effect of Fluid–Structure Interaction. *Energies* **2022**, *15*, 670. https://doi.org/10.3390/en15020670

Academic Editor: Mofazzal Hossain

Received: 17 November 2021
Accepted: 14 January 2022
Published: 17 January 2022

Publisher's Note: MDPI stays neutral with regard to jurisdictional claims in published maps and institutional affiliations.

1. Introduction

With the continuous development of high-temperature, high-pressure, and high-yield gas wells, the difficulty of natural gas exploitation is increasing, which puts forward higher requirements for wellbore integrity [1,2]. Tubing is an important part of natural gas development. It is in a special environment and the load is complex and changeable [3,4]. The tubing will be affected by the impact of unsteady fluid and other dynamic loads, which easily cause vibration and even resonance of the tubing string [5,6]. This will not only affect the mechanical properties of the tubing, but accelerate the fatigue failure and affect the integrity of the wellbore and safe production [7].

Many domestic and foreign scholars have conducted research on the modal analysis of pipes or tubing strings. Lenwoue [8] established a poro-elasto-plastic model using nonlinear finite element software, which aimed at analyzing the influence of drill string vibration cyclic loads on the development of the wellbore natural fracture. Xu [9] established an FEM of tubing erosion based on erosion mode, fluid–solid coupling, and the impact of sand. The erosion rule and failure region on the tubing were obtained. Zhang [10] studied the dynamical modeling, multipulse orbits, and chaotic dynamics of cantilevered pipe conveying pulsating fluid with harmonic external force using the energy-phase method.

Jiang [11] analyzed the modal characteristics of the simplified U-shaped tubing structure under fluid–structure interaction using ANSYS Workbench software. Wróbel [12] undertook experimental investigations in order to identify the influence of hydraulic pressure on the natural frequencies and mode shapes of a basic hydraulic line model. Xin [13] analyzed the two typical tubings (straight pipe and curved pipe) using the finite element method (FEM) in ADINA software, in order to solve the problem whereby the vibration of a hydraulic pipe poses a severe safety concern in aircraft. Lubecki [14] presented experimental studies of dynamic changes in the length of a microhydraulic hose under the influence of step pressure and flow load. Chai [15] proposed a simplified bilinear stiffness and damping model according to hysteresis loops of the clamp obtained from experimental testing. Then, the finite element model of an L-type tubing system with clamps was established using Timoshenko beam theory in combination with the aforementioned stiffness-damping model. Sekacheva [16] proposed a method for determining the probability of the occurrence of increased vibrations using modal analysis in the ANSYS Workbench software package. Wu [17] used the multiple scale method to obtain an equivalent nonlinear wave equation from the complicated nonlinear governing equation describing the fluid conveyed in a pipe. Guo [18] has established a nonlinear flow-induced vibration model of tubing string by using the micro-element method and energy method with the Hamiltonian variational principle, in order to address the vibration failure problem of the tubing string induced by high-speed fluid flow in the tubing of HPHT oil and gas well. Nan [19] comprehensively analysed the dynamic failure modes and safety criteria of buried tubings based on the study of the dynamic response characteristics of tubings.

However, when analyzing the mechanical properties of the tubing, it is not rigorous to only conduct dry modal analysis and consider less the influence of the fluid–structure interaction effect on the modal characteristics of the tubing. It is necessary to consider the characteristics of fluid and tubing and the interaction between them at the same time, which not only effectively saves the analysis time and cost, but also ensures that the simulation results are closer to the laws of physical phenomena [20,21]. However, it also increases the complexity of tubing mechanics analysis [22].

Based on the mathematical principles of fluid mechanics, solid mechanics, and vibration mechanics, the modal changes of the tubing under dry mode and wet mode are simulated and analyzed by using the finite element method, and the effects of fluid–structure interaction, inlet pressure, and temperature in the tubing on the calculation results are discussed.

2. Vibration Mechanics Model of Tubing

Structural mode mostly refers to the mode in vacuum without considering the influence of a surrounding fluid. This mode is called "dry mode" [23]. There is fluid in the tubing, and some tubings are even placed in the fluid, so that their inner and outer walls are surrounded by the fluid. The complex changes of fluid density and pressure have a great impact on the tubing mode, which cannot be ignored. This inherent mode taking into account the influence of surrounding fluid is called "wet mode".

The fluid–structure interaction should be considered in the stress analysis of the tubing. The coupling mode adopted is bidirectional fluid–structure interaction [24,25], that is, the data exchange is mutual. Both the fluid analysis results are transmitted to the solid structure, and the solid analysis results (such as displacement, velocity and acceleration) are transmitted back to the fluid. The tubing and the fluid are discussed separately. The vibration equation of the tubing is:

$$[M_t]\{\ddot{x}\} + [C_t]\{\dot{x}\} + [K_t]\{x\} = \left\{F_f\right\} + \{F\} \tag{1}$$

where $[M_t]$, $[C_t]$, and $[K_t]$ are the mass matrix, damping matrix, and stiffness matrix of tubing; $\left\{F_f\right\}$ is the excitation of fluid to the tubing; $\{F\}$ is the external excitation of the tubing; $\{x\}$ is the displacement column vector of the tubing, $\{\dot{x}\}$ is the velocity column vector of the tubing; $\{\ddot{x}\}$ is the acceleration column vector of the tubing.

The wave equation of the fluid is [26]:

$$\left[M_f\right]\{\ddot{p}\} + \left[C_f\right]\{\dot{p}\} + \left[K_f\right]\{p\} = \{F_t\} \tag{2}$$

where $[M_f]$, $[C_f]$, and $[K_f]$ are the mass matrix, damping matrix, and stiffness matrix of the fluid; $\{F_t\}$ is the excitation of the tubing to the fluid; $\{p\}$ is the pressure column vector of fluid, and $\{\dot{p}\}$ and $\{\ddot{p}\}$ are the first-order derivative column vector and the second-order derivative column vector of fluid pressure with respect to time, respectively.

The excitation of fluid to tubing $\left\{F_f\right\}$ and the excitation of tubing to fluid $\{F_t\}$ can be expressed as follows:

$$\{F_t\} = -\rho_f \left[I_{tp}\right]\{\ddot{x}\} \tag{3}$$

$$\left\{F_f\right\} = \left[I_{tp}\right]\{\ddot{p}\} \tag{4}$$

where ρ_f is the fluid density and $[I_{tp}]$ is the inertia matrix of the tubing cross section.

The simultaneous Equations (1)–(4) can be used:

$$\begin{cases} [M_t]\{\ddot{x}\} + [C_t]\{\dot{x}\} + [K_t]\{x\} - \left[I_{tp}\right]^{\mathrm{T}}\{\ddot{p}\} - \{F\} = 0 \\ \left[M_f\right]\{\ddot{p}\} + \left[C_f\right]\{\dot{p}\} + \left[K_f\right]\{p\} + \rho_f\left[I_{tp}\right]\{\ddot{x}\} = 0 \end{cases} \tag{5}$$

Standardize Equation (5):

$$[M]\{\ddot{\theta}\} + [C]\{\dot{\theta}\} + [K]\{\theta\} - \{F\} = 0 \tag{6}$$

where $[M]$, $[C]$, and $[K]$ are the mass coupling matrix, damping coupling matrix, and stiffness coupling matrix of tubing and fluid, respectively; $\{\theta\}$ is the parameter column vector, while $\{\dot{\theta}\}$ and $\{\ddot{\theta}\}$ are the first-order derivative column vector and the second-order derivative column vector of parameter with respect to time, respectively.

$$[M] = \begin{bmatrix} M_f & \rho_f I_{tp} \\ -I_{tp} & M_t \end{bmatrix} \tag{7}$$

$$[C] = \begin{bmatrix} C_f & 0 \\ 0 & C_t \end{bmatrix} \tag{8}$$

$$[K] = \begin{bmatrix} K_f & 0 \\ 0 & K_t \end{bmatrix} \tag{9}$$

$$\{\ddot{\theta}\} = \left\{ \begin{array}{c} \ddot{p} \\ \ddot{x} \end{array} \right\}, \{\dot{\theta}\} = \left\{ \begin{array}{c} \dot{p} \\ \dot{x} \end{array} \right\}, \{\theta\} = \left\{ \begin{array}{c} p \\ x \end{array} \right\} \tag{10}$$

The wet modal analysis of the tubing can be carried out by combining Equations (6)–(10).

If the influence of damping and external excitation of the tubing is ignored, i.e., $[C_t] = 0$, $\{F\} = 0$, the simultaneous Equations (1)–(4) can be used:

$$\begin{cases} [M_t]\{\ddot{x}\} + [K_t]\{x\} - \left[I_{tp}\right]^{\mathrm{T}}\{\ddot{p}\} = 0 \\ \left[M_f\right]\{\ddot{p}\} + \left[K_f\right]\{p\} + \rho_f\left[I_{tp}\right]\{\ddot{x}\} = 0 \end{cases} \tag{11}$$

Standardize Equation (5):

$$[M]\{\ddot{\theta}\} + [K]\{\theta\} = 0 \tag{12}$$

The wet modal analysis ignoring the influence of damping and external excitation of the tubing can be carried out by combining Equations (7), (9), (10), and (12).

3. Dry Mode Finite Element Analysis of Tubing

3.1. Establishment of Finite Element Mechanical Mode

As gas storage wells often face extreme conditions of "large injection and large production" and "strong injection and strong production", it is more convincing to use tubing calculations for gas storage wells. It is difficult to establish the full-scale model of the tubing due to the length. Therefore, a single tubing with a length of 9.6 m is taken as an example to establish the finite element model. The outer diameter is 177.8 mm (some gas storage wells have larger tubing sizes), the wall thickness is 10.36 mm, the elastic modulus of the tubing is 206 GPa, the Poisson's ratio is 0.3, and the density is 7850 kg/m^3. The tubing model is discretized with eight node hexahedron elements. The number of nodes is 167,060 and the number of elements is 24,966. The tubing mesh is shown in Figure 1. Taking any hexahedral element on the tubing as an example, its dimensions are dx, dy, and dz. The normal stress is σ_x, σ_y, σ_z, and the shear stress is τ_{xy}, τ_{yz}, τ_{zx}. The corresponding three-dimensional stress diagram is shown in Figure 2.

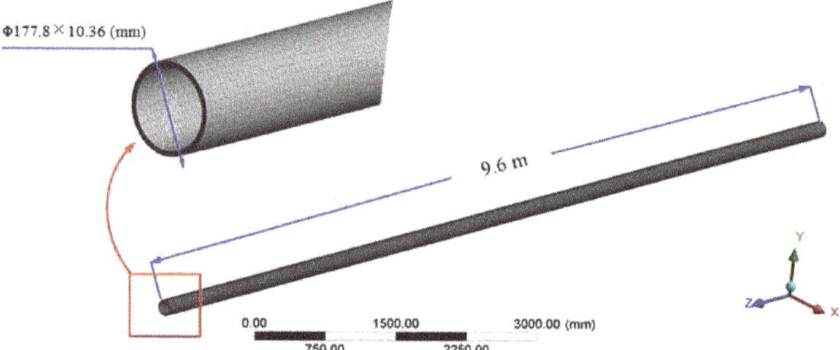

Figure 1. Mesh of single tubing model.

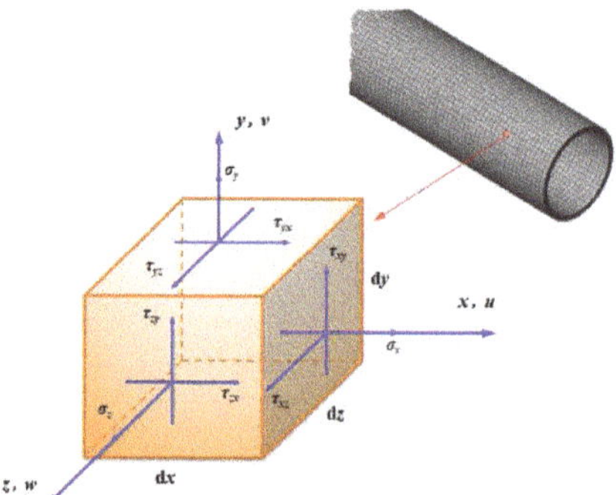

Figure 2. Three dimensional stress diagram of hexahedral element of tubing.

The external load of the tubing mainly includes the pressure of the fluid, the gravity of the tubing, and temperature. Before the wet mode simulation, the vibration mode of the tubing in vacuum should be determined, and the influence of fluid flow on the tubing mode, i.e., dry mode analysis, should not be considered temporarily. Because at least one end of the tubing is fixed, the constrained mode of the tubing is simulated directly.

3.2. Simulation

ANSYS workbench finite element software was used to solve the modal analysis of the tubing.

After the finite element mechanics model is established, constraints and loads are applied to it in the "Mechanical" module of the workbench solver. The constraint conditions for the tubing are generally divided into no constraint, one end constraint and both ends constraint. Since the solver cannot solve unconstrained conditions, two settings of one end constraint and both ends of the tubing are set, and the constraint mode is fixed. When performing load sensitivity analysis, it is also necessary to apply initial pressure and temperature loads.

Set the initial conditions of the modal analysis in the "Modal" module. The default maximum number of modes for the solver is 6. These modes are the rigid body modes that each object has, that is, the translational modes in 3 directions and the rotational modes in 3 directions, so the maximum number of modes needs to be increased (only the first 6 modes are considered for tubing modal analysis) [27] to find the corresponding modal at more frequencies. Here, the maximum number of modes is increased to 20. Then, it can be solved.

3.3. Dry Modal Analysis of Tubing

Under the boundary conditions of constraining only one end and both ends of the tubing, the modal analysis is carried out. The simulation results of the natural frequencies of each order of the oil tubing in both cases are shown in Figure 3. It can be seen from the figure that after constraining both ends of the tubing, the natural frequency of the tubing body is higher than when only one end is constrained. The reason for this is that after the two ends of the tubing are fixed, the stiffness of the tubing increases, that is, the excitation that promotes the resonance of the tubing increases, and the tubing is not prone to vibration.

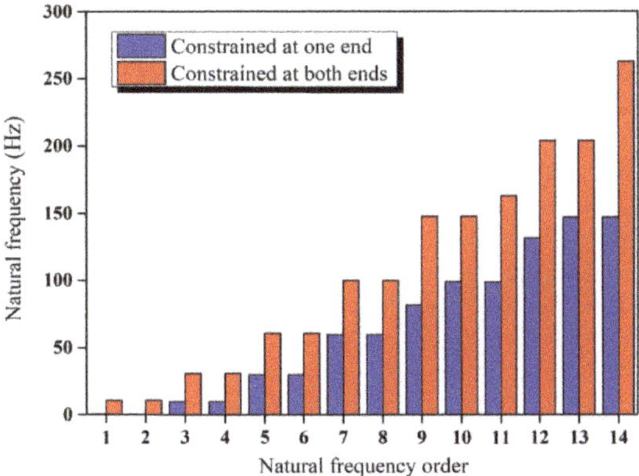

Figure 3. Comparison of natural frequencies of tubing under two constraints.

The number of modes depends on the node degrees of freedom, which should be distinguished from the model. It is the element node rather than the overall model. After the element is divided, each node corresponds to a degree of freedom. The tubing model has 167,060 nodes, but it is not necessary to solve all modes, which is also impractical. The low-frequency modal amplitude is the largest and most dangerous, the high-frequency modal amplitude is very small, and the modal research with high frequency is meaningless. Therefore, only the first few modes at low frequencies need to be studied.

As shown in Figures 4 and 5, when only one end of the tubing is constrained, the seventh mode of the tubing is the first-order Y-direction bending mode. That is, it swings back and forth along the radial Y direction of the tubing, and the vibration frequency is 59.556 Hz at this time. There are three bends in the tubing. The maximum deformation occurs at the unconstrained end of the tubing, and the value is 3.152 mm. Similarly, at this frequency, the eighth mode of the tubing is the first-order X-direction bending mode. That is, it swings back and forth along the radial X direction of the tubing. The bending part and the maximum deformation part are similar to the seventh mode.

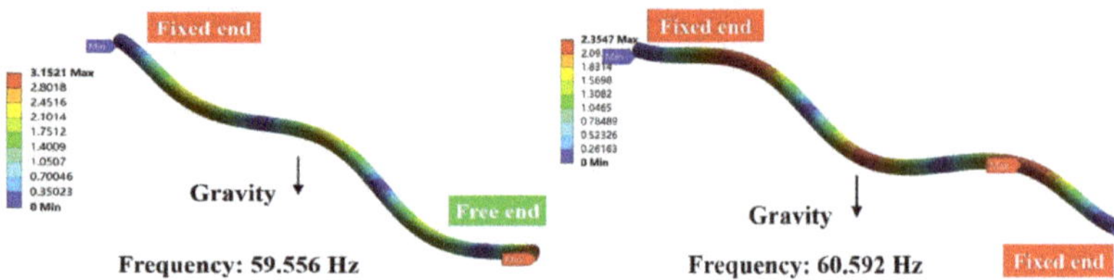

Figure 4. First-order Y-direction bending state of tubing under two constraints.

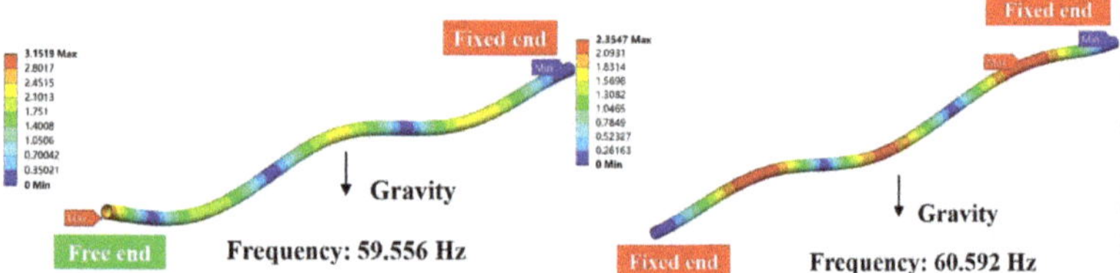

Figure 5. First-order X-direction bending state of tubing under two constraints.

When the two ends of the tubing are constrained, the fifth mode is the first-order Y-direction bending mode, and the vibration frequency is 60.592 Hz. Different from the tubing that only constrains one end, although the tubing also has three bends, the large deformation appears in the three bends. Moreover, due to the action of gravity, the maximum deformation occurs at the lowest bending point (in the direction of gravity acceleration) of the three bends, and the value is about 2.355 mm. At this vibration frequency, the sixth mode of the tubing is the first-order X-direction bending mode, and the bending part and maximum deformation part are similar to the fifth mode.

It can be seen from the above results that when only one end of the tubing is constrained, the other end is in a free state, and the deformation of the whole tubing will be transmitted to the free end. Therefore, for the tubing fixed at the wellhead, an important reason for installing packer near the bottom of the well is to prevent the bottom of the tubing from swinging greatly under the excitation of a specific frequency. However, large deformation will appear in the middle of the tubing after both ends are restrained, which easily causes buckling of the tubing [28]. Therefore, some effective measures need to be taken to deal with such problems, such as thickening the tubing wall thickness, placing centralizers [29], adding titanium alloy tubing or expansion joints, etc.

As shown in Figure 6, the ninth mode of the tubing is the first-order torsional mode when only one end is constrained. That is, the tubing rotates back and forth around its own axis, and the vibration frequency is 81.48 Hz. The maximum deformation of the tubing occurs at the bottom section and the value is 2.335 mm. The 11th mode of the tubing is the first-order torsional mode when the two ends are constrained, and the vibration frequency is 162.98 Hz. The maximum deformation of the tubing occurs in the middle, and the value is basically the same as the displacement of the tubing when only one end is constrained. In this case, the overall displacement of the tubing is symmetrically distributed from the middle to both ends.

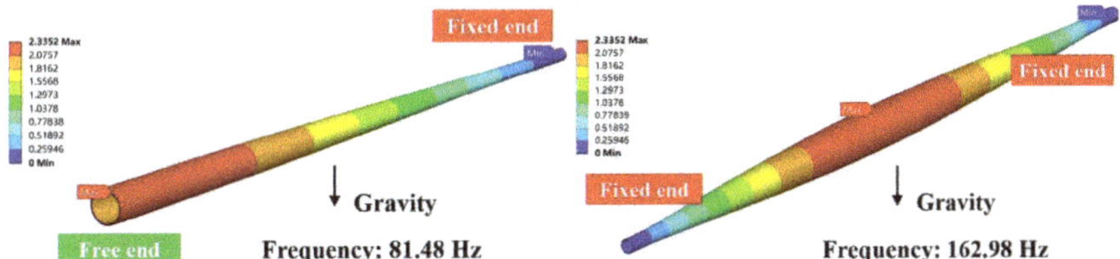

Figure 6. First-order torsional mode of tubing under two constraints.

As shown in Figures 7 and 8, the second-order and third-order X-direction bending modes of the tubing under two constraints are shown. It can be seen that the arch bending number of the tubing increases with the increase of the order, but the vibration mode and large deformation position of the tubing are similar to the first-order. It can also be found that the maximum displacement of the tubing under the two constraints of the second-order X-direction bending modes is 3.127 mm and 2.354 mm, respectively, and the maximum displacement of the tubing under the two constraints of the third-order X-direction bending modes is 3.109 mm and 2.354 mm, respectively. Compared with the first-order displacement, it can be seen that the amplitude of the low-frequency mode is large and the amplitude of the high-frequency mode is small, which verifies the previous conclusion.

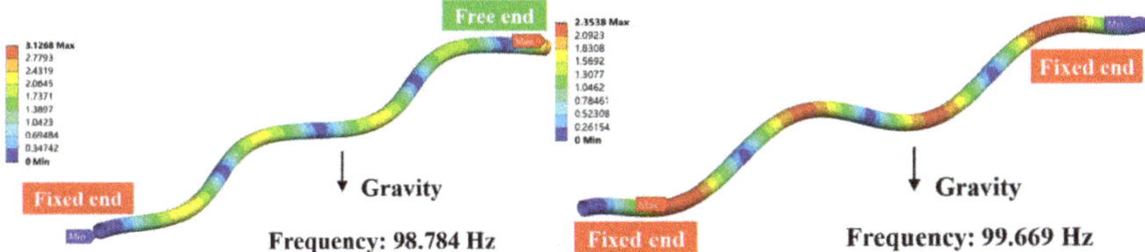

Figure 7. Second-order X-direction bending modes of tubing under two constraints.

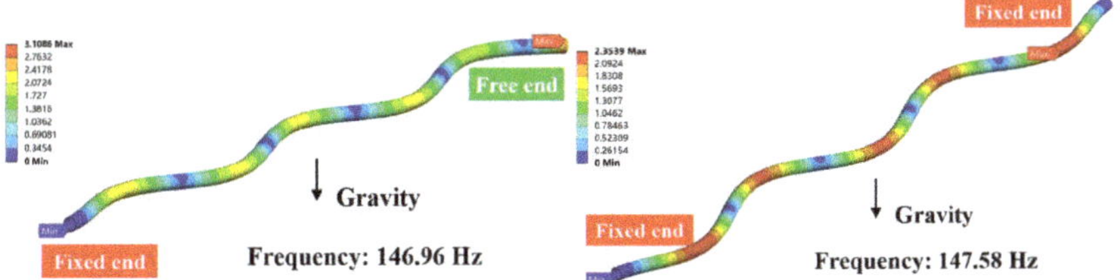

Figure 8. Third-order X-direction bending modes of tubing under two constraints.

4. Wet Modal Finite Element Analysis of Tubing

The influence of fluid flow should be considered in the wet modal analysis of tubing, so the fluid area needs to be established separately in the calculation. Considering the fluid–structure interaction effect, the modal analysis results of the tubing will be more accurate. When the pressure load in the tubing is applied, the imported pressure data are the real flow data of the fluid. Similarly, the influence of deformation on fluid flow in the tubing can also be studied inversely through the calculated tubing deformation data, that is, the bidirectional fluid–structure interaction effect. The established finite element model of the tubing and the fluid in the tubing are shown in Figure 9 (a is the tubing model and b is the fluid model).

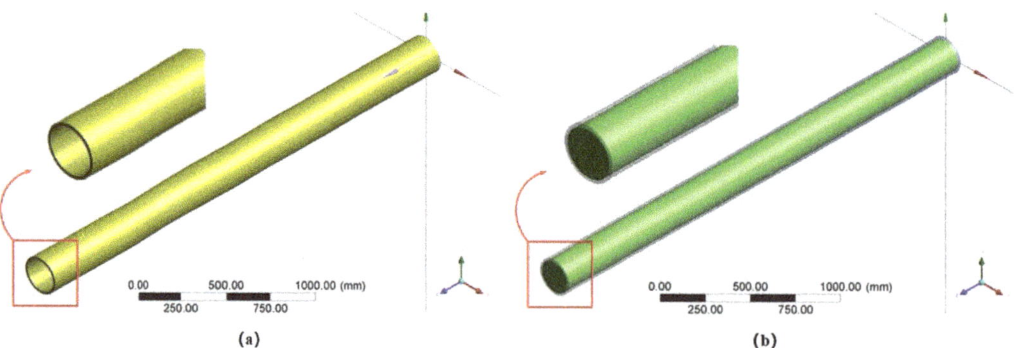

(a)   **(b)**

Figure 9. Finite element model of the tubing and the fluid in the tubing. ((**a**) is the tubing model and (**b**) is the fluid model).

The eight-node hexahedron element is used to mesh the tubing and the fluid model in the tubing. The mesh model is shown in Figure 10. The external wall of all tubing is defined as boundary wall, the surface on the fluid side of the fluid–structure interface is defined as boundary interface$_f$, and the surface on the tubing side of the fluid–structure interface is defined as boundary interface$_s$. The two surfaces form a fluid–structure interaction interface to realize the data transfer between the tubing and the fluid in the tubing. This is also the reason why the model is divided by a hexahedral eight node element. If tetrahedral mesh is used, the calculation results may not converge, resulting in inaccurate modal analysis.

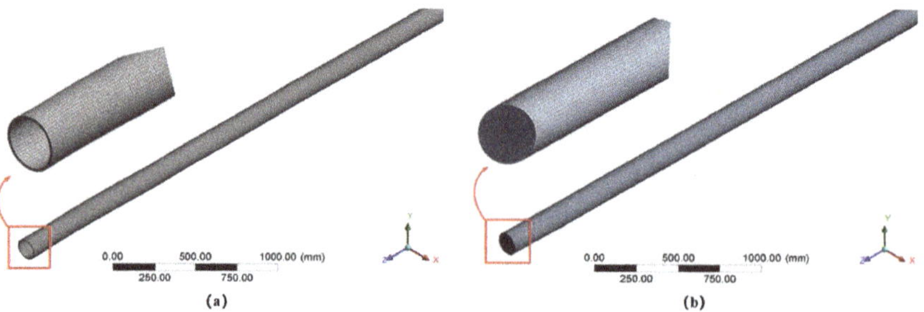

(a)   **(b)**

Figure 10. Mesh generation of the tubing and the fluid in the tubing ((**a**) is the tubing mesh and (**b**) is the fluid mesh).

The calculation process is divided into two steps. Firstly, the fluid flow in the tubing should be simulated, that is, after the boundary conditions of inlet and outlet of the tubing are determined, the pressure and temperature distribution in the tubing under steady-state conditions should be calculated. Secondly, these load data need to be applied to the inner wall of the tubing, as shown in Figure 11. At this time, the load in the tubing is no longer a two-dimensional constant load or a uniformly varying linear load, but a load formed by the real flow of fluid acting on the inner wall of the tubing. The temperature and pressure values of each element divided by the tubing are different, which is difficult to obtained by the analytical method.

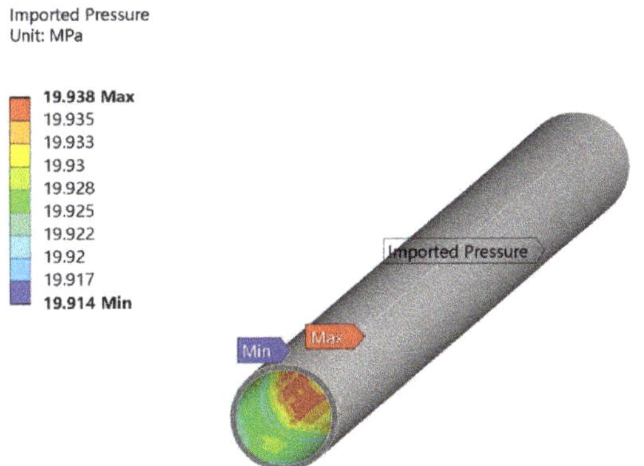

Figure 11. Pressure load caused by fluid flow in tubing inner wall.

5. Analysis of Influencing Factors of Tubing Mode

5.1. Influence of Fluid–Structure Interaction Effect on Tubing Mode

Assuming that the tubing is constrained at both ends, the wet modal analysis of the tubing is carried out, as shown in Figure 12a. It can be seen from the figure that the natural frequency of the tubing considering the fluid–structure interaction effect is small. The reason for this is that after considering the fluid–structure interaction effect, the research object includes not only the tubing, but also the high-temperature and high-pressure natural gas flowing in the tubing. The overall mass increases obviously, so the natural frequency decreases.

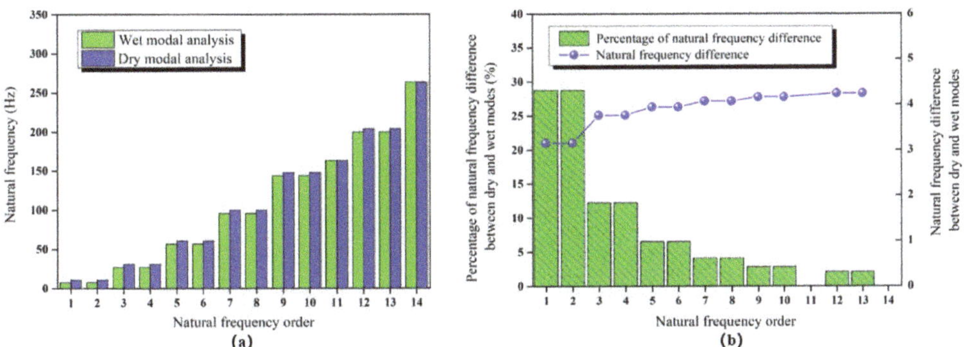

Figure 12. Natural frequencies of tubing under two modal analysis methods ((**a**) is natural frequencies of tubing under two modal analysis methods, (**b**) is difference of natural frequency of tubing under two modal analysis methods).

As can be seen from Figure 12b, with the increase of order, the difference of tubing natural frequency gradually increases, but the difference proportion gradually decreases. This shows that in the process of increasing the order, the deformation parts of the tubing increase, and the influence of fluid–structure interaction on the vibration frequency of the tubing decreases gradually. However, in the actual gas production process, high-order frequency rarely occurs. For the frequency within the 14th order, the influence of fluid–structure interaction effect on tubing mode cannot be ignored.

Figure 13 shows the maximum displacement of the tubing under two modal analysis methods. It can be seen from Figure 13 that the maximum displacement of the tubing is large when the natural frequency is small, so the maximum displacement of the tubing in wet mode is slightly larger. The first and second-order modes of the tubing are single, and there is only one large deformation part. All deformations are concentrated in the center of the tubing since both ends are constrained, and the displacement decreases gradually along both ends of the tubing. For the third-order modes and above, the large deformation parts gradually increase and will not be concentrated in one place, so the maximum displacement decreases significantly. The 11th order mode of the tubing is the first-order torsional mode. At this time, the tubing rotates back and forth around its own axis. The transverse deformation is not obvious as other orders, and the main deformation part is in the middle of the tubing. The 14th order mode of the tubing is tensile mode. The tensile displacement of the tubing at low frequency is not large, and the end of the tubing is fixed, so the maximum displacement is much smaller than other orders. The frequencies within the 11th and 14th orders are small deformation modes, which show the characteristics of similar frequency and small displacement. In the actual production process, in the case of inevitable resonance, if it can be induced to reach the corresponding torsional or tensile mode, the buckling and deformation damage of the tubing can be reduced as much as possible.

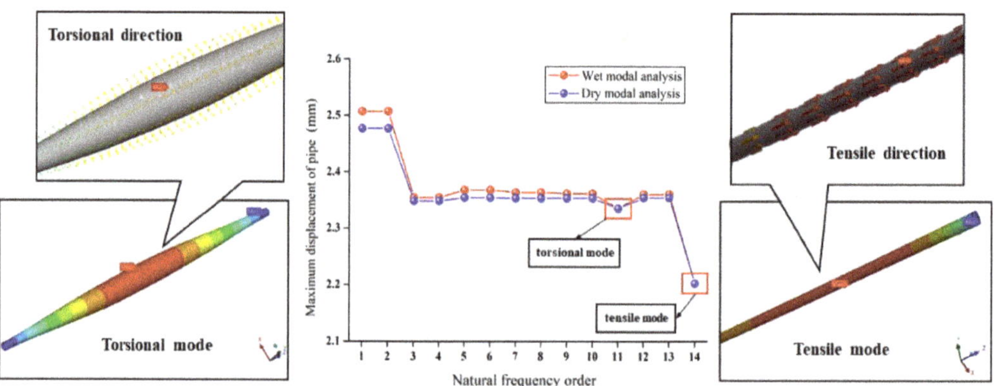

Figure 13. Maximum displacement of tubing under two modal analysis methods.

5.2. Influence of Inlet Pressure on Tubing Mode

The tubing is a three-dimensional cylindrical space and the pressure of flow acting on each point of the inner wall is inconsistent, so the inlet pressure of the tubing is taken as the boundary condition. Both ends of the tubing are constrained, and the tubing modes under the two analysis methods are simulated and calculated. The natural frequencies of the tubing are obtained when the inlet pressure is 10 MPa, 20 MPa, 30 MPa, 40 MPa, 50 MPa, and 60 MPa respectively. The influence of temperature is not considered for the time being. The natural frequencies of the tubing under different inlet pressures of the first 14 orders are extracted, as shown in Tables 1 and 2. The difference curve of tubing natural frequency affected by the fluid–structure interaction effect under different inlet pressure is drawn, as shown in Figure 14.

Table 1. Natural frequencies of tubing under different inlet pressures (considering the fluid-structure interaction effect).

Order	10 MPa	20 MPa	30 MPa	40 MPa	50 MPa	60 MPa
1	10.891	10.604	10.308	10.003	9.6872	9.3599
2	10.891	10.604	10.308	10.003	9.6872	9.3599
3	30.721	30.346	29.966	29.581	29.19	28.793
4	30.721	30.346	29.966	29.581	29.19	28.793
5	60.477	60.071	59.662	59.250	58.836	58.418
6	60.477	60.071	59.662	59.251	58.836	58.418
7	99.550	99.124	98.697	98.268	97.837	97.404
8	99.550	99.125	98.697	98.268	97.837	97.404
9	147.46	147.02	146.58	146.14	145.69	145.25
10	147.46	147.02	146.58	146.14	145.69	145.25
11	163.00	162.97	162.95	162.92	162.89	162.86
12	203.67	203.22	202.76	202.31	201.86	201.40
13	203.67	203.22	202.76	202.31	201.86	201.40
14	263.03	263.03	263.04	263.04	263.05	263.05

Table 2. Natural frequencies of tubing under different inlet pressures (fluid-structure interaction effect is not considered).

Order	10 MPa	20 MPa	30 MPa	40 MPa	50 MPa	60 MPa
1	11.249	10.973	10.688	10.395	10.093	9.7807
2	11.249	10.973	10.688	10.395	10.093	9.7807
3	31.198	30.829	30.455	30.077	29.693	29.303
4	31.198	30.829	30.455	30.077	29.693	29.303
5	60.994	60.592	60.187	59.779	59.368	58.954
6	60.994	60.592	60.187	59.779	59.368	58.954
7	100.09	99.669	99.244	98.817	98.388	97.958
8	100.09	99.669	99.244	98.817	98.388	97.958
9	148.02	147.58	147.14	146.70	146.26	145.82
10	148.02	147.58	147.14	146.70	146.26	145.82
11	163.01	162.98	162.95	162.93	162.90	162.87
12	204.24	203.79	203.34	202.89	202.43	201.98
13	204.24	203.79	203.34	202.89	202.43	201.98
14	263.01	263.02	263.02	263.03	263.04	263.04

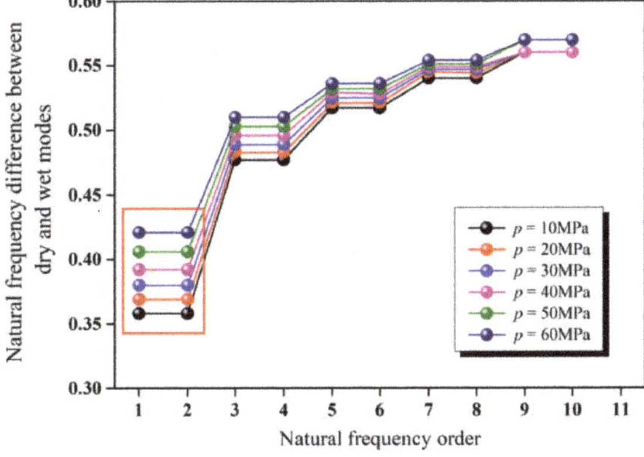

Figure 14. Influence of fluid-structure interaction effect on natural frequency of tubing under different inlet pressure.

According to the data in Tables 1 and 2, the natural frequency of the tubing calculated after considering the fluid–structure interaction effect is small under the same inlet pressure. As shown in Figure 14, in the low-order mode, the difference of tubing natural frequency increases with the gradual increase of inlet pressure. With the increase of order, the difference of tubing natural frequency under different pressures gradually tends to be consistent, indicating that the influence of inlet pressure on tubing natural frequency decreases with the increase of order. When the tubing is in torsional and tensile modes, the natural frequency of the tubing basically does not change.

Figure 15 shows the maximum displacement of the tubing under different inlet pressures. It can be seen that the maximum displacement of tubing corresponding to each order also increases with the gradual increase of inlet pressure. The 11th order is torsional mode and the tubing is twisted along its own axis. The displacement is small, which is independent of the value of inlet pressure. If the fluid–structure interaction effect is not considered, the overall mass of the research object will be reduced, and the maximum displacement of the tubing will be small.

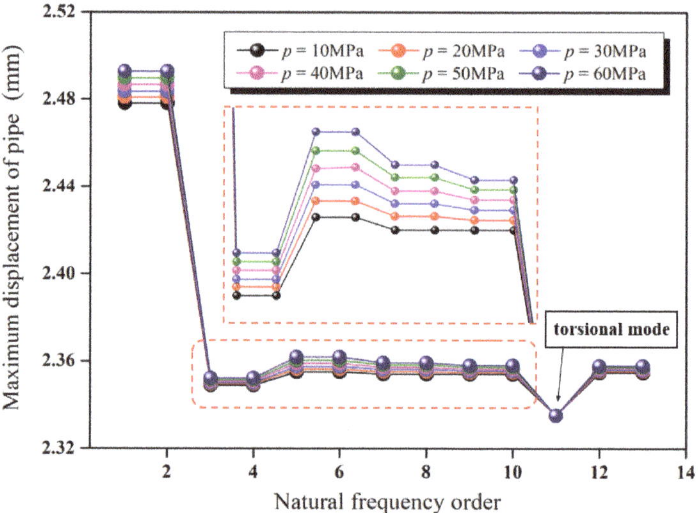

Figure 15. Maximum displacement of tubing under different inlet pressure.

5.3. Effect of Temperature on Tubing Mode

Both ends of the tubing are constrained, and the natural frequencies of the tubing are obtained when the ambient temperature is 30 °C, 40 °C, 50 °C, 60 °C, 70 °C, 80 °C, and 90 °C, respectively. The influence of pressure is not considered for the time being. The natural frequencies of the tubing under different ambient temperatures of the first 14 orders are extracted, as shown in Table 3. The difference curve of tubing natural frequency under different inlet pressure is drawn, as shown in Figure 16.

Table 3. Natural frequencies of tubing under different inlet pressures (considering the fluid-structure interaction effect).

Order	30 °C	40 °C	50 °C	60 °C	70 °C	80 °C	90 °C
1	11.152	10.677	10.178	9.6502	9.0894	8.4884	7.8379
2	11.152	10.677	10.178	9.6502	9.0894	8.4884	7.8379
3	31.068	30.443	29.803	29.149	28.478	27.790	27.083
4	31.068	30.443	29.803	29.149	28.478	27.790	27.084
5	60.854	60.177	59.492	58.799	58.098	57.388	56.669
6	60.854	60.177	59.492	58.800	58.098	57.388	56.669
7	99.946	99.238	98.525	97.807	97.084	96.354	95.620
8	99.946	99.239	98.525	97.807	97.084	96.355	95.620
9	147.87	147.14	146.41	145.67	144.93	144.19	143.44
10	147.87	147.14	146.41	145.67	144.93	144.19	143.44
11	163.03	163.02	163.00	162.99	162.98	162.97	162.95
12	204.09	203.34	202.60	201.84	201.09	200.34	199.58
13	204.09	203.34	202.60	201.85	201.09	200.34	199.58
14	263.02	263.04	263.05	263.07	263.08	263.10	263.12

According to the data in Table 3, the influence trend of temperature on the natural frequency of tubing is similar to that of inlet pressure. The higher the temperature, the smaller the natural frequency of the tubing, which easily causes vibration [30]. With the increase of ambient temperature, the maximum displacement of tubing corresponding to each order increases. The large deformation parts of the tubing corresponding to the first and second orders are concentrated, so the maximum displacement is significantly greater than the following orders. In the 3rd to 14th orders, the maximum displacement of the tubing corresponding to 5th and 6th orders is significantly higher than that of other orders. The vibration modes of these two orders of tubing are first-order bending and there are few overall deformation parts, so the displacement of buckling parts is large. After that, the deformation part of the tubing increases with the increase of the order, and the displacement of each buckling part decreases accordingly. The 11th order is torsional mode, and the tubing displacement is small and independent of the value of ambient temperature.

Temperature has a great impact on the strength of the tubing. The stress intensity and strain of the tubing at different temperatures are studied and analyzed, and the results are shown in Figure 17. When the ambient temperature increases from 30 to 90 °C, the stress intensity and elastic strain will increase with the difference of variation law. The increase of elastic strain shows a linear trend. According to the calculated results, the tubing elastic strain function considering the effect of temperature is constructed, and the back calculation error is less than 5%. The increasing trend of stress intensity decreases slowly with the increase of temperature. Therefore, it is inferred that the attenuation rate will gradually accelerate with the continuous increase of temperature.

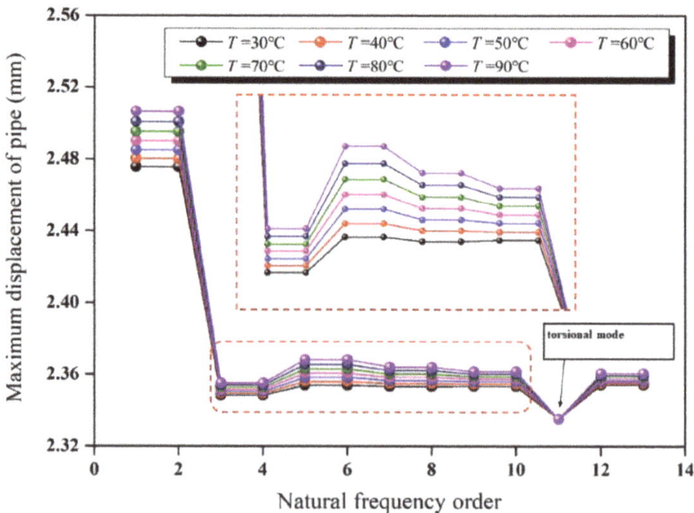

Figure 16. Maximum displacement of tubing under different temperatures.

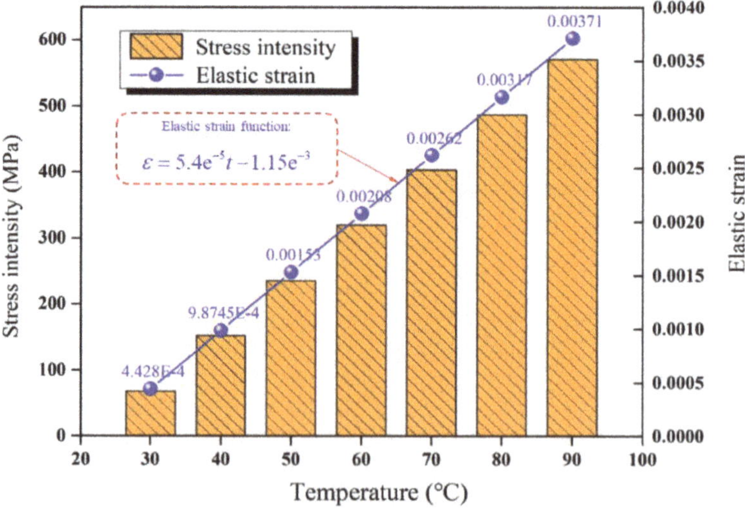

$$\varepsilon = 5.4e^{-5}t - 1.15e^{-3}$$

Figure 17. Stress intensity and strain of tubing at different temperatures.

6. Conclusions

In this paper, the modal changes of tubing under dry mode and wet mode are simulated and compared, and the effects of fluid–structure interaction effect, inlet pressure, and ambient temperature on the tubing modal are discussed. The following conclusions can be drawn:

1. After considering the fluid–solid coupling effect, the research objects include tubings and natural gas in the tubing. The mass increases, so the natural frequency decreases, but the displacement is slightly larger. In the actual production process, the buckling and deformation damage of the tubing can be reduced by inducing the corresponding torsional or tensile mode.

2. The greater the pressure in the tubing, the greater the equivalent stress on the tubing body, so the tubing is more prone to vibration, i.e., the natural frequency is lower. Furthermore, after considering the fluid–solid coupling effect, the pressure in the tubing is the true pulsating pressure of the fluid. The prestress applied to the tubing wall changes with time, and the pressures at different parts are different. At this time, the tubing is changed at different frequencies. Vibration is prone to occur, i.e., the natural frequency is smaller than the dry mode.

3. The higher the temperature, the lower the rigidity of the tubing body and the faster the strength attenuation, so the tubing is more prone to vibration, i.e., the natural frequency is lower. Both the stress intensity and the elastic strain increase with the temperature, so the displacement of the tubing also increases. The influence of temperature on the tubing modal is slightly greater than that of pressure.

Author Contributions: Conceptualization, methodology, software, C.L.; validation, formal analysis, investigation, J.J.; writing—original draft preparation, writing—review and editing, J.D. All authors have read and agreed to the published version of the manuscript.

Funding: This research received no external funding.

Conflicts of Interest: The authors declare no conflict of interest.

References

1. Alkhamis, M.; Imqam, A. A Simple Classification of Wellbore Integrity Problems Related to Fluids Migration. *Arab. J. Sci. Eng.* **2021**, *46*, 6131–6141. [CrossRef]
2. Zhang, L.; Yan, X.; Yang, X.; Zhao, X. Evaluation of wellbore integrity for HTHP gas wells under solid-temperature coupling using a new analytical model. *J. Nat. Gas Sci. Eng.* **2015**, *25*, 347–358. [CrossRef]
3. Zhang, Z.; Wang, J.; Li, Y.; Ming, L.; Zhang, C. Effects of instantaneous shut-in of high production gas well on fluid flow in tubing. *Pet. Explor. Dev.* **2020**, *47*, 642–650. [CrossRef]
4. Kou, Z.; Wang, H. Transient Pressure Analysis of a Multiple Fractured Well in a Stress-Sensitive Coal Seam Gas Reservoir. *Energies* **2020**, *13*, 3849. [CrossRef]
5. Jun, L.; Guo, X.; Wang, G.; Liu, Q.; Fang, D.; Huang, L.; Mao, L. Bi-nonlinear vibration model of tubing string in oil & gas well and its experimental verification. *Appl. Math. Model.* **2020**, *81*, 50–69.
6. Liu, J.; Zeng, L.; Guo, X.; Dai, L.; Huang, X.; Cai, L. Nonlinear flow-induced vibration response characteristics of leaching tubing in salt cavern underground gas storage. *J. Energy Storage* **2021**, *41*, 102909. [CrossRef]
7. Zhang, Z.; Wang, J.; Li, Y.; Liu, H.; Meng, W.; Li, L. Research on the influence of production fluctuation of high-production gas well on service security of tubing string. *Oil Gas Sci. Technol. Revue d'IFP Energies Nouv.* **2021**, *76*, 54. [CrossRef]
8. Lenwoue, A.; Deng, J.; Feng, Y.; Li, H.; Oloruntoba, A.; Selabi, N.B.S.; Marembo, M.; Sun, Y. Numerical Investigation of the Influence of the Drill String Vibration Cyclic Loads on the Development of the Wellbore Natural Fracture. *Energies* **2021**, *14*, 2015. [CrossRef]
9. Xu, J.; Mou, Y.; Xue, C.; Ding, L.; Wang, R.; Ma, D. The study on erosion of buckling tubing string in HTHP ultra-deep wells considering fluid–solid coupling. *Energy Rep.* **2021**, *7*, 3011–3022. [CrossRef]
10. Zhang, Y.F.; Yao, M.H.; Zhang, W.; Wen, B.C. Dynamical modeling and multi-pulse chaotic dynamics of cantilevered pipe conveying pulsating fluid in parametric resonance. *Aerosp. Sci. Technol.* **2017**, *68*, 441–453. [CrossRef]
11. Jiang, Y.; Zhu, L. Modal analysis of liquid-filled tubing under fluid-structure interaction by simulation and experiment methods. *Vibroeng. Procedia* **2018**, *21*, 42–47.
12. Wróbel, J.; Blaut, J. Influence of Pressure Inside a Hydraulic Line on Its Natural Frequencies and Mode Shapes. In Proceedings of the International Scientific-Technical Conference on Hydraulic and Pneumatic Drives and Control, Staniszow, Poland, 21–23 October 2020; Springer: Berlin/Heidelberg, Germany, 2020.
13. Xin, L.; Wang, S.; Ran, L. Modal Analysis of Two Typical Fluid-filled Pipes in Aircraft. In Proceedings of the 2011 International Conference on Fluid Power and Mechatronics, Beijing, China, 17–20 August 2011; pp. 469–473.
14. Lubecki, M.; Stosiak, M.; Bocian, M.; Urbanowicz, K. Analysis of selected dynamic properties of the composite hydraulic microhose. *Eng. Fail. Anal.* **2021**, *125*, 105431. [CrossRef]
15. Chai, Q.; Zeng, J.; Ma, H.; Li, K.; Han, Q. A dynamic modeling approach for nonlinear vibration analysis of the L-type tubing system with clamps. *Chin. J. Aeronaut.* **2020**, *33*, 3253–3265. [CrossRef]
16. Sekacheva, A.A.; Pastukhova, L.G.; Alekhin, V.N.; Noskov, A.S. Natural frequencies of a vertical tubing element. In Proceedings of the IV International Conference on Safety Problems of Civil Engineering Critical Infrastructures, Ekaterinburg, Russia, 4–5 October 2019; p. 481.
17. Wu, Q.; Qi, G. Global dynamics of a pipe conveying pulsating fluid in primary parametrical resonance: Analytical and numerical results from the nonlinear wave equation. *Phys. Lett. A* **2019**, *383*, 1555–1562. [CrossRef]

18. Guo, X.; Liu, J.; Wang, G.; Dai, L.; Fang, D.; Huang, L.; He, Y. Nonlinear flow-induced vibration response characteristics of a tubing string in HPHT oil & gas well. *Appl. Ocean Res.* **2020**, *106*, 102468.

19. Jiang, N.; Zhu, B.; Zhou, C.; Li, H.; Wu, B.; Yao, Y.; Wu, T. Blasting vibration effect on the buried tubing: A brief overview. *Eng. Fail. Anal.* **2021**, *129*, 105709. [CrossRef]

20. Tian, X.; Zhou, S.; Hu, L.; Li, W.; He, W. Vibration Fatigue Characteristic of High-Pressure Fuel Pipe Based on Fluid-Solid Coupling Model. *J. Vib. Meas. Diagn.* **2018**, *38*, 1234–1239.

21. Xu, Y.; Zhang, H.; Guan, Z. Dynamic Characteristics of Downhole Bit Load and Analysis of Conversion Efficiency of Drill String Vibration Energy. *Energies* **2021**, *14*, 229. [CrossRef]

22. Luo, J.; Zhang, K.; Liu, J.; Mu, L.; Wang, F. The Study of Tubing Vibration Mechanism in High Pressure Gas Well. *World J. Eng. Technol.* **2021**, *9*, 128–137. [CrossRef]

23. Shi, L.; Zhu, J.; Wang, L.; Chu, S.; Tang, F.P.; Jiang, Y.; Xu, T.; Yu, J. Analysis of Strength and Modal Characteristics of a Full Tubular Pump Impeller Based on Fluid-Structure Interaction. *Energies* **2021**, *14*, 6395. [CrossRef]

24. Zheng, X.; Wang, Z.; Triantafyllou, M.S.; Karniadakis, G.E. Fluid-structure interactions in a flexible pipe conveying two-phase flow. *Int. J. Multiph. Flow* **2021**, *141*, 103667. [CrossRef]

25. Mohmmed, A.O.; Al-Kayiem, H.H.; Osman, A.B.; Sabir, O. One-way coupled fluid-structure interaction of gas-liquid slug flow in a horizontal pipe: Experiments and simulations. *J. Fluids Struct.* **2020**, *97*, 103083. [CrossRef]

26. Huang, Z.; Li, L.; Hu, G. *Fatigue Life Prediction of Tubing Strings in Natural Gas Wells*; Chongqing University Press: Chongqing, China, 2012.

27. Lee, G.; Jhung, M.; Bae, J.; Kang, S. Numerical Study on the Cavitation Flow and Its Effect on the Structural Integrity of Multi-Stage Orifice. *Energies* **2021**, *14*, 1518. [CrossRef]

28. Lian, Z.; Mou, Y.; Liu, Y.; Xu, D. Buckling behaviors of tubing strings in HTHP ultra-deep wells. *Nat. Gas Ind.* **2018**, *38*, 89–94.

29. Mou, Y.; Lian, Z.; Zhang, Q.; Lu, Z.; Li, J. Study on influence of centralizer on buckling behavior of tubing string in ultra-deep gas well. *J. Saf. Sci. Technol.* **2018**, *14*, 109–114.

30. Bęben, D. The Influence of Temperature on Degradation of Oil and Gas Tubing Made of L80-1 Steel. *Energies* **2021**, *14*, 6855. [CrossRef]

Article

Establishment and Solution of Four Variable Water Hammer Mathematical Model for Conveying Pipe

Jiehao Duan [1,2], Changjun Li [1,2,*] and Jin Jin [3]

[1] College of Petroleum Engineering, Southwest Petroleum University, Chengdu 610500, China; 201611000075@stu.swpu.edu.cn
[2] CNPC Key Laboratory of Oil & Gas Storage and Transportation, Southwest Petroleum University, Chengdu 610500, China
[3] CCDC Downhole Service Company, Chengdu 610500, China; tt190710@126.com
* Correspondence: lichangjunemail@sina.com; Tel.: +86-18728495286

Abstract: Transient flow in pipe is a much debated topic in the field of hydrodynamics. The water hammer effect caused by instantaneous valve closing is an important branch of transient flow. At present, the fluid density is regarded as a constant in the study of the water hammer effect in pipe. When there is gas in the pipe, the variation range of density is large, and the pressure-wave velocity should also change continuously along the pipe. This study considers the interaction between pipeline fluid motion and water hammer wave propagation based on the essence of water hammer, with the pressure, velocity, density and overflow area set as variables. A new set of water hammer calculation equations was deduced and solved numerically. The effects of different valve closing time, flow rate and gas content on pressure distribution and the water hammer effect were studied. It was found that with the increase in valve closing time, the maximum fluctuating pressure at the pipe end decreased, and the time of peak value also lagged behind. When the valve closing time increased from 5 s to 25 s, the difference in water hammer pressure was 0.72 MPa, and the difference in velocity fluctuation amplitude was 0.076 m/s. The findings confirm: the greater the flow, the greater the pressure change at the pipe end; the faster the speed change, the more obvious the water hammer effect. High-volume flows were greatly disturbed by instantaneous obstacles such as valve closing. With the increase of time, the pressure fluctuation gradually attenuated along the pipe length. The place with the greatest water hammer effect was near the valve. Under the coupling effect of time and tube length, the shorter the time and the shorter the tube length, the more obvious the pressure fluctuation. Findings also confirm: the larger the gas content, the smaller the fluctuation peak of pipe end pressure; the longer the water hammer cycle, the smaller the pressure-wave velocity. The actual pressure fluctuation value was obviously lower than that without gas, and the size of the pressure wave mainly depended on the gas content. When the gas content increased from 1% to 9%, the difference of water hammer pressure was 0.41 MPa.

Keywords: pipeline; water hammer; gas-liquid two-phase flow; pressure; velocity

Citation: Duan, J.; Li, C.; Jin, J. Establishment and Solution of Four Variable Water Hammer Mathematical Model for Conveying Pipe. *Energies* **2022**, *15*, 1387. https://doi.org/10.3390/en15041387

Academic Editors: Min Wang, Sheng-Qi Yang, Kun Du, Chun Zhu, Qi Wang, Wen Zhang and Antonio Crespo

Received: 7 January 2022
Accepted: 11 February 2022
Published: 14 February 2022

Publisher's Note: MDPI stays neutral with regard to jurisdictional claims in published maps and institutional affiliations.

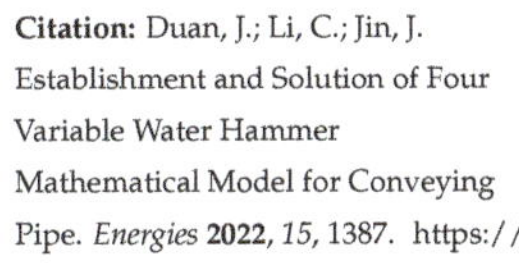

1. Introduction

Since the mid nineteenth century, the unsteady flow of fluid has been a much discussed topic for scholars. In steady flow, the parameter at any point does not change with time; while in unsteady flow, the parameter at any point is a function of time and changes with it [1,2]. When boundary conditions at the end of a pipeline change, the steady flow in the pipeline will also change accordingly, manifested as a sudden change in the flow rate and pressure of the fluid, which causes momentum conversion. The occurrence time of this phenomenon is generally very short, so it is called transient flow, also known as water hammer or water hammer [3,4]. At present, the most representative work on water hammer is "Transient Flow" written by Professor Streeter, an American hydraulicsist [5]. This monograph systematically explains the types of water hammer process, the mechanism and derivation of water hammer, the applicable conditions of the model and the corresponding solution method, etc. Today, the method of characteristics (MOC), first proposed in the monograph, are widely used to characterize the water hammer process.

The classical transient flow models of pressurized pipe flow all use a reservoir-pipe-valve system as the research object and close the valve as the research condition [6,7]. The mathematical model includes the motion equation and the continuity equation. The pipeline element is selected as the control body. The momentum equation is established by the momentum change of the fluid entering and exiting the control body, and the continuity equation is established by the mass conservation of the fluid entering and exiting the control body. When the valve is closed, a short section of fluid at the valve stops flowing instantaneously, and the next section of fluid next to this section of fluid and in the upstream direction is blocked, decelerating to a standstill. By analogy, the fluid in the tube gradually stops moving along the upstream direction. Since the kinetic energy of the fluid is reduced and converted into pressure energy, the pressure in the tube gradually increases along the upstream direction. That is, a pressure wave propagates upstream at a wave velocity a [8,9]. Figure 1 shows the movement of the liquid in a horizontal pipeline after the valve is suddenly closed at the end of a pipeline, ignoring the friction and local loss.

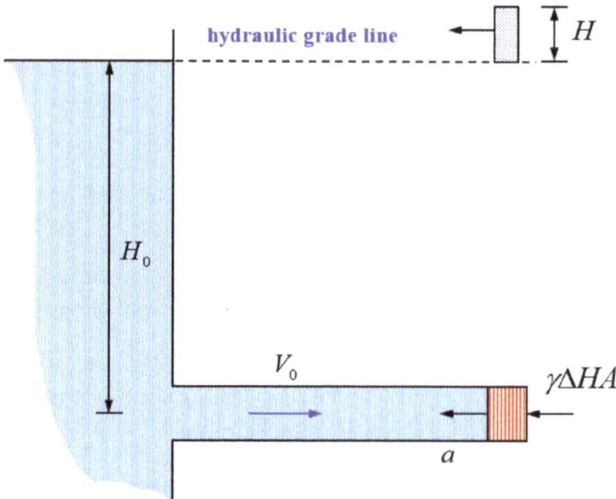

Figure 1. The fluid movement state in the pipe after the valve is suddenly closed (ignoring friction) [10].

In recent years, many scholars have performed research on the transient flow in fluid pipelines. Baba [11] considered the numerical calculation of the transient flow model of a hydrogen-natural gas mixture in pipelines with different types of valves. The influence of the valve function was analyzed, and the calculation results under different mass ratios and gas mixture flow parameters were given. Wahba [12] used one-dimensional and two-dimensional water hammer models to numerically study the attenuation of turbulence in pipelines. The one-dimensional model was solved by the characteristic line method, and the two-dimensional model was solved by the semidiscretization method and the fourth-order Runge—Kutta method. Kandil [13] used the characteristic line method to numerically solve the water hammer equation and studied pressure fluctuations by changing the elastic modulus of the pipeline and Poisson's ratio. Hossam [14] first used the wave characteristic line method to numerically simulate the transient flow in a viscoelastic pipe, re-derived the Joukowsky equation of the elastic pipe, and derived the quadratic equation of energy conservation in the viscoelastic pipe. Khamoushi [15] proposed a one-dimensional model of transient flow in non-Newtonian fluids, using Zielke's unsteady friction solution for power law fluids and cross fluids. The time increment updated the weight function of the Zielke model. Abdeldayem [16] studied and explored the performance of different unsteady friction models in terms of accuracy, efficiency, and reliability to determine the most suitable model for engineering practice. Through comparison, it was found that Vítkovský's unsteady friction model is the most suitable. Taking the reservoir-pipe-valve system as an example, the effectiveness of the method was verified by experimental data. Lashkarbolok [17] used radial-basis functions and least-squares optimization technology to solve the numerical solution of one-dimensional equations governing transient pipe flow. The method can deal with geometrically nonuniform pipes by employing an arbitrary distribution of scattering nodes in the space domain. Aliabadi [18] studied fluid-structure interaction (FSI) during water hammer in a viscoelastic (VE) pipe in the frequency domain. The main aim was to investigate pressure and stress waves using the transfer matrix method (TMM) in a typical reservoir-VE pipe-valve (RPV) system. Both major coupling mechanisms, namely Poisson and junction, are taken into account. Firouzi [19] proposed a time-varying reliability method to predict the failure risk of pipeline strength reduction due to the combination of water hammer pressure and corrosion and proposed the coupling model of pitting corrosion and water hammer pressure. It was found that with an increase in the average incidence of water hammer, the risk of pipeline failure increases, and the sudden closure of pipeline valves will have a destructive impact on the safety of a pipe network. Henclik [20] proposed an original concept of applying a shock response spectrum (SRS) method to water hammer events. The method is shortly presented, its numerical approach developed, and its application to water hammer loads described. SRS computations and analysis of water hammer experimental results measured at a laboratory pipeline are performed and concluded. Chen [21] constructed a fluid-solid coupling 4 equation model applied to high-temperature and high-pressure oil and gas based on the fluid-solid coupling theory and selected the best coupling method and combined the characteristic line method to obtain the pressure-wave velocity. Zhu [22,23] conducted a transient flow experiment using a sudden valve closing of a gas-liquid two-phase pressurized pipe flow and used the experiment to verify that the pressure-wave velocity of a pressurized pipe flow transient flow model was corrected by the gas content rate. At the same time, considering the nonconstant friction and the viscoelasticity of the pipe, the influence on gas flow and pressure-wave dissipation was studied, and the relevant parameters were checked. Fu [24,25] used the split-phase flow model to simulate the liquid column separation caused by transient flow in the pressure pipe flow and also considered the influence of the pipe stress wave on the pressure wave. Due to the existence of nonconservative terms in the model and to ensure the convergence of the results, this model needed to be solved separately while relaxation parameters were introduced. Han [26] examined bridging the water hammer in long-distance pipelines. He based his study on classic water hammer theory, considering the growth and reduction of cavities, the air content in the pipeline, and the water hammer after the instantaneous change in the pressure in the corrugated tube, among other factors. The process was described in semianalytical mathematics, and the influence of the growth and reduction time of cavities, the size of the gas content, and the relative elevation of the highest point of the pipeline to the horizontal section on bridging the water hammer pressure was analyzed.

After investigation, it was found that a large number of domestic and foreign literatures regard the fluid density in the conveying pipe as a constant. When the fluid in the tube is all liquid, the error caused by the fluid density as a constant is within the engineering allowable range. However, when there is gas (natural gas, etc.) in the tube, the gas has different degrees of compressibility and the density varies widely. Treating it as a constant violates the actual process of fluid flow in the tube. The density is closely related to the pressure-wave speed, which should also be constantly changing along the pipeline.

Furthermore, when calculating water hammer pressure, earlier researchers did not consider the change and inconsistency of the research object at the beginning and end of the period due to fluid inflow and outflow in the dt period. Based on the essence of water hammer, the interaction between pipeline fluid motion and water hammer wave propagation is considered in this paper, and the pressure, velocity, density and overflow area are set as variables. A new set of water hammer calculation equations is deduced and solved numerically. The effects of different valve closing time, flow rate and gas content on pressure distribution and water hammer effect are studied. Such calculations of the propagation process of the water hammer wave are more reasonable and reliable.

2. Mathematical Model

2.1. Fluid Motion Equation

Taking the water body in the flow section of length x as the research object, this paper tracks it to study the water body in the flow section. The analysis diagram of the interaction between the pipe water movement and the water hammer wave propagation is shown in Figure 2. The initial length of the stream segment is set to x. The liquid in the flow section between section 1-1 and section 2-2 at the initial time is taken as the research object. Taking into account the original flow of the liquid, after the dt period, the flow segment 1-2 moves to a new position $1'$-$2'$, and the crest of the water hammer wave also propagates to the position of section $1'$-$1'$.

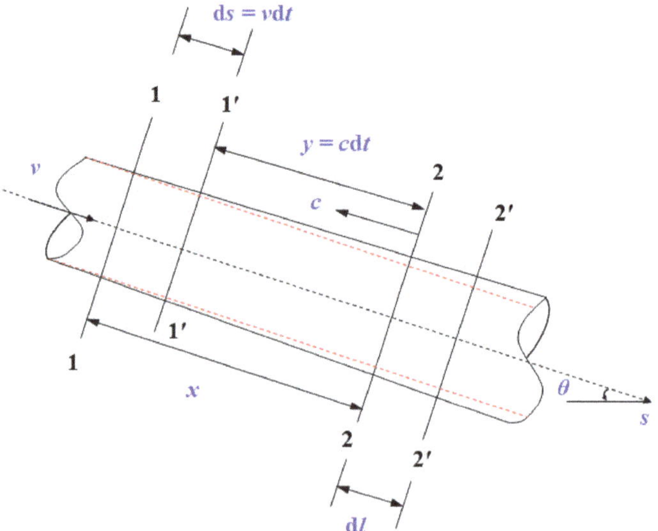

Figure 2. Schematic diagram of water hammer wave propagation.

As shown in Figure 2, after the dt period, section 1-1 of the studied control body moves to the position of section $1'$-1. The movement distance is ds. The water hammer wave moves from section 2-2 to Section $1'$-$1'$ at wave speed c. The movement distance is y, and section 2-2 moves to Section $2'$-$2'$.

At the beginning of dt, the momentum of the liquid in the 1-2 flow section is:

$$\left(\rho + \frac{\partial\rho}{\partial s}\frac{x}{2}\right)\left(A + \frac{\partial A}{\partial s}\frac{x}{2}\right)\left(v + \frac{\partial v}{\partial s}\frac{x}{2}\right)x = \rho Avx + \frac{1}{2}\frac{\partial(\rho Av)}{\partial s}x^2$$
$$+ \frac{1}{4}\left(\rho\frac{\partial A}{\partial s}\frac{\partial v}{\partial s} + A\frac{\partial\rho}{\partial s}\frac{\partial v}{\partial s} + v\frac{\partial\rho}{\partial s}\frac{\partial A}{\partial s}\right)x^3 + \frac{1}{8}\frac{\partial\rho}{\partial s}\frac{\partial A}{\partial s}\frac{\partial v}{\partial s}x^4 \approx \rho Avx + \frac{1}{2}\frac{\partial(\rho Av)}{\partial s}x^2 \qquad (1)$$

At the end of time dt, the average density of the fluid in the $1'$-$2'$ flow section is:

$$\tilde{\rho} = \frac{1}{2}\left\{\rho + \frac{\partial\rho}{\partial s}\mathrm{d}s + \frac{\partial}{\partial t}\left(\rho + \frac{\partial\rho}{\partial s}\mathrm{d}s\right)\mathrm{d}t + \rho + \frac{\partial\rho}{\partial s}(x + \mathrm{d}l) + \frac{\partial}{\partial t}\left[\rho + \frac{\partial\rho}{\partial s}(x + \mathrm{d}l)\right]\mathrm{d}t\right\}$$
$$\approx \rho + \frac{\partial\rho}{\partial s}\mathrm{d}s + \frac{\partial\rho}{\partial t}\mathrm{d}t + \frac{1}{2}\frac{\partial\rho}{\partial s}x \qquad (2)$$

In the same way, at the end of time dt, the average area of the fluid in the 1'-2' flow section is:

$$\tilde{A} \approx A + \frac{\partial A}{\partial s}ds + \frac{\partial A}{\partial t}dt + \frac{1}{2}\frac{\partial A}{\partial s}x \tag{3}$$

The average velocity of the fluid in the 1'-2' stream section is:

$$\tilde{v} \approx v + \frac{\partial v}{\partial s}ds + \frac{\partial v}{\partial t}dt + \frac{1}{2}\frac{\partial v}{\partial s}x \tag{4}$$

Therefore, at the end of time dt, the fluid momentum in the 1'-2' flow section is:

$$
\begin{aligned}
mo &= \left(\rho + \frac{\partial \rho}{\partial s}ds + \frac{\partial \rho}{\partial t}dt + \frac{1}{2}\frac{\partial \rho}{\partial s}x\right)\left(A + \frac{\partial A}{\partial s}ds + \frac{\partial A}{\partial t}dt + \frac{1}{2}\frac{\partial A}{\partial s}x\right)\left(v + \frac{\partial v}{\partial s}ds + \frac{\partial v}{\partial t}dt + \frac{1}{2}\frac{\partial v}{\partial s}x\right)(y + dl) \\
&= \rho Avx + \frac{\partial(\rho Av)}{\partial s}xds + \frac{\partial(\rho Av)}{\partial t}xdt + \frac{1}{2}\frac{\partial(\rho Av)}{\partial s}x^2 + \rho Avx\frac{\partial v}{\partial s}dt
\end{aligned}
\tag{5}
$$

At the time of dt, the increase of momentum in the studied control body is:

$$
\begin{aligned}
mo =\ & \left[\rho Avx + \frac{\partial(\rho Av)}{\partial s}xds + \frac{\partial(\rho Av)}{\partial t}xdt + \frac{1}{2}\frac{\partial(\rho Av)}{\partial s}x^2 + \rho Avx\frac{\partial v}{\partial s}dt\right] - \\
& \left[\rho Avx + \frac{1}{2}\frac{\partial(\rho Av)}{\partial s}x^2\right] = \frac{\partial(\rho Av)}{\partial s}xds + \frac{\partial(\rho Av)}{\partial t}xdt + \frac{1}{2}\frac{\partial(\rho Av)}{\partial s}x^2
\end{aligned}
\tag{6}
$$

The external force acting on the control body includes the hydrodynamic pressure of the cross-section at both ends, the hydrodynamic pressure of the side, gravity and the frictional resistance of the pipe wall. Because the distance of the control body movement in the dt period is very small, when analyzing the combined external force it receives, the analysis of the 1-2 flow section is equivalent to the 1'-2' analysis. Therefore, at the time of dt, the impulse of the external force received by the control body is:

$$\frac{\partial(\rho Av)}{\partial s}xds + \frac{\partial(\rho Av)}{\partial t}xdt + \rho Avx\frac{\partial v}{\partial s}dt = \left[-\frac{\partial(pA)}{\partial s}x + p\frac{\partial A}{\partial s}x + \rho gAx\sin\theta - \tau_0\chi x\right]dt \tag{7}$$

After sorting, the formula can be obtained:

$$\left(\frac{1}{\rho}\frac{d\rho}{dp} + \frac{1}{A}\frac{dA}{dp}\right)\frac{dp}{dt} + \frac{1}{v}\frac{dv}{dt} + \frac{\partial v}{\partial s} + \frac{1}{\rho v}\frac{\partial p}{\partial s} - \frac{g}{v}\sin\theta + \frac{\tau_0\chi}{\rho Av} = 0 \tag{8}$$

Through the simultaneous establishment of a one-dimensional unsteady continuity equation:

$$\frac{1}{v}\frac{\partial v}{\partial t} + \frac{\partial v}{\partial s} + \frac{1}{\rho v}\frac{\partial p}{\partial s} - \frac{g}{v}\sin\theta + \frac{\tau_0\chi}{\rho Av} = 0 \tag{9}$$

Because of $\frac{\partial z}{\partial s} = -\sin\theta$, substitute the above formula to obtain:

$$\frac{1}{g}\frac{\partial v}{\partial t} + \frac{v}{g}\frac{\partial v}{\partial s} + \frac{1}{\gamma}\frac{\partial p}{\partial s} + \frac{\partial z}{\partial s} + \frac{\tau_0\chi}{\gamma A} = 0 \tag{10}$$

Equation (10) is the water hammer motion equation.

2.2. Fluid Continuity Equation

The derivation of a water hammer continuity equation uses the law of mass conservation and takes the fluid in the flow section with length x as the research object, as shown in Figure 2. Both ends of the pipe section are section 1-1 and section 2-2, respectively.

The difference in the quality of the liquid flowing into and out of the control body during the dt period is:

$$\frac{1}{g}\frac{\partial v}{\partial t} + \frac{v}{g}\frac{\partial v}{\partial s} + \frac{1}{\gamma}\frac{\partial p}{\partial s} + \frac{\partial z}{\partial s} + \frac{\tau_0\chi}{\gamma A} = 0 \tag{11}$$

At the beginning of the dt period, the fluid quality in flow section 1-2 is:

$$\left(\rho + \frac{\partial \rho}{\partial s}\frac{x}{2}\right)\left(A + \frac{\partial A}{\partial s}\frac{x}{2}\right)x \approx \rho A x + \frac{1}{2}\frac{\partial(\rho A)}{\partial s}x^2 \tag{12}$$

At the end of the dt period, the fluid quality in flow section 1-2 is:

$$\left(\rho + \frac{\partial \rho}{\partial s}\frac{x}{2} + \frac{\partial \rho}{\partial t}dt\right)\left(A + \frac{\partial A}{\partial s}\frac{x}{2} + \frac{\partial A}{\partial t}dt\right)x \approx \rho A x + \frac{\partial(\rho A)}{\partial s}\frac{x^2}{2} + \frac{\partial(\rho A)}{\partial t}x dt \tag{13}$$

Therefore, the fluid mass increase in flow segment 1-2 during the dt period is:

$$\left[\rho A x + \frac{\partial(\rho A)}{\partial s}\frac{x^2}{2} + \frac{\partial(\rho A)}{\partial t}x dt\right] - \left[\rho A x + \frac{\partial(\rho A)}{\partial s}\frac{x^2}{2}\right] = \frac{\partial(\rho A)}{\partial t}x dt \tag{14}$$

According to the law of conservation of mass, during the dt period, the difference between the mass of fluid flowing into and out of the control body should be equal to the increase in the mass of the fluid in the control body during the same period, that is:

$$\frac{\partial(\rho A)}{\partial t} + \frac{\partial(\rho A v)}{\partial s} = 0 \tag{15}$$

Equation (15) is the water hammer continuity equation derived in this paper. Therefore, the water hammer calculation equations can be obtained as:

$$\begin{cases} \frac{\partial(\rho A)}{\partial t} + \frac{\partial(\rho A v)}{\partial s} = 0 \\ \frac{1}{g}\frac{\partial v}{\partial t} + \frac{v}{g}\frac{\partial v}{\partial s} + \frac{1}{\gamma}\frac{\partial p}{\partial s} + \frac{\partial z}{\partial s} + \frac{\tau_0 \chi}{\gamma A} = 0 \end{cases} \tag{16}$$

2.3. Complete Equations of Water Hammer

Generally in engineering, it is customary to use the piezometer head H to reflect the pressure p. Therefore, substituting $p = \rho g(H - z)$ into the equation of motion, we can get:

$$\frac{dp}{dt} = \frac{\partial p}{\partial t} + v\frac{\partial p}{\partial s} = \frac{\partial[\rho g(H - z)]}{\partial t} + v\frac{\partial[\rho g(H - z)]}{\partial s} = \rho g\frac{d(H - z)}{dt} + g(H - z)\frac{d\rho}{dt} \tag{17}$$

where, $\frac{\partial z}{\partial t} = 0$, $\frac{\partial z}{\partial s} = -\sin\theta$, θ is the included angle between the pipe axis and the horizontal direction.

The above formula considers the variation of liquid density ρ with time t and distance s during water hammer, which is different from the traditional water hammer equations. Since:

$$\frac{\partial p}{\partial s} = \frac{\partial[\rho g(H - z)]}{\partial s} = \gamma\frac{\partial H}{\partial s} - \gamma\frac{\partial z}{\partial s} + g(H - z)\frac{\partial \rho}{\partial s} \tag{18}$$

Then substitute $\gamma = \rho g$, $\tau_0 = \frac{1}{8}f\rho v^2$ and $\chi_0 = \pi D = 2\sqrt{\pi A}$ into the equation of motion to obtain:

$$\frac{\partial v}{\partial t} + v\frac{\partial v}{\partial s} + g\frac{\partial H}{\partial s} + \frac{(H - z)g}{\rho}\frac{\partial \rho}{\partial s} + \frac{fv|v|}{4}\sqrt{\frac{\pi}{A}} = 0 \tag{19}$$

Since the fluid elasticity equation is:

$$\frac{1}{\rho}\frac{d\rho}{dp} = \frac{1}{K} \tag{20}$$

The elastic equation of the pipe is:

$$\frac{1}{A}\frac{dA}{dp} = \frac{D}{Ee} \tag{21}$$

The simultaneous hydroelasticity Equations (20) and (19) can be obtained:

$$\frac{\partial \rho}{\partial t} + v \frac{\partial \rho}{\partial s} - \frac{\rho}{\left[\frac{K}{\rho g} - (H - z)\right]} \left(\frac{\partial H}{\partial t} + v \frac{\partial H}{\partial s} + v \sin \theta\right) = 0 \tag{22}$$

The simultaneous elastic equation of pipe (21) and Equation (19) can be obtained:

$$\frac{\partial A}{\partial t} + v \frac{\partial A}{\partial s} - \frac{\frac{2AK}{Ee}\sqrt{\frac{A}{\pi}}}{\left[\frac{K}{\rho g} - (H - z)\right]} \left(\frac{\partial H}{\partial t} + v \frac{\partial H}{\partial s} + v \sin \theta\right) = 0 \tag{23}$$

After sorting, the formula can be obtained:

$$\frac{\partial H}{\partial t} + v \frac{\partial H}{\partial s} + v \sin \theta + \frac{\frac{1}{\rho g} - \frac{(H-z)}{K}}{\frac{2}{Ee}\sqrt{\frac{A}{\pi}} + \frac{1}{K}} \frac{\partial v}{\partial s} = 0 \tag{24}$$

Thus, the water hammer closure equations are:

$$\begin{cases} \frac{\partial v}{\partial t} + v \frac{\partial v}{\partial s} + g \frac{\partial H}{\partial s} + \frac{(H-z)g}{\rho} \frac{\partial \rho}{\partial s} + \frac{fv|v|}{4}\sqrt{\frac{\pi}{A}} = 0 \\[2ex] \frac{\partial H}{\partial t} + v \frac{\partial H}{\partial s} + v \sin \theta + \frac{\frac{1}{\rho g} - \frac{(H-z)}{K}}{\frac{2}{Ee}\sqrt{\frac{A}{\pi}} + \frac{1}{K}} \frac{\partial v}{\partial s} = 0 \\[2ex] \frac{\partial \rho}{\partial t} + v \frac{\partial \rho}{\partial s} - \frac{\rho}{\left[\frac{K}{\rho g} - (H-z)\right]} \left(\frac{\partial H}{\partial t} + v \frac{\partial H}{\partial s} + v \sin \theta\right) = 0 \\[2ex] \frac{\partial A}{\partial t} + v \frac{\partial A}{\partial s} - \frac{\frac{2AK}{Ee}\sqrt{\frac{A}{\pi}}}{\left[\frac{K}{\rho g} - (H-z)\right]} \left(\frac{\partial H}{\partial t} + v \frac{\partial H}{\partial s} + v \sin \theta\right) = 0 \end{cases} \tag{25}$$

3. Model Solving and Boundary Conditions

3.1. Difference Equation

In the calculation of water hammer equations, the more common methods are the characteristic line method and the direct difference method based on the finite difference method [27,28]. Because their deformation form cannot be transformed into characteristic equations, the characteristic line method cannot be used for calculation of the Gestalt equations proposed in this paper. Instead, the direct difference method is selected for calculation.

Compatibility, convergence and stability must be considered in difference calculations [28–30]. Change the differential equation into a difference equation. If the computational grid is reduced, the difference equation tends to be consistent with the differential equation. For example, the solution of the difference equation converges to the solution of the differential equation, which is convergence. If the rounding error and initial error in the calculation are always controlled within a limited range, and the calculated value is close to the true solution, then this scheme is stable. For the diffusion difference scheme adopted in this paper, the stability condition is the Courant condition. Since the water hammer equation is a hyperbolic equation, its stability condition should still meet the following requirements even though the concept of water hammer wave velocity is not introduced:

$$\frac{\Delta s}{\Delta t} \geq v \text{ OR } \frac{\Delta s}{\Delta t} \geq \left| v \pm \frac{1}{\sqrt{\rho\left(\frac{1}{K} + \frac{D}{Ee}\right)}} \right|_{\max} \tag{26}$$

In order to solve this model, the basic equation needs to be discretized and the differential equation needs to be transformed into a differential algebraic equation. Here, the diffusion difference scheme is used to solve the equation. Figure 3 shows the *L-t* plane of characteristic grid of the four variable water hammer model. Use *X* to represent the objective function head *H*, velocity *V* and density ρ and area *A*. Then the partial differential of *X* to time and position can be expressed as:

$$\frac{\partial X}{\partial t} = \frac{X_i^k - \left[\xi X_i^{k-1} + \frac{1-\xi}{2}\left(X_{i+1}^{k-1} + X_{i-1}^{k-1}\right)\right]}{\Delta t} \tag{27}$$

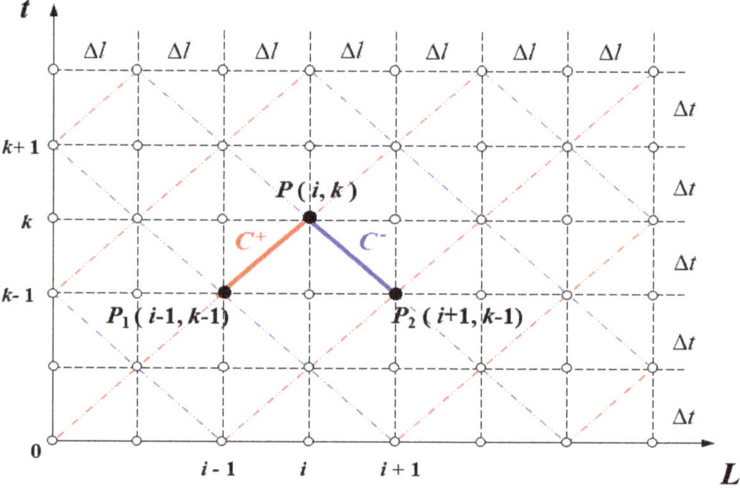

Figure 3. *L-t* plane of characteristic grid.

$$\frac{\partial X}{\partial z} = \frac{X_{i+1}^{k-1} - X_{i-1}^{k-1}}{2\Delta z} \tag{28}$$

By introducing Equations (27) and (28) into the water hammer closure calculation model, the following difference equations can be obtained:

$$\frac{v_i^k - \left[\xi v_i^{k-1} + \frac{1-\xi}{2}\left(v_{i+1}^{k-1} + v_{i-1}^{k-1}\right)\right]}{\Delta t} + v_i^{k-1}\frac{v_{i+1}^{k-1} - v_{i-1}^{k-1}}{2\Delta z}$$
$$+ g\frac{H_{i+1}^{k-1} - H_{i-1}^{k-1}}{2\Delta z} + \frac{H_i^{k-1}g}{\rho_i^{k-1}}\frac{\rho_{i+1}^{k-1} - \rho_{i-1}^{k-1}}{2\Delta z} + \frac{fv_i^{k-1}\left|v_i^{k-1}\right|}{2D} = 0 \tag{29}$$

$$\frac{H_i^k - \left[\xi H_i^{k-1} + \frac{1-\xi}{2}\left(H_{i+1}^{k-1} + H_{i-1}^{k-1}\right)\right]}{\Delta t} + v_i^{k-1}\frac{H_{i+1}^{k-1} - H_{i-1}^{k-1}}{2\Delta z}$$
$$+ v_i^{k-1}\sin\theta + \frac{\frac{1}{\rho g} - \frac{(H-z)}{K}}{\frac{2}{Ee}\sqrt{\frac{A}{\pi}} + \frac{1}{K}}\frac{v_{i+1}^{k-1} - v_{i-1}^{k-1}}{2\Delta z} = 0 \tag{30}$$

$$\frac{\rho_i^k - \left[\xi\rho_i^{k-1} + \frac{1-\xi}{2}\left(\rho_{i+1}^{k-1} + \rho_{i-1}^{k-1}\right)\right]}{\Delta t} + v_i^{k-1}\frac{\rho_{i+1}^{k-1} - \rho_{i-1}^{k-1}}{2\Delta z} - \frac{\rho_i^{k-1}}{\frac{K}{\rho_i^{k-1}g} - \left(H_i^{k-1} - z_i\right)}\cdot$$
$$\left\{\frac{H_i^k - \left[\xi H_i^{k-1} + \frac{1-\xi}{2}\left(H_{i+1}^{k-1} + H_{i-1}^{k-1}\right)\right]}{\Delta t} + v_i^{k-1}\left(\frac{H_{i+1}^{k-1} - H_{i-1}^{k-1}}{2\Delta z} + \sin\theta\right)\right\} = 0 \tag{31}$$

$$\frac{A_i^k - \left[\xi A_i^{k-1} + \frac{1-\xi}{2}\left(A_{i+1}^{k-1} + A_{i-1}^{k-1}\right)\right]}{\Delta t} + v_i^{k-1}\frac{A_{i+1}^{k-1} - A_{i-1}^{k-1}}{2\Delta z} - \frac{2KA_i^{k-1}}{\rho_i^{k-1}Ee}\sqrt{\frac{A_i^{k-1}}{\pi}}.$$

$$\left\{\frac{\rho_i^k - \left[\xi\rho_i^{k-1} + \frac{1-\xi}{2}\left(\rho_{i+1}^{k-1} + \rho_{i-1}^{k-1}\right)\right]}{\Delta t} + v_i^{k-1}\frac{\rho_{i+1}^{k-1} - \rho_{i-1}^{k-1}}{2\Delta z}\right\} = 0 \tag{32}$$

Make the following assumptions:

$$\begin{cases} X_{H1} = \xi H_i^{k-1} + \frac{1-\xi}{2}\left(H_{i+1}^{k-1} + H_{i-1}^{k-1}\right), & X_{H2} = \left(H_{i+1}^{k-1} - H_{i-1}^{k-1}\right)\frac{\Delta t}{\Delta z} \\[2mm] X_{v1} = \xi v_i^{k-1} + \frac{1-\xi}{2}\left(v_{i+1}^{k-1} + v_{i-1}^{k-1}\right), & X_{v2} = \left(v_{i+1}^{k-1} - v_{i-1}^{k-1}\right)\frac{\Delta t}{\Delta z} \\[2mm] X_{\rho 1} = \xi\rho_i^{k-1} + \frac{1-\xi}{2}\left(\rho_{i+1}^{k-1} + \rho_{i-1}^{k-1}\right), & X_{\rho 2} = \left(\rho_{i+1}^{k-1} - \rho_{i-1}^{k-1}\right)\frac{\Delta t}{\Delta z} \\[2mm] X_{A1} = \xi A_i^{k-1} + \frac{1-\xi}{2}\left(A_{i+1}^{k-1} + A_{i-1}^{k-1}\right), & X_{A2} = \left(A_{i+1}^{k-1} - A_{i-1}^{k-1}\right)\frac{\Delta t}{\Delta z} \end{cases} \tag{33}$$

After sorting, the expressions of the four unknowns are as follows:

$$\begin{cases} H_i^k = X_{H1} - v_i^{k-1}\left(\frac{X_{H2}}{2} + \Delta t\sin\theta\right) - X_{v2}\left[\frac{\frac{1}{\rho_i^{k-1}g} - \frac{\left(H_i^{k-1}-z_i\right)}{K}}{\frac{4}{Ee}\sqrt{\frac{A_i^{k-1}}{\pi}} + \frac{1}{K}}\right] \\[4mm] v_i^k = X_{v1} - \frac{v_i^{k-1}}{2}X_{v2} - \frac{g}{2}X_{H2} - \frac{\left(H_i^{k-1}-z_i\right)g}{2\rho_i^{k-1}}X_{\rho 2} - \frac{f\Delta t}{2D}v_i^{k-1}\left|v_i^{k-1}\right| = 0 \\[4mm] \rho_i^k = X_{\rho 1} - \frac{v_i^{k-1}}{2}X_{\rho 2} + \frac{\rho_i^{k-1}}{\left[\frac{K}{\rho_i^{k-1}g} - \left(H_i^{k-1}-z_i\right)\right]}\left[H_i^k - X_{H1} + v_i^{k-1}\left(\frac{X_{H2}}{2} + \Delta t\sin\theta\right)\right] \\[4mm] A_i^k = X_{A1} - \frac{v_i^{k-1}}{2}X_{A2} + \frac{2A_i^{k-1}K}{\rho_i^{k-1}Ee}\sqrt{\frac{A_i^{k-1}}{\pi}}\left(\rho_i^k - X_{\rho 1} + \frac{v_i^{k-1}}{2}X_{\rho 2}\right) \end{cases} \tag{34}$$

The fluid in the oil pipeline is not all liquid in the field. What happens when there is a small amount of gas in the oil pipeline? Assuming that there are only gas-liquid two phases in the pipe, the density of the gas-liquid mixture is [31]:

$$\rho_m = \rho_g C_g + \rho_l(1 - C_g) \tag{35}$$

where, C_g is the gas content.

Since the density in this paper is considered a variable, according to the difference principle, we can get:

$$\rho_{mi}^k = \rho_{gi}^k C_{gi}^k + \rho_{li}^k\left(1 - C_{gi}^k\right) \tag{36}$$

$$\rho_{mi}^{k-1} = \rho_{gi}^{k-1} C_{gi}^{k-1} + \rho_{li}^{k-1}\left(1 - C_{gi}^{k-1}\right) \tag{37}$$

$$\rho_{mi+1}^{k-1} = \rho_{gi+1}^{k-1} C_{gi+1}^{k-1} + \rho_{li+1}^{k-1}\left(1 - C_{gi+1}^{k-1}\right) \tag{38}$$

$$\rho_{mi-1}^{k-1} = \rho_{gi-1}^{k-1} C_{gi-1}^{k-1} + \rho_{li-1}^{k-1}\left(1 - C_{gi-1}^{k-1}\right) \tag{39}$$

Accordingly, the flow rate also becomes the mixing flow rate, and its expression is:

$$v_m = \frac{Q_g + Q_l}{A} \tag{40}$$

Then according to the difference principle, we can get:

$$v_{\mathrm{m}i}^{k} = \frac{Q_{\mathrm{g}i}^{k} + Q_{\mathrm{l}i}^{k}}{A_i^k} \tag{41}$$

$$v_{\mathrm{m}i}^{k-1} = \frac{Q_{\mathrm{g}i}^{k-1} + Q_{\mathrm{l}i}^{k-1}}{A_i^{k-1}} \tag{42}$$

$$v_{\mathrm{m}i+1}^{k-1} = \frac{Q_{\mathrm{g}i+1}^{k-1} + Q_{\mathrm{l}i+1}^{k-1}}{A_{i+1}^{k-1}} \tag{43}$$

$$v_{\mathrm{m}i-1}^{k-1} = \frac{Q_{\mathrm{g}i-1}^{k-1} + Q_{\mathrm{l}i-1}^{k-1}}{A_{i-1}^{k-1}} \tag{44}$$

Thus, formulas (29)–(32) are deformed as:

$$\frac{v_{\mathrm{m}i}^{k} - \left[\xi v_{\mathrm{m}i}^{k-1} + \frac{1-\xi}{2}\left(v_{\mathrm{m}i+1}^{k-1} + v_{\mathrm{m}i-1}^{k-1}\right)\right]}{\Delta t} + v_{\mathrm{m}i}^{k-1}\frac{v_{\mathrm{m}i+1}^{k-1} - v_{\mathrm{m}i-1}^{k-1}}{2\Delta z}$$
$$+ g\frac{H_{i+1}^{k-1} - H_{i-1}^{k-1}}{2\Delta z} + \frac{H_i^{k-1}g}{\rho_{\mathrm{m}i}^{k-1}}\frac{\rho_{\mathrm{m}i+1}^{k-1} - \rho_{\mathrm{m}i-1}^{k-1}}{2\Delta z} + \frac{f v_{\mathrm{m}i}^{k-1}\left|v_{\mathrm{m}i}^{k-1}\right|}{2D} = 0 \tag{45}$$

$$\frac{H_i^{k} - \left[\xi H_i^{k-1} + \frac{1-\xi}{2}\left(H_{i+1}^{k-1} + H_{i-1}^{k-1}\right)\right]}{\Delta t} + v_{\mathrm{m}i}^{k-1}\frac{H_{i+1}^{k-1} - H_{i-1}^{k-1}}{2\Delta z}$$
$$+ v_{\mathrm{m}i}^{k-1}\sin\theta + \frac{\frac{1}{\rho_{\mathrm{m}i}^{k-1}g} - \frac{\left(H_i^{k-1}-z\right)}{K}}{\frac{2}{Ee}\sqrt{\frac{A_i^{k-1}}{\pi} + \frac{1}{K}}}\frac{v_{\mathrm{m}i+1}^{k-1} - v_{\mathrm{m}i-1}^{k-1}}{2\Delta z} = 0 \tag{46}$$

$$\frac{\rho_{\mathrm{m}i}^{k} - \left[\xi\rho_{\mathrm{m}i}^{k-1} + \frac{1-\xi}{2}\left(\rho_{\mathrm{m}i+1}^{k-1} + \rho_{\mathrm{m}i-1}^{k-1}\right)\right]}{\Delta t} + v_i^{k-1}\frac{\rho_{\mathrm{m}i+1}^{k-1} - \rho_{\mathrm{m}i-1}^{k-1}}{2\Delta z} - \frac{\rho_{\mathrm{m}i}^{k-1}}{\frac{K}{\rho_i^{k-1}g} - \left(H_i^{k-1}-z_i\right)} \cdot$$
$$\left\{\frac{H_i^{k} - \left[\xi H_i^{k-1} + \frac{1-\xi}{2}\left(H_{i+1}^{k-1} + H_{i-1}^{k-1}\right)\right]}{\Delta t} + v_i^{k-1}\left(\frac{H_{i+1}^{k-1} - H_{i-1}^{k-1}}{2\Delta z} + \sin\theta\right)\right\} = 0 \tag{47}$$

$$\frac{A_i^{k} - \left[\xi A_i^{k-1} + \frac{1-\xi}{2}\left(A_{i+1}^{k-1} + A_{i-1}^{k-1}\right)\right]}{\Delta t} + v_i^{k-1}\frac{A_{i+1}^{k-1} - A_{i-1}^{k-1}}{2\Delta z} - \frac{2KA_i^{k-1}}{\rho_i^{k-1}Ee}\sqrt{\frac{A_i^{k-1}}{\pi}} \cdot$$
$$\left\{\frac{\rho_{\mathrm{m}i}^{k} - \left[\xi\rho_{\mathrm{m}i}^{k-1} + \frac{1-\xi}{2}\left(\rho_{\mathrm{m}i+1}^{k-1} + \rho_{\mathrm{m}i-1}^{k-1}\right)\right]}{\Delta t} + v_i^{k-1}\frac{\rho_{\mathrm{m}i+1}^{k-1} - \rho_{\mathrm{m}i-1}^{k-1}}{2\Delta z}\right\} = 0 \tag{48}$$

The corresponding assumptions will also be changed accordingly:

$$\begin{cases} X_{v1}^{\mathrm{m}} = \xi v_{\mathrm{m}i}^{k-1} + \frac{1-\xi}{2}\left(v_{\mathrm{m}i+1}^{k-1} + v_{\mathrm{m}i-1}^{k-1}\right), & X_{v2}^{\mathrm{m}} = \left(v_{\mathrm{m}i+1}^{k-1} - v_{\mathrm{m}i-1}^{k-1}\right)\frac{\Delta t}{\Delta z} \\[2mm] X_{\rho1}^{\mathrm{m}} = \xi\rho_{\mathrm{m}i}^{k-1} + \frac{1-\xi}{2}\left(\rho_{\mathrm{m}i+1}^{k-1} + \rho_{\mathrm{m}i-1}^{k-1}\right), & X_{\rho2}^{\mathrm{m}} = \left(\rho_{\mathrm{m}i+1}^{k-1} - \rho_{\mathrm{m}i-1}^{k-1}\right)\frac{\Delta t}{\Delta z} \end{cases} \tag{49}$$

After sorting, the expressions of four unknowns considering gas-liquid two-phase are obtained as follows:

$$\begin{cases} H_i^k = X_{H1} - v_{m_i}^{k-1}\left(\frac{X_{H2}}{2} + \Delta t \sin\theta\right) - X_{v2}^m\left[\dfrac{\frac{1}{\rho m_i^{k-1}g} - \frac{\left(H_i^{k-1} - z_i\right)}{K}}{\frac{4}{Ee}\sqrt{\frac{A_i^{k-1}}{\pi} + \frac{1}{K}}}\right] \\[2em] v_{m_i}^k = X_{v1}^m - \frac{v_{m_i}^{k-1}}{2}X_{v2}^m - \frac{g}{2}X_{H2} - \frac{\left(H_i^{k-1} - z_i\right)g}{2\rho m_i^{k-1}}X_{\rho2}^m - \frac{f\Delta t}{2D}v_{m_i}^{k-1}\left|v_{m_i}^{k-1}\right| = 0 \\[2em] \rho m_i^k = X_{\rho1}^m - \frac{v_{m_i}^{k-1}}{2}X_{\rho2}^m + \dfrac{\rho m_i^{k-1}}{\left[\frac{K}{\rho m_i^{k-1}g} - \left(H_i^{k-1} - z_i\right)\right]}\left[H_i^k - X_{H1} + v_{m_i}^{k-1}\left(\frac{X_{H2}}{2} + \Delta t \sin\theta\right)\right] \\[2em] A_i^k = X_{A1} - \frac{v_{m_i}^{k-1}}{2}X_{A2} + \frac{2A_i^{k-1}K}{\rho m_i^{k-1}Ee}\sqrt{\frac{A_i^{k-1}}{\pi}}\left(\rho m_i^k - X_{\rho1}^m + \frac{v_{m_i}^{k-1}}{2}X_{\rho2}^m\right) \end{cases} \tag{50}$$

3.2. Boundary Conditions

The boundary point is different from the inner node. It has an equation with boundary conditions, and the number of equations is more than the number of unknowns. For the previous solutions, one is to synthesize the continuity equation and motion equation into a partial differential equation and then transform it into a difference equation. The second is not to establish the continuity equation and the motion equation, but to introduce the characteristic equation. For example, the inverse characteristic line equation can be introduced for the upstream and the along characteristic line equation can be introduced for the downstream [32,33]. For the equation form of four variables in this paper, neither method is applicable. The solution obtained by method one does not converge and cannot meet the stability condition. However, if the continuity Equation (24) is abandoned, the ideal results can be obtained by using the equation of motion (19) and Equations (20) and (21) together with the upstream boundary conditions. The downstream boundary of a simple pipeline can be treated by method one, and the results are stable and convergent.

The equations of motion (19)–(21) are used to solve the upstream boundary conditions. When the upstream is a reservoir, the reservoir water level can be regarded as constant, and the boundary condition is a constant, that is $\frac{\partial H}{\partial t} = 0$, from the motion equation:

$$v_0^k = v_0^{k-1}\left(1 - X'_{v2}\right) - g\left[X'_{H2} + \frac{\left(H_0^{k-1} - z_0\right)}{\rho_0^{k-1}}X'_{\rho2} - \frac{f\Delta t}{2gD}v_0^{k-1}\left|v_0^{k-1}\right|\right] \tag{51}$$

$$\rho_0^k = \rho_0^{k-1} - v_0^{k-1}X'_{\rho2} + \frac{\rho_0^{k-1}}{\frac{K}{\rho_0^{k-1}g} - \left(H_0^{k-1} - z_0\right)}v_0^{k-1}\left(X'_{A2} + \Delta t \sin\theta\right) \tag{52}$$

Combine formulas (20) and (21) and substitute the calculated value to obtain:

$$A_0^k = A_0^{k-1} - v_0^{k-1}X'_{A2} + \frac{2A_0^{k-1}K}{\rho_0^{k-1}Ee}\sqrt{\frac{A_0^{k-1}}{\pi}}\left(\rho_0^k - \rho_0^{k-1} + v_0^{k-1}X'_{\rho2}\right) \tag{53}$$

where in:

$$\begin{cases} X'_{H2} = \left(H_1^{k-1} - H_0^{k-1}\right)\frac{\Delta t}{\Delta z} \qquad X'_{v2} = \left(v_1^{k-1} - v_0^{k-1}\right)\frac{\Delta t}{\Delta z} \\[1em] X'_{\rho2} = \left(\rho_1^{k-1} - \rho_0^{k-1}\right)\frac{\Delta t}{\Delta z} \qquad X'_{A2} = \left(A_1^{k-1} - A_0^{k-1}\right)\frac{\Delta t}{\Delta z} \end{cases} \tag{54}$$

Here, the equation of motion (19) and the continuity Equation (24) are integrated into a partial differential equation and then transformed into a difference equation. The specific form is:

$$\frac{\partial H}{\partial t} + (v + v_0)\frac{\partial H}{\partial s} + \frac{v_0}{g}\frac{\partial v}{\partial t} + \left[\frac{\frac{1}{\rho g} - \frac{(H-z)}{K}}{\frac{2}{Ee}\sqrt{\frac{A}{\pi} + \frac{1}{K}}} + \frac{v_0 v}{g}\right]\frac{\partial v}{\partial s} + \frac{(H-z)v_0}{\rho}\frac{\partial \rho}{\partial s} + v\sin\theta + v_0\frac{fv|v|}{2gD} = 0 \tag{55}$$

When the downstream end of the pipeline is a valve, it can be regarded as an orifice to obtain [34]:

$$v_t^A = v_m \tau_k \sqrt{\frac{H_t^A}{H_0}} \tag{56}$$

where H_0 is the initial head pressure of the downstream section. For downstream N section, rewrite it as:

$$v_N^k = v_m \tau_k \sqrt{\frac{H_N^k}{H_N^0}} \tag{57}$$

Solve the above equation with Equations (20), (21) and (39):

$$v_N^k = -Y_1 Y_2 + \sqrt{(Y_1 Y_2)^2 - 2Y_2 Y_3} \tag{58}$$

When the flow rate is obtained, the following formula (37) can be obtained:

$$H_N^k = -Y_1 v_N^k - Y_3 \tag{59}$$

Substitute into Equations (20) and (21) to obtain:

$$\rho_N^k = \rho_N^{k-1} - v_N^{k-1} X''_{\rho 2} + \frac{\rho_N^{k-1}}{\frac{K}{\rho_N^{k-1}g} - \left(H_N^{k-1} - z_0\right)}\left[H_N^k - H_N^{k-1} + v_N^{k-1}(X''_{A2} + \Delta t\sin\theta)\right] \tag{60}$$

$$A_N^k = A_N^{k-1} - v_N^{k-1} X''_{A2} + \frac{2A_N^{k-1}K}{\rho_N^{k-1}Ee}\sqrt{\frac{A_N^{k-1}}{\pi}}\left(\rho_N^k - \rho_N^{k-1} + v_N^{k-1} X''_{\rho 2}\right) \tag{61}$$

where v_m is the initial flow velocity at the downstream section of the pipeline and τ_k is the opening of the downstream valve.

$$Y_1 = \frac{v_0}{g} \tag{62}$$

$$Y_2 = \frac{(v_m \tau_k)^2}{2H_N^0} \tag{63}$$

$$Y_3 = -H_N^{k-1} + \left(v_N^{k-1} + v_0\right)X''_{H2} - v_0\left[\frac{v_N^{k-1}}{g} - \frac{(Hv_N^{k-1} - z_0)}{\rho v_N^{k-1}}X''_{\rho 2} - \frac{f\Delta t}{2gD}v_N^{k-1}|v_N^{k-1}|\right]$$

$$+ \left[\frac{\frac{1}{\rho_N^{k-1}g} - \frac{(H_N^{k-1} - z_N)}{K}}{\frac{2}{Ee}\sqrt{\frac{A_N^{k-1}}{\pi} + \frac{1}{K}}} + \frac{v_0 v_N^{k-1}}{g}\right]X''_{v2} + v_N^{k-1}\cdot\Delta t\sin\theta \tag{64}$$

$$\begin{cases} X''_{H2} = \left(H_N^{k-1} - H_{N-1}^{k-1}\right)\frac{\Delta t}{\Delta z} \\ X''_{\rho 2} = \left(\rho_N^{k-1} - \rho_{N-1}^{k-1}\right)\frac{\Delta t}{\Delta z} \end{cases} \qquad \begin{aligned} X''_{v2} &= \left(v_N^{k-1} - v_{N-1}^{k-1}\right)\frac{\Delta t}{\Delta z} \\ X''_{A2} &= \left(A_N^{k-1} - A_{N-1}^{k-1}\right)\frac{\Delta t}{\Delta z} \end{aligned} \tag{65}$$

The above is the treatment of the boundary problem of a simple pipeline. Figure 4 shows the calculation flowchart.

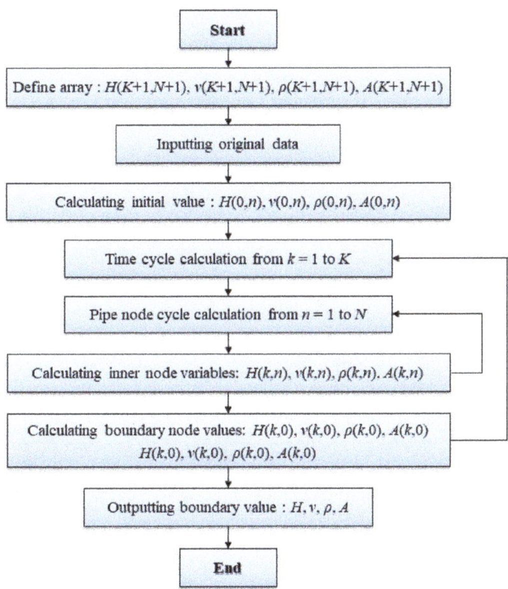

Figure 4. Flowchart of calculation of water hammer four variable closed model.

3.3. Model Validation

To verify the accuracy of the water hammer model, a water hammer calculation was carried out on a case of a fluid pipeline in the literature, and the calculation results were compared with those in the literature [35].

As shown in Figure 5, the calculation results of water hammer are compared. The red is the calculation result of the literature, and the blue is the calculation result of the article model. It can be seen from the figure that there are certain deviations in the results, which may be caused by different mathematical models since the model in this paper considers the four variables of pressure, flow rate, density and flow area.

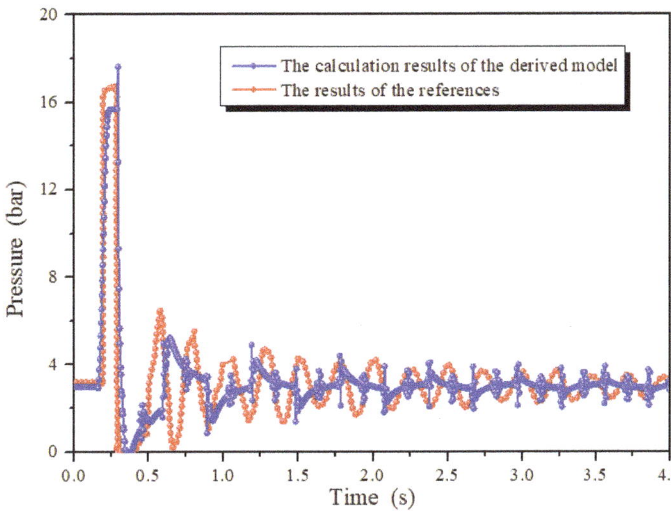

Figure 5. Comparison between calculation results of the present model and literature results.

4. Field Application and Result Discussion

4.1. Sample Pipeline Foundation Parameters

Taking an oil pipeline X as an example, this paper simulates the pressure fluctuation in the pipeline after the valve at the end is closed and the effects of different factors such as valve closing time, flow and gas content on water hammer pressure. The buried depth of the pipeline is 1.8 m, the pipe diameter is 300 mm, and the wall thickness is 6 mm. The physical parameters of oil, soil, anticorrosion coating and pipeline are shown in Table 1. The atmospheric temperature is 20 °C, the soil temperature is 5 °C, and the oil temperature is 22 °C.

Table 1. Physical parameters of oil, soil, anticorrosion coating and pipe.

Parameter	Value	Parameter	Value
Density of oil	841 kg/cm^3	Specific heat of soil	2225 J/(K·kg)
Thermal diffusivity of oil	5.91 mm^2/s	Specific heat of pipe	465 J/(K·kg)
Thermal conductivity of oil	0.162 W/(m·K)	Specific heat of anticorrosive coating	2000 J/(K·kg)
Coefficient of thermal expansion of oil	8.3×10^{-4} K^{-1}	Specific heat of oil	1900 J/(K·kg)
Density of soil	1455 kg/cm^3	Thermal conductivity of soil	1.4 W/(m·K)
Density of pipe	7850 kg/cm^3	Thermal conductivity of pipe	48 W/(m·K)
Density of anticorrosive coating	1200 kg/cm^3	Thermal conductivity of anticorrosive coating	0.15 W/(m·K)

4.2. Result and Discussion

Assuming that the valve closing time is about 5 s, the valve closing index m is 1.0 (linear valve closing), and the calculation time is 200 s. Figure 6 is the change of pressure at the pipe end after valve closing. It can be seen from the figure that at the moment of valve closing, the fluid flow rate at the pipe end suddenly changes to 0, the pressure increases rapidly, reaches the peak value of 4.95 MPa at 5.08 s, then starts to decline, reaching the valley value of 3.37 MPa at 12.80 s. Due to friction along the way, the fluctuating pressure will decay quickly and finally reach the equilibrium pressure.

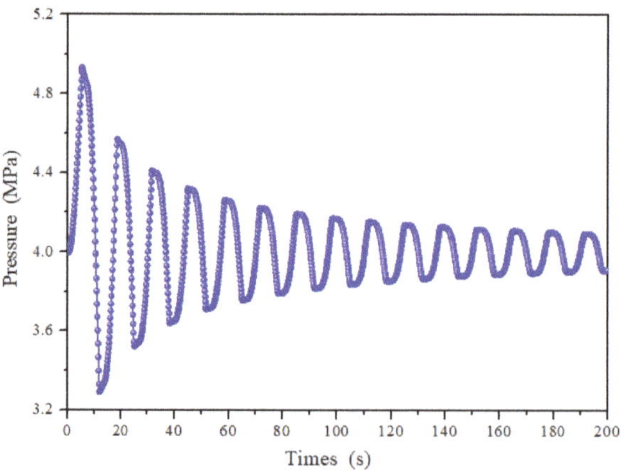

Figure 6. Pressure fluctuation at the end of the pipeline.

Take the section close to the pipe end as the research plane and calculate the fluctuation of velocity in this plane as shown in Figure 7. Before closing the valve, the initial flow velocity of the section is 2.85 m/s. When the valve is closed, the flow rate drops rapidly and fluctuates around 0. According to the principle of the Bernoulli equation, the fluctuation will become weaker and weaker over time. Finally, the fluid pressure in the pipe will be a certain value and the flow rate will be 0.

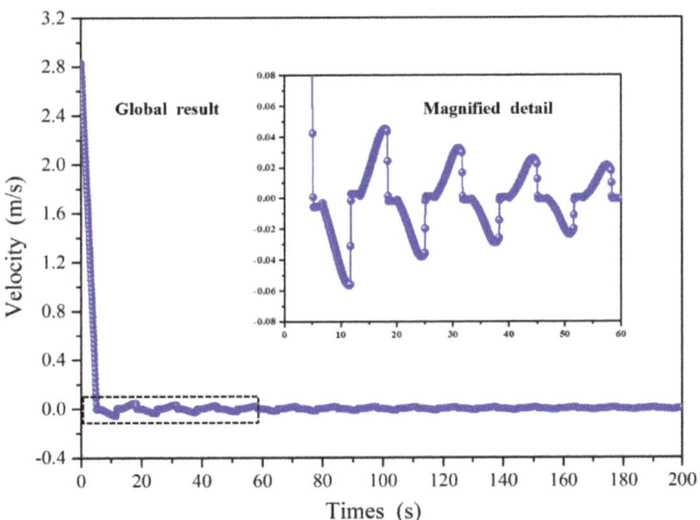

Figure 7. Difference of natural frequency of tubing under two model analysis methods.

The effects of different valve closing time on the fluctuation of pressure and velocity at the pipe end are studied by the model. As shown in Figure 8a, the simulation results of fluctuating pressure at the pipe end under five different valve closing times are shown. According to the calculation, with the gradual increase of valve closing time, the maximum fluctuating pressure at the pipe end decreases, and the time of peak value lags behind. Figure 8b shows the simulation results of fluctuation velocity near the pipe end under five different valve closing times. The shorter the valve closing time, the earlier the flow rate drops to 0 and fluctuates around 0, and the greater the fluctuation range.

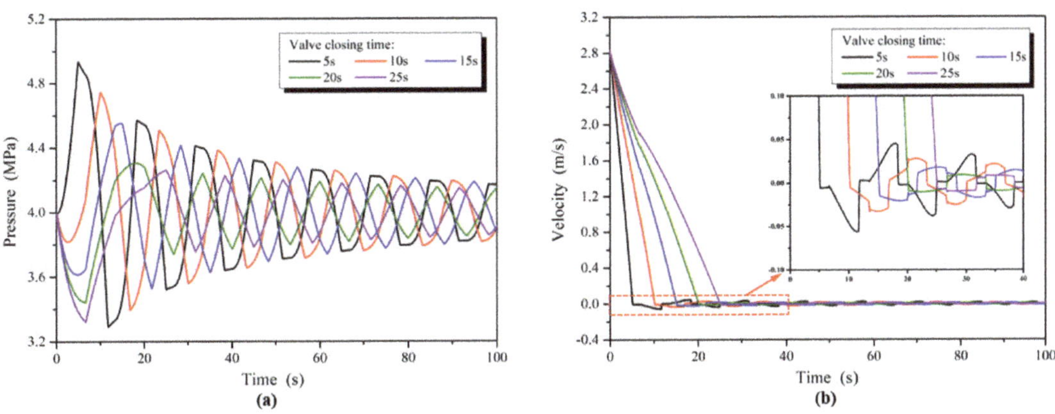

Figure 8. Influence of valve closing time on pressure and velocity fluctuation at the end of a pipeline: (**a**) represents pressure response; (**b**) represents velocity response.

Figure 9 shows the influence of different flow rates on pipe end pressure. It can be seen from Figure 9a that the pressure at the pipe end first decreases and then increases at the moment of valve closing. When the flow in the pipe is 0.4 kg/s, the decline amplitude is the largest, but the rise speed is also the fastest and the peak value is the highest. Similarly, when the flow rate is 0.4 kg/s, the decline rate of pipe end pressure from peak to valley is the fastest and the valley is the smallest. That is, the greater the flow, the greater the change of pipe end pressure, the faster the speed change, and the more obvious the water hammer effect.

Figure 9. Influence of flow rate on pressure and velocity fluctuation at the end of a pipeline: (**a**) represents pressure response; (**b**) represents velocity response.

As Figure 9b shows, the greater the flow rate, the greater the initial flow velocity in the pipe and the fastest the decline rate of flow velocity. According to the enlarged view shown in Figure 9b, when the flow rate is large (e.g., flow rate is 0.4 kg/s), flow rate fluctuation is more intense. Moreover, there will be small fluctuations at inflection points such as falling from peak value and rising from valley value. The larger the flow, the more obvious the small fluctuations. It shows that the flow with large flow is greatly disturbed by instantaneous obstacles such as valve closing. However, with the increase of time, the impact of flow will be less and less obvious, whether it is pressure or flow rate.

After the fluctuating pressure is generated at the pipe end, it will propagate upstream along the pipeline to form a pressurization wave surface. To intuitively obtain the propagation of pressure wave in all calculation time, the three-dimensional diagram of pressure-wave propagation with pipe length and time after valve closing is drawn (Figure 10).

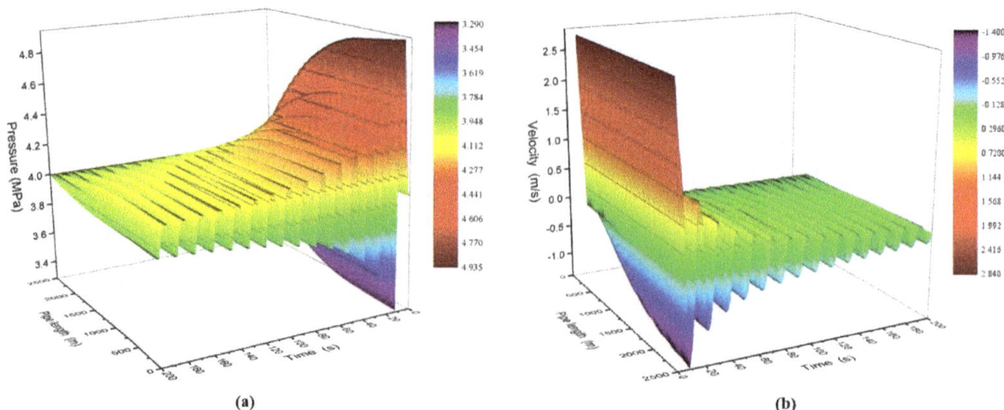

Figure 10. Three-dimensional diagram of pressure-wave propagation with pipe length and time after valve closing: (**a**) represents pressure response; (**b**) represents velocity response.

The effects of different pipe length and time on the fluctuating pressure in the pipe are studied, as shown in Figure 11. Figure 11a shows the pressure change chart in a 500 m section at 40 s. Pressure changes with pipe length when the blue line is 40 s. The red line shows the change of pressure with time at 500 m. By analogy, Figure 11b–d are the pressure change charts of a 1000 m section at 80 s, 1500 m section at 120 s and 2000 m section at 160 s after valve closing. The meaning of the blue line and the red line is the same as that in Figure 11a. It can be seen from the figure that the pressure fluctuation will gradually attenuate along the pipe length with the increase of time. The place with the greatest water hammer effect is near the valve. Under the coupling effect of time and tube length, the shorter the time and the shorter the tube length, the more obvious the pressure fluctuation; the longer the time and the longer the tube, the weaker the pressure fluctuation.

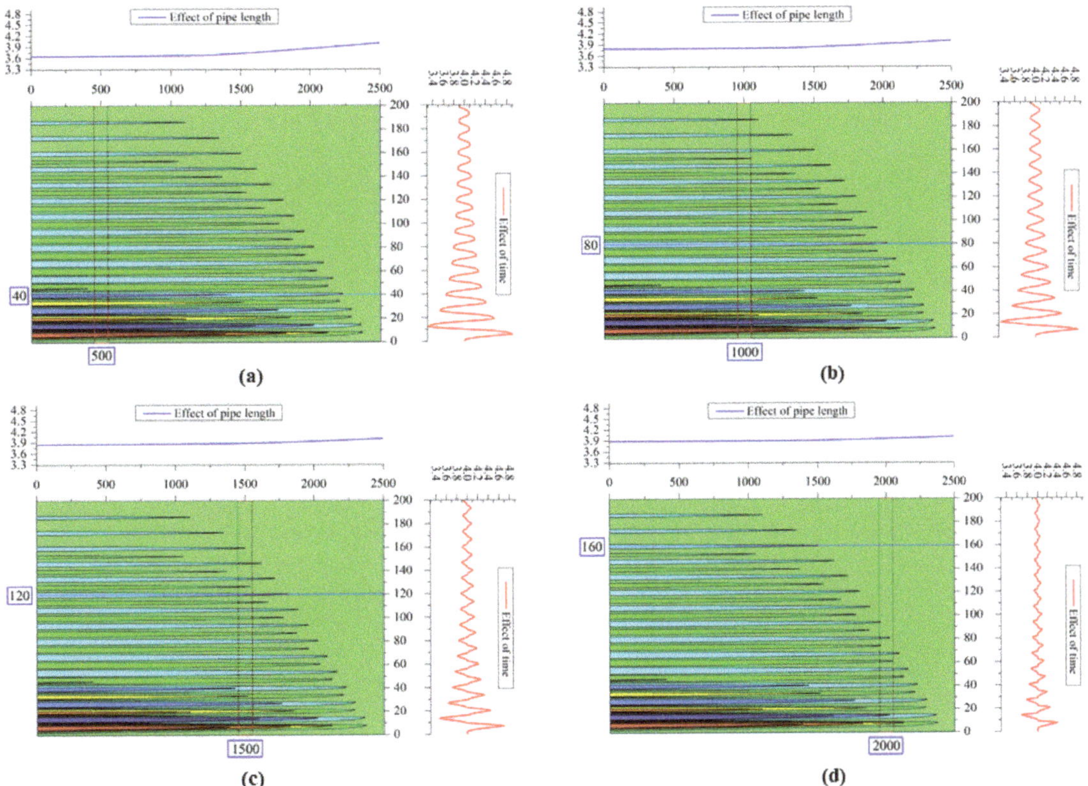

Figure 11. Effect of different pipe length and time on fluctuating pressure in pipe: (**a**) represents the pressure change chart in a 500 m section at 40 s; (**b**) represents the pressure change chart in a 1000 m section at 80 s; (**c**) represents the pressure change chart in a 1500 m section at 120 s; (**d**) represents the pressure change chart in a 2000 m section at 160 s.

Through the water hammer model established in this paper, the changes of pipe end pressure and flow velocity under five conditions with gas content of 1%, 3%, 5%, 7% and 9% were calculated, respectively (Figure 12). Figure 12a shows that the greater the gas content, the smaller the fluctuation peak of pipe end pressure. The time of the first peak of pressure wave is basically the same, but the time of the first valley value begins to be different. The greater the gas content, the later the minimum pressure appears. With the increase of time, this difference becomes more and more obvious, which is reflected in a longer water hammer cycle. Since the water hammer period is the ratio of pipe length to pressure-wave velocity, the greater the void fraction, the smaller the pressure-wave velocity. Slightly different is that there is a difference in the time when the first fluctuation value of velocity wave appears. The greater the void fraction, the longer the time lag of flow rate fluctuation, as shown in Figure 12b. According to Bernoulli's theory, pressure and velocity should be in a trade-off relationship and as the former increases, the latter decreases. Therefore, when the gas content is large, although the velocity fluctuation lags, the absolute value of velocity is large.

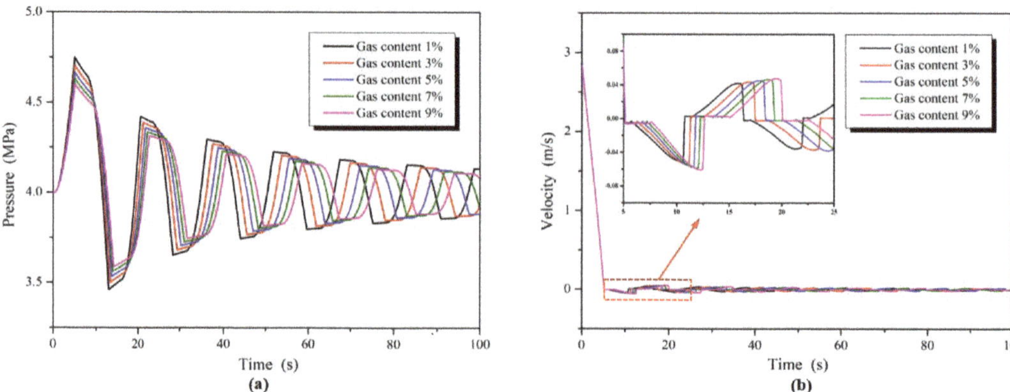

Figure 12. Influence of gas content on pressure and velocity fluctuation at the end of a pipeline: (**a**) represents pressure response; (**b**) represents velocity response.

When there is free gas in the oil pipe, the gas will form uneven bubbles in the pipe. When the bubble is free in the oil, it can be approximately regarded as an elastic deformable body. When there is pressure fluctuation in the oil, the bubbles will be forced to compress and increase their internal energy. At this time, the bubbles are not stable, and work will be done to the surrounding oil to release energy, so that the momentum of the oil increases and compresses the surrounding bubbles. Due to energy dissipation, the increase of oil momentum is not equal to the increase of internal energy after bubble forcing, and the energy of bubble forcing comes from pressure fluctuation. Therefore, the actual pressure fluctuation value is obviously lower than that without gas, and the size of the pressure wave mainly depends on the gas content [36–38].

5. Conclusions

Based on the essence of water hammer, the interaction between pipeline fluid motion and water hammer wave propagation is considered, and the pressure, velocity, density and overflow area are set as variables. A new set of water hammer calculation equations is deduced and solved numerically. The effects of different valve closing time, flow rate and gas content on pressure distribution and water hammer effect were studied. The following conclusions can be drawn:

With the increase of valve closing time, the maximum fluctuating pressure at the pipe end decreases, and the time of peak value also lags behind. The shorter the valve closing time, the earlier the flow rate drops to zero and fluctuates around zero, and the greater the fluctuation range. When the valve closing time increases from 5 s to 25 s, the difference of water hammer pressure is 0.72 MPa and the difference of velocity fluctuation amplitude is 0.076 m/s.

With greater flow, the pressure change at the pipe end is greater. With faster speed changes, the water hammer effect becomes more obvious. With larger flow, flow velocity fluctuations increase, and there will be a small amplitude fluctuation. Our calculations show that the flow with large flow is greatly disturbed by instantaneous obstacles such as valve closing. However, with the increase of time, the impact of flow will be less and less obvious, whether it is pressure or flow rate.

With the increase of time, the pressure fluctuation will gradually attenuate along the pipe length. The place with the greatest water hammer effect is near the valve. Under the coupling effect of time and tube length, the shorter the time and the shorter the tube length, the more obvious the pressure fluctuation.

Our calculations show that the larger the gas content is, the smaller the fluctuation peak of pipe end pressure is, the longer the water hammer cycle is, and the smaller the pressure-wave velocity is. The actual pressure fluctuation value is obviously lower than that without gas, and the size of the pressure wave mainly depends on the gas content. When the gas content increases from 1% to 9%, the difference in water hammer pressure is 0.41 MPa. When the gas content is large, although the velocity fluctuation lags, the absolute value of velocity is large.

Author Contributions: Conceptualization, methodology, software, C.L.; investigation, J.J.; validation, formal analysis, writing—original draft preparation, writing—review and editing, J.D. All authors have read and agreed to the published version of the manuscript.

Funding: This research received no external funding.

Conflicts of Interest: The authors declare no conflict of interest.

Nomenclature

A	Flow channel area	m^2
a	Pressure-wave velocity	m/s
C_g	Gas content	%
D	Pipe diameter	m
e	Pipe wall thickness	m
E	Elastic modulus of pipes	MPa
f	Friction coefficient	dimensionless
g	Gravitational acceleration	$\mathrm{m/s}^2$
H	Water head	m
i	Time node number	dimensionless
k	Spatial node number	dimensionless
K	Fluid compressibility coefficient	dimensionless
mo	Fluid momentum	kg·m/s
p	Pressure	MPa
Q_g	Gas flow	kg/s
Q_l	Liquid flow	kg/s
Δs	Displacement step	m
s	Location	m
Δt	Time step	s
t	Time	s
v	Flow velocity	m/s
v_m	Gas-liquid mixing flow velocity	m/s
x	Length of flow section	m
y	Moving distance	m
z	Position of reference point	m
ρ	Fluid density	$\mathrm{kg/m}^3$
ρ_g	Gas density	$\mathrm{kg/m}^3$
ρ_l	Liquid density	$\mathrm{kg/m}^3$
ρ_m	Multiphase fluid density	$\mathrm{kg/m}^3$
θ	Included angle between pipeline axis and horizontal direction	°
γ	Specific gravity	dimensionless
ζ	Fitting coefficient	dimensionless
χ	Wet perimeter	m
τ_0	Pipe inner wall shear stress	MPa

References

1. Ye, J.; Zeng, W.; Zhao, Z.; Yang, J.; Yang, J. Optimization of Pump Turbine Closing Operation to Minimize Water Hammer and Pulsating Pressures During Load Rejection. *Energies* **2020**, *13*, 1000. [CrossRef]
2. Chen, S.; Zhang, J.; Li, G.; Yu, X. Influence Mechanism of Geometric Characteristics of Water Conveyance System on Extreme Water Hammer during Load Rejection in Pumped Storage Plants. *Energies* **2019**, *12*, 2854. [CrossRef]
3. Kandil, M.; Kamal, A.M.; El-Sayed, T.A. Study the effect of pipematerials properties on the water hammer considering the fluid-structure interaction, frictionless model. *Int. J. Press. Vessel. Pip.* **2021**, 104550. [CrossRef]
4. Mery, H.O.; Hassan, J.M.; Ekaid, A.L. Viscoelastic loop strategy for water hammer control in a pressurized steel pipeline. *Mater. Today Proc.* 2021; *in press.* [CrossRef]
5. Wylie, S. *Transient Flow*; Water Conservancy and Electric Power Press: Beijing, China, 1983.
6. Garg, R.K.; Kumar, A. Experimental and numerical investigations of water hammer analysis in pipeline with two different materials and their combined configuration. *Int. J. Press. Vessel. Pip.* **2020**, *188*, 104219. [CrossRef]
7. Behroozi, A.M.; Vaghefi, M. Numerical simulation of water hammer using implicit Crank-Nicolson Local Multiquadric Based Differential Quadrature. *Int. J. Press. Vessel. Pip.* **2020**, *181*, 104078. [CrossRef]
8. Yang, J.; Lv, Y.; Liu, D.; Wang, Z. Pressure Analysis in the Draft Tube of a Pump-Turbine under Steady and Transient Conditions. *Energies* **2021**, *14*, 4732. [CrossRef]
9. Zhang, L.; Zhang, J.; Yu, X.; Lv, J.; Zhang, X. Transient Simulation for a Pumped Storage Power Plant Considering Pressure Pulsation Based on Field Test. *Energies* **2019**, *12*, 2498. [CrossRef]
10. Slawomir, H. Numerical modeling of water hammer with fluid–structure interaction in a pipeline with viscoelastic supports. *J. Fluids Struct.* **2018**, *76*, 469–487.
11. Baba, G.A.; Norsarahaida, A. The Effect of Water Hammer on Pressure Oscillation of Hydrogen Natural Gas Transient Flow. *Appl. Mech. Mater.* **2014**, *554*, 251–255. [CrossRef]
12. Wahba, E.M. On the Propagation and Attenuation of Turbulent Fluid Transients in Circular Pipes. *J. Fluids Eng.* **2015**, *138*, 031106. [CrossRef]
13. Kandil, M.; Kamal, A.M.; El-Sayed, T.A. Effect pipes material on water hammer. *Int. J. Press. Vessel. Pip.* **2019**, *179*, 103996. [CrossRef]
14. Abdel-Gawad, H.A.A.; Djebedjian, B. Modeling water hammer in viscoelastic pipes using the wave characteristic method. *Appl. Math. Model.* **2020**, *83*, 322–341. [CrossRef]
15. Khamoushi, A.; Keramat, A.; Majd, A. One-Dimensional Simulation of Transient Flows in Non-Newtonian Fluids. *J. Pipeline Syst. Eng. Pract.* **2020**, *11*, 04020019. [CrossRef]
16. Abdeldayem, O.M.; Ferràs, D.; van der Zwan, S.; Kennedy, M. Analysis of Unsteady Friction Models Used in Engineering Software for Water Hammer Analysis: Implementation Case in WANDA. *Water* **2021**, *13*, 495. [CrossRef]
17. Lashkarbolok, M.; Tijsseling, A.S. Numerical simulation of water-hammer in tapered pipes using an implicit least-squares approach. *Int. J. Press. Vessel. Pip.* **2020**, *187*, 104161. [CrossRef]
18. Aliabadi, H.K.; Ahmadi, A.; Keramat, A. Frequency response of water hammer with fluid-structure interaction in a viscoelastic pipe. *Mech. Syst. Signal Process.* **2020**, *144*, 106848. [CrossRef]
19. Firouzi, A.; Yang, W.; Shi, W.; Li, C.Q. Failure of corrosion affected buried cast iron pipes subject to water hammer. *Eng. Fail. Anal.* **2021**, *120*, 104993. [CrossRef]
20. Henclik, S. Application of the shock response spectrum method to severity assessment of water hammer loads. *Mech. Syst. Signal Process.* **2021**, *157*, 107649. [CrossRef]
21. Chen, T.; Su, Z.; Zhu, J.; Li, M. Analysis and improvement of 4-equation based on FSI. *Chin. J. Appl. Mech.* **2016**, *33*, 565–569.
22. Zhu, Y.; Wu, C.; Yuan, Y.; Shi, Z. Experimental and modeling study on gas-liquid two-phase transient flow in viscoelastic pipes. *J. Harbin Inst. Technol.* **2018**, *50*, 89–93+172.
23. Zhu, Y.; Wu, C.; Yuan, Y.; Shi, Z. Analysis of pressure damping in air-water transient flow in viscoelastic pipes. *J. Hydraul. Eng.* **2018**, *49*, 303–312.
24. Fu, Y.; Jiang, J.; Li, Y.; Ying, R. Calculation of pipe water hammer pressure with liquid column separation by improved two-fluid model. *Trans. Chin. Soc. Agric. Eng.* **2018**, *34*, 58–65.
25. Fu, Y.; Jiang, J.; Li, Y.; Ying, R. Water hammer simulated method based on two-fluid model in two-phase flow. *J. Huazhong Univ. Sci. Technol.* **2018**, *46*, 126–132.
26. Han, K.; Ding, F.; Mao, Z. Solving water column separation and cavity collapse for pipelines by semi-analytical method. *Trans. Chin. Soc. Agric. Eng.* **2019**, *35*, 33–39.
27. Twyman, J. Transient flow analysis using the method of characteristics MOC with five-point interpolation scheme. *Obras Proy.* **2018**, *24*, 62–70. [CrossRef]
28. Fan, L.; Wang, F.; Li, H. Characteristic-line-based model predictive control for hyperbolic distributed parameter systems and its application. *Control. Theory Appl.* **2013**, *30*, 1329–1334.
29. Napolitano, M. Computational fluid dynamics in Europe, a personal view. *Acta Aerodyn. Sin.* **2016**, *34*, 139–156.
30. Zhang, D. Discrete integrable systems: Multidimensional consistency. *Acta Phys. Sin.* **2020**, *69*, 010202. [CrossRef]
31. Zhang, Z.; Wang, J.; Li, Y.; Liu, H.; Meng, W.; Li, L. Research on the Influence of Production Fluctuation of High-Production Gas Well on Service Security of Tubing String. *Oil Gas Sci. Technol.* **2021**, *76*, 54. [CrossRef]

32. Nerella, R.; Rathnam, E.V. Fluid Transients and Wave Propagation in Pressurized Conduits Due to Valve Closure. *Procedia Eng.* **2015**, *127*, 1158–1164. [CrossRef]
33. Henclik, S. A Numerical Approach to the Standard Model of Water Hammer with Fluid-Structure Interaction. *J. Teor. Appl. Mech.* **2015**, *53*, 543–555. [CrossRef]
34. Tomaszewski, A.; Przybylinski, T.; Lackowski, M. Experimental and Numerical Analysis of Multi-Hole Orifice Flow Meter: Investigation of the Relationship between Pressure Drop and Mass Flow Rate. *Sensors* **2020**, *20*, 7281. [CrossRef] [PubMed]
35. Ismaier, A.; Schluecker, E. Fluid dynamic interaction between water hammer and centrifugal pumps. *Nucl. Eng. Des.* **2009**, *239*, 3151–3154. [CrossRef]
36. Zolfaghary, A.H.; Naghashzadegan, M.; Shokri, V. Comparison of Numerical Methods for Two-Fluid Model for Gas–Liquid Transient Flow Regime and Its Application in Slug Modeling Initiation. *Iran. J. Sci. Technol. Trans. Mech. Eng.* **2018**, *43*, 663–673. [CrossRef]
37. Zhai, L.; Yang, J.; Wang, Y. An investigation of transition processes from transient gas–liquid plug to slug flow in horizontal pipe: Experiment and Cost-based recurrence analysis. *Nucl. Eng. Des.* **2021**, *379*, 111253. [CrossRef]
38. Sondermann, C.N.; Freitas, R.; Rachid, F.; Bodstein, G. A suitability analysis of transient one-dimensional two-fluid numerical models for simulating two-phase gas-liquid flows based on benchmark problems. *Comput. Fluids* **2021**, *229*, 105070. [CrossRef]

 *energies*

MDPI

Article

Numerical Study and Force Chain Network Analysis of Sand Production Process Using Coupled LBM-DEM

Tian Xia [1,2], Qihong Feng [1,2,*], Sen Wang [1,2], Jiyuan Zhang [1,2], Wei Zhang [1,2] and Xianmin Zhang [1,2]

[1] School of Petroleum Engineering, China University of Petroleum (East China), Qingdao 266580, China; b15020061@s.upc.edu.cn (T.X.); wangsen@upc.edu.cn (S.W.); 20180075@upc.edu.cn (J.Z.); b17020072@s.upc.edu.cn (W.Z.); happym@upc.edu.cn (X.Z.)
[2] Key Laboratory of Unconventional Oil & Gas Development, China University of Petroleum (East China), Ministry of Education, Qingdao 266580, China
* Correspondence: fqihong@upc.edu.cn

Abstract: Sand production has caused many serious problems in weakly consolidated reservoirs. Therefore, it is very urgent to find out the mechanism for this process. This paper employs a coupled lattice Boltzmann method and discrete element method (LBM-DEM) to study the sand production process of the porous media. Simulation of the sand production process is conducted and the force chain network evolvement is analyzed. Absolute and relative permeability changes before and after the sand production process are studied. The effect of injection flow rate, cementation strength, and confining pressure are investigated. During the simulation, strong force chain rupture and force chain reorganization can be identified. The mean shortest-path distance of the porous media reduces gradually after an initial sharp decrease while the mean degree and clustering coefficient increase in the same way. Furthermore, the degree of preferential wettability for water increases after the sand production process. Moreover, a critical flow rate below which porous media can reach a steady state exists. Results also show that porous media under higher confining pressure will be more stable due to the higher friction resistance between particles to prevent sand production.

Keywords: lattice Boltzmann method; discrete element method; sand production; force chain network analysis

Citation: Xia, T.; Feng, Q.; Wang, S.; Zhang, J.; Zhang, W.; Zhang, X. Numerical Study and Force Chain Network Analysis of Sand Production Process Using Coupled LBM-DEM. *Energies* **2022**, *15*, 1788. https://doi.org/10.3390/en15051788

Academic Editor: Michele La Rocca

Received: 29 January 2022
Accepted: 23 February 2022
Published: 28 February 2022

Publisher's Note: MDPI stays neutral with regard to jurisdictional claims in published maps and institutional affiliations.

1. Introduction

Sand production is one of the common problems in the oil field during the production process, especially in weakly consolidated or unconsolidated sandstone reservoirs. The stress and pore pressure near the perforation cavity would be redistributed and the rock would change from an equilibrium state to another after the perforation process. The stress state around the perforation cavities region changed due to the considerable force exerted on the rock by the fluid as the fluids flow into the wellbore. Thus, rock failure will occur in the weakly consolidated or unconsolidated reservoir consequently. In this condition, sand particles will be produced along with the reservoir fluids, which lead to the evolution of perforation cavities and deteriorate the formation condition. A lot of money was spent to deal with problems including borehole instability, casing collapse, and well cleaning. Research shows that almost 30% of these sandstone reservoirs may produce sand easily [1]. Besides this, it is reported that the weakly consolidated or unconsolidated reservoir contains 70% of oil and gas in the world [2,3]. However, the sand production prediction is full of uncertainty. Therefore, accurate simulation and prediction of the sand production process are very important and helpful in sand control and enhancing oil recovery.

The commonly used methods to predict sand production consist of experiments and numerical simulations. Terzaghi firstly used a box with a movable lid at the bottom to conduct the sand production experiment and laid a foundation for the later laboratory

---

experiments [4]. Up to now, experiments under more complex conditions such as uni-axial or tri-axial stress environments with axial and multiphase fluid injections can be conducted [5–8]. In these experiments, different categories of influencing factors including the characteristics of fluids, the state of the formation, the effect of confining pressure, and the effect of outlet size are discussed. However, the sand production experiment needs some assumptions since the scale is small and its result cannot be used in field practice due to the boundary effects. Besides this, it is hard to obtain the in-situ rock sample in some reservoirs that are weakly consolidated or unconsolidated.

Various numerical methods considering the plasticity and micromechanics of the rock, erosion effect, and the mechanical mechanisms due to waterflooding have been developed to accurately predict sand production. These methods consist of continuum method and discrete method. The continuum method considers the rock a continuous material and is widely used in sand production prediction because of its simplicity. It includes models based on plasticity theory, mixture theory, and the micromechanics approach. Methods based on plasticity theory considered the yielding behavior of the rock as the result of tensile and shear failures due to the depletion of well pressure [9]. The effect of hydrostatic stresses due to fluid flow is considered and many models are proposed to predict sand production [10]. Methods based on mixture theory considered the effect of hydrodynamic forces on the sand production process [11]. In these methods, the erosion phenomenon was studied and considered as the cause of the rock failure [12]. More influencing factors including flow rate, viscosity, density, grain size, and cavity height are employed to propose the model [13]. The micromechanics approach combined the poro-mechanical and erosion models extensively rather than employing mechanical yielding models and the erosion models independently. Two semi-analytical models, namely SPAM and SLAM model, were presented based on this concept and study the sand production due to the sheer pressure on the rock [14]. However, the continuum methods cannot reflect the real behavior of the rock such as nonlinear characteristics of mechanical response due to its simplicity. The discrete method treats the rock as a combination of many individual sand particles that have their characteristics. O'Connor et al. firstly used the DEM method to simulate sand production [15]. Based on their research, different shapes of particles such as the n-sided polygon were considered [16]. The growth of a micro-failure leading to rock mass failure is concluded by employing PFC2D for simulating the sanding process in hollow cylinder tests with fluid flow [17]. The coupled LBM-DEM method precisely presented the solid-fluid interface and enhanced the reliability of the prediction of sand production and the lattice Boltzmann method (LBM) has been proved to be the most appropriate method in investigating the fluid flow mechanism in the porous media [18–22]. However, these methods are mostly developed on two dimensions and the bond between the particles is not fully considered in the simulation. Besides this, the characteristic of force chain network in the sand production process is not studied in these research works.

Compared with particle methods, it is more appropriate to use the fluid-solid coupling method since sand production is a process in which fluid and solid interact with each other. We adopted the coupled lattice Boltzmann method and discrete element method (LBM-DEM) in this paper since it is effective to deal with the complicated boundary conditions and it is more feasible in pore-scale (mesoscale) fluid flow. Besides this, the CFD-DEM method is computationally expensive when dealing with fine mesh and solving the nonlinear Navier–Stokes equation in the pore scale. Then we validate the method by simulating the sedimentation of a single particle. Finally, we conduct the simulation of the sand production process and analyze the force network evolvement. Absolute and relative permeability before and after the sand production process are calculated and the effect of injection flow rate, cementation strength, and confining pressure are investigated. Compared with previous studies, we conducted our simulation on three dimensions and fully considered the bond between particles. Besides this, the complex network analysis of the force chain network evolution in the sand production process, which was not studied in previous research, is fully discussed in this paper.

2. Methodology

The coupled LBM-DEM method was firstly proposed by Cook, et al. [23]. This method was further improved by a great number of researchers subsequently [24–27]. The basic principles of each method and how the two methods are coupled with each other are given as follows.

2.1. Bonded DEM

DEM was developed to deal with rock mechanics problems in the 1970s [28]. It has been regarded as a useful tool for analyzing the behavior of materials with a discontinuous characteristic. In this method, the porous media are considered a combination of separate particulates. Due to the simplicity and numerical efficiency, we often use spherical particles.

In this method, Newton's second law is used to describe the motion of each particle. The equation that governs the translation motion of a particle can be described as [28]

$$m\frac{d\mathbf{v}}{dt} = \mathbf{F}_b + \mathbf{F}_c + \mathbf{F}_h \tag{1}$$

where m and $\mathbf{v}$ denote the mass and the velocity of the particle, respectively. $\mathbf{F}_b$ indicates the body force, $\mathbf{F}_c$ refers to the total contact force which is calculated by summing up all the contact force on the particle. $\mathbf{F}_h$ represents the hydrodynamic force exerted by the flowing fluid. The equation that governs the rotation of a particle is [28]

$$I\frac{d\boldsymbol{\omega}}{dt} = \mathbf{T}_c + \mathbf{T}_h \tag{2}$$

where I is the moment of inertia of the particle, and $\boldsymbol{\omega}$ is the angular of the particle. $\mathbf{T}_c$ refers to the total torque generated by the contact force and $\mathbf{T}_h$ indicates the hydrodynamic torque exerted by the flowing fluid.

In the DEM simulation, particles moved freely in the domain. We defined that contact exists when the distance between two particles centers is shorter than the summation of the particle radius. Figure 1 illustrates two particles in contact. The difference between the summation of the radius and the particle center distance was called overlapping length. By using the Hertz contact model, the contact force can be calculated by [28]

$$F_n = k_n \delta_n^{3/2} \tag{3}$$

where k_n denotes normal stiffness, which is influenced by the elastic modulus of the material. δ_n refers to the overlapping length.

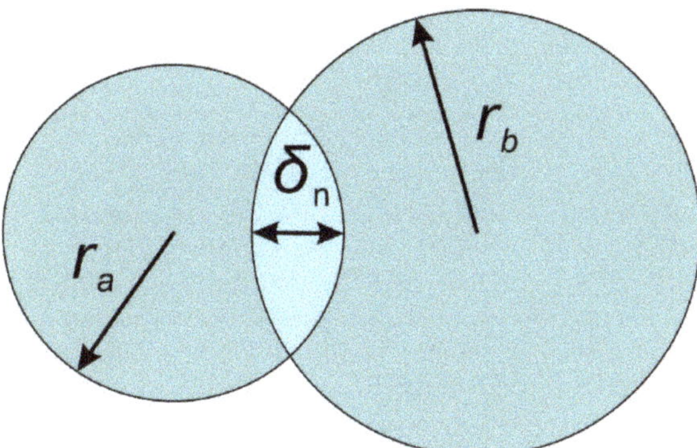

Figure 1. The sketch of two contact particles.

For each particle at each time step, the accelerations can be computed by using Equations (1) and (2). Particle velocity and position would be calculated by employing equations of motion consequently. In this way, the information of any single particle at any time step can be obtained and it is convenient for us to do further analyses.

For the cementation of the rock, we used the contact bond model to simulate the bond characteristics [29,30]. In this model, the grain particles are approximated as spherical particles and the cementation bonds are considered as sticks joining the two adjacent particles. The bond element can be regarded as two springs with normal and shear stiffness acting between the contacting particles. The contact bond model can be shown as [31]

$$F_n^b = \begin{cases} k_n^b \delta \ (F_n^b \leq F_{\max}) \\ 0 \ (F_n^b > F_{\max}) \end{cases} \tag{4}$$

where k_n^b is the normal stiffness of the cement, $F_{\max}$ is the critical tensile force. The magnitude of the critical force can be calculated as

$$F_{\max} = \sigma_{\max} S_c \tag{5}$$

where $\sigma_{\max}$ is the bond strength and the S_c is the cross-sectional area of the bond element. The bond strength of the bond element can be determined by conducting a uniaxial tensile test [32,33]. The bond element can be regarded as a beam and the radius of the cross-section $r_b = (r_1 + r_2)/2$, where r_1 and r_2 are the radii of the bonded particles and the cross-sectional area can be calculated subsequently [34]. If the tensile force acting on the cement exceeds or equals the normal bond tensile strength, the bond breaks and the normal contact force should be zero.

2.2. Lattice Boltzmann Method

Fluid flow is often modeled by Navier–Stokes equations, and LBM is a potential tool for the solution of these equations. The BGK model is most widely used in LBM modeling and streaming and collision is given by [35]

$$f_a(\mathbf{x} + \mathbf{e}_a \Delta t, t + \Delta t) - f_a(\mathbf{x}, t) = -\frac{\Delta t}{\tau} \left[f_a(\mathbf{x}, t) - f_a^{eq}(\mathbf{x}, t) \right] \tag{6}$$

where $\mathbf{x}$ is the position in the lattice space, τ is the collision relaxation time, Δt is the time interval, $\mathbf{e}_a$ refers to the lattice velocity, a indicates the discretized direction, f and f^{eq} denote the local particle distribution function and equilibrium distribution function. In the equation, the left side is the streaming part and the right side is the collision part.

Collision of the fluid particles is regarded as a relaxation towards a local equilibrium and the equilibrium distribution function f^{eq} is defined as [35]

$$f_a^{eq}(\mathbf{x}) = w_a \rho(\mathbf{x}) \left[1 + 3\frac{\mathbf{e}_a \cdot \mathbf{u}}{c^2} + \frac{9}{2} \frac{(\mathbf{e}_a \cdot \mathbf{u})^2}{c^4} - \frac{3}{2} \frac{\mathbf{u}^2}{c^2} \right] \tag{7}$$

where the weights w_a are $1/3$ for the rest particles ($a = 0$), $1/18$ for $a = 1\sim6$, and $1/36$ for $a = 7\sim18$, and c is the basic speed on the lattice given by $c = \Delta x / \Delta t$, Δx denotes the lattice spacing. The configuration of D3Q19 lattice cell is presented in Figure 2.

The distribution function represents the frequency of occurrence in a certain direction and the summation of frequencies of all directions can be considered as fluid densities. Therefore, the macroscopic fluid density is [35]

$$\rho = \sum_{a=0}^{18} f_a \tag{8}$$

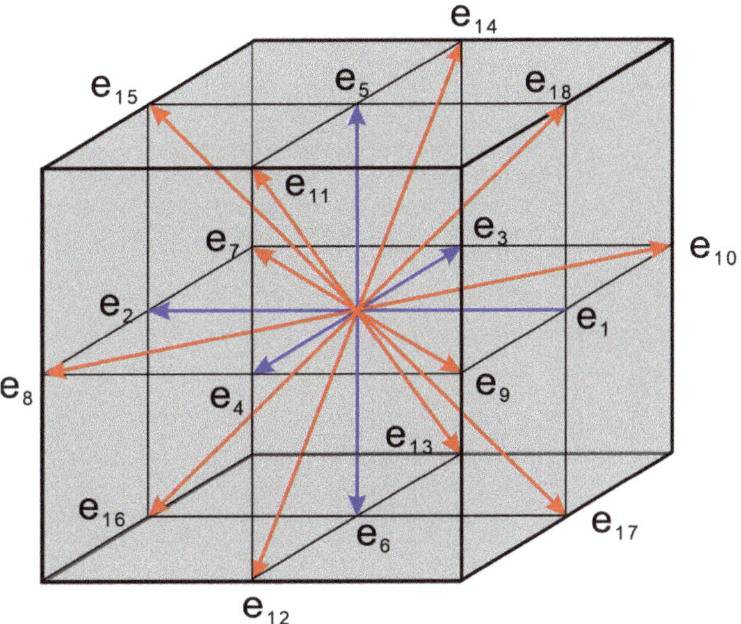

Figure 2. The configuration of D3Q19 showing the 19 discrete velocities.

The macroscopic velocity **u** is an average of the microscopic velocities $\mathbf{e}_a$ weighted by the directional densities f_a and can be expressed as [35]

$$\mathbf{u} = \frac{1}{\rho} \sum_{a=0}^{18} f_a \mathbf{e}_a \tag{9}$$

Equation (9) allows us to connect the discrete microscopic velocities in LBM with the continuous macroscopic velocities representing the fluid's motion.

Viscosity and pressure can be described by [35]

$$\nu = c_s^2 (\tau - \frac{1}{2})\Delta t \tag{10}$$

$$p = \rho c_s^2 \tag{11}$$

where c_s is the sound velocity given by [36]

$$c_s = \sqrt{\frac{c^2}{3}} \tag{12}$$

Shan-Chen model, also known as the pseudo-potential model, is one of the most widely used multiphase Lattice Boltzmann models. In this paper, we employed this method to simulate two-phase flow in the porous media and obtain the relative permeabilities in different stages of sand production.

When the fluid consists of multiple components, a pseudo-potential model can be established by incorporating interaction force between fluid components. For two-phase flow, two distribution functions represent the flow of wetting phase and non-wetting phase and the evolution function of the σth component can be described as [37]

$$f_a^\sigma(\mathbf{x} + \mathbf{e}_a \Delta t, t + \Delta t) - f_a^\sigma(\mathbf{x}, t) = -\frac{\Delta t}{\tau_\sigma}[f_a^\sigma(\mathbf{x}, t) - f_a^{\sigma(eq)}(\mathbf{x}, t)] \tag{13}$$

where f_a^σ denote the local particle distribution function of the σth component, τ_σ is the relaxation time of the σth component and the corresponding kinematic viscosity can be expressed as [37]

$$\nu_\sigma = c_s^2(\tau_\sigma - \frac{1}{2})\Delta t \tag{14}$$

In order to simulate two-phase flow in porous media, many researchers proposed different forms to incorporate the interaction force between two fluids in the fluid flow [38–42]. In this paper, we incorporate the interaction force in the collision process by adding a term into the velocity field and the macroscopic velocity of the σth component can be calculated by [43]

$$\mathbf{u}_\sigma^{eq} = \tilde{\mathbf{u}} + \frac{\tau_\sigma \mathbf{F}_\sigma}{\rho_\sigma} \tag{15}$$

$$\tilde{\mathbf{u}} = \frac{\sum_\sigma \frac{\rho_\sigma \mathbf{u}_\sigma}{\tau_\sigma}}{\sum_\sigma \frac{\rho_\sigma}{\tau_\sigma}} \tag{16}$$

where $\tilde{\mathbf{u}}$ is the compound velocity of all components without considering additional forces. $\mathbf{F}_\sigma$ denotes the total force exerted on the σth component, which includes the fluid–fluid interaction $\mathbf{F}_f$, the fluid–solid interaction $\mathbf{F}_{ads}$, and the external force $\mathbf{F}_e$ [44].

Therefore, the whole fluid velocity can be defined as [45,46]

$$\mathbf{u} = \frac{1}{\rho}\left(\sum_\sigma \rho_\sigma \mathbf{u}_\sigma + \frac{1}{2}\sum_\sigma \mathbf{F}_\sigma\right) \tag{17}$$

where ρ is the total density of all the fluid components and can be expressed as [45]

$$\rho = \sum_\sigma \rho_\sigma \tag{18}$$

The fluid–fluid interaction only considers the interaction between the nearest neighbors and can be expressed by [44]

$$\mathbf{F}_f^\sigma(x,t) = -\psi^\sigma(\mathbf{x},t)\sum_{\sigma'}^{S} G_{\sigma\sigma'}\sum_a \omega_i \psi^{\sigma'}(\mathbf{x}+\mathbf{e}_a\Delta t,t)\mathbf{e}_a \tag{19}$$

where $G_{\sigma\sigma'}$ denotes the Green function reflecting the strength of the intercomponent potential between the σth component and the σ'th component, $\psi^\sigma(\mathbf{x},t)$ is the effective density of the σth component and is commonly taken as the fluid density.

At the fluid-solid interface, the solid wall can be treated as a component with constant density. Therefore, the interaction force between the fluid and solid can be written as [44]

$$\mathbf{F}_{ads}^\sigma(\mathbf{x},t) = -\psi^\sigma(\mathbf{x},t)\sum_\sigma^{S} G_s^\sigma \sum_a \omega_i \psi(\rho_w)s(\mathbf{x}+\mathbf{e}_a\Delta t,t)\mathbf{e}_a \tag{20}$$

where G_s^σ denotes the Green function reflecting the strength of potential between the fluid and the solid. $s(\mathbf{x}+\mathbf{e}_a\Delta t,t)$ is a function that is equal to 1 when the node is solid and zero when the node is fluid. ρ_w is the density of the solid wall.

2.3. LBM-DEM Coupling

The moving boundary condition is very important for the coupled LBM and DEM methods. Researchers have come up with numerous moving boundary conditions for coupling. In this study, we employ a method proposed by Noble and Torczynski [25,26,47].

In the coupling method, the collision operator in the LBM equation is expressed with the solid fraction γ in each node (Figure 3). Thus, Equation (6) can be described as [18]

$$f_a(\mathbf{x} + \mathbf{e}_a \Delta t, t + \Delta t) \ = \ f_a(\mathbf{x}, t) - \frac{1}{\tau}(1 - \beta)\left[f_a(\mathbf{x}, t) - f_a^{eq}(\mathbf{x}, t)\right] + \beta f_a^m \tag{21}$$

where β is a weighting function which depends on the solid fraction γ in each node [18],

$$\beta \ = \ \frac{\gamma(\tau - 0.5)}{(1 - \gamma) + (\tau - 0.5)} \tag{22}$$

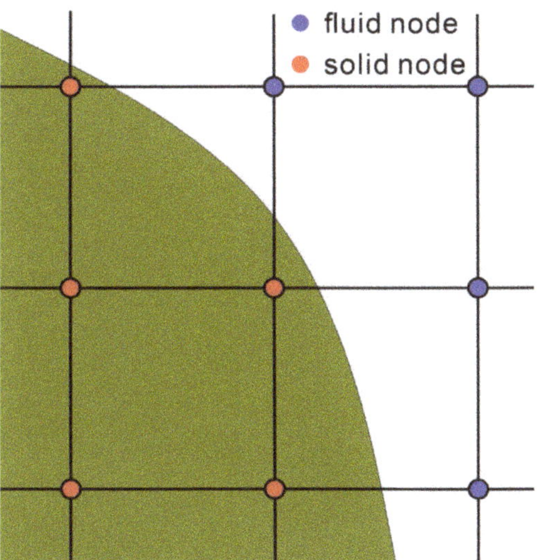

Figure 3. Noble and Torczynski's scheme.

f_a^m is the collision term that represents the bounce-back of the non-equilibrium part of the distribution function. It can be described by [18]

$$f_a^m \ = \ f_{-a}(\mathbf{x}, t) - f_a(\mathbf{x}, t) + f_a^{eq}(\rho, \mathbf{v}_b) - f_{-a}^{eq}(\rho, \mathbf{v}) \tag{23}$$

where $-a$ is the direction opposite of a. The hydrodynamic forces and torque acted on a particle can be calculated as [18]

$$\mathbf{F}_{fluid} \ = \ \frac{h^2}{\Delta t}\left[\sum_n \left(\beta_n \sum_a f_a^m \mathbf{e}_a\right)\right] \tag{24}$$

$$\mathbf{T}_{fluid} \ = \ \frac{h^2}{\Delta t}\left[\sum_n (\mathbf{x}_n - \mathbf{x}_c) \times \sum_n \left(\beta_n \sum_a f_a^m \mathbf{e}_a\right)\right] \tag{25}$$

where n is the number of nodes covered by a particle. The process of coupled LBM and DEM is shown in Figure 4.

2.4. Method Validation

In this section, we will verify the boundary condition for a solid that moves across the lattice by simulating a single sphere settling into a low particle Reynold number. For this case, it has a specific analytical expression for both the force and the torque [3]. The force for this case is given by

$$F = \frac{-6\pi\rho\nu r v}{1 - 0.6526(r/l) + 0.1475(r/l)^3 - 0.131(r/l)^4 - 0.0644(r/l)^5} \tag{26}$$

where ρ is the fluid density, ν is the fluid kinetic viscosity, v is the velocity of the particle, r denotes the particle radius, and l represents the distance between the particle center and the wall.

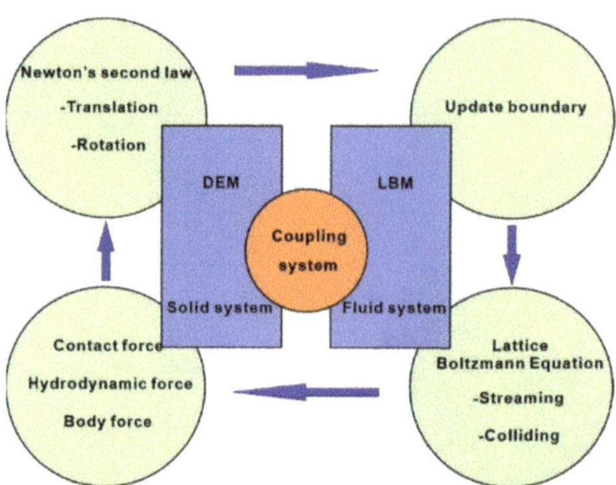

Figure 4. The cycle of computation for coupled LBM and DEM.

Our simulation was set up similarly to that by Strack et al. (2007). Two fixed walls $2l$ distance apart were employed in the transverse direction. The width and depth of the computational domain were set to a relatively larger value of $8l$ to decrease the dependence of the simulation results on the domain size. The walls were modeled with a bounce-back boundary condition, and periodic boundary conditions were adopted in the other two directions. A particle was initially located in the computational domain with a distance of l to both the left and the top. The particle diameter was covered by 10 lattices, so the computational domain was $40 \times 160 \times 160$ lattices in the simulation. No body force was applied to the fluid and the particle settled under the influence of gravity. Due to the buoyancy force exerted on the particle, the gravitational force on the particle was considered based on the density difference between the particle and the fluid. The number of CPU cores is 4 in our simulation.

Figure 5a depicted the velocity of the particle and the force acting on the particle in the simulation. The velocity was normalized by the final settling velocity. The final setting velocity denotes the particle velocity in the steady state, which is defined by the relative error of settling velocity between n and n − 1 time steps smaller than the convergence limit 1×10^{-6}. The force was normalized by the analytical values based on the final settling velocity and the vertical displacements were normalized by the lattice spacing. From the figure, we can see that the velocity of the particle and the force exerted on the particle changed synchronously. When the particle reached the final settling velocity, the force exerted on the particle did not change anymore as well. The particle reached an equilibrium state and kept settling at a constant speed.

We studied the computational error of this method and the cumulative error of simulations under different Reynold numbers and lattice spacing can be seen from Figure 5b. The cumulative error was calculated by

$$\varepsilon = \sum \left| \frac{F_{sim} - F_a}{F_a} \right| \tag{27}$$

where F_{sim} is the simulation results calculated by the LBM-DEM coupling method, Fa denotes the analytical method. As we can see from the figure, the simulation results become more and more accurate with the decrease of the lattice spacing. This is because small lattice spacing means more lattices covered on the particle, which will reduce the oscillation. It also can be spotted that the cumulative computational error is larger when the Reynold number is larger. A larger Reynold number indicates larger particle velocity and computational error is accumulated due to the update of the particle position and recalculation of the solid fraction. Therefore, it should be noticed the displacement of each step for particles should be shorter than the lattice spacing when employing this method to ensure the accuracy and stability of the computation.

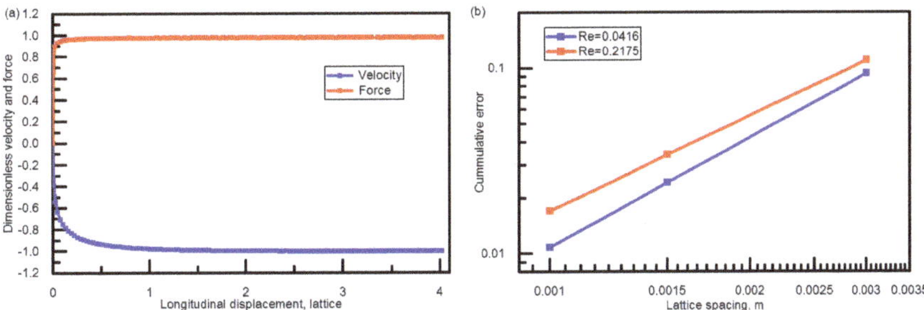

Figure 5. (**a**) Normalized velocity and Force for a spherical particle settling between infinite walls. (**b**) Cumulative computational error under different Reynold numbers with different lattice spacing.

2.5. Model Setup

In this section, we used a 3D LBM-DEM approach to simulate the sand production process. The cross-section of our model for the simulation of the sand production process from a cemented reservoir is shown in Figure 6. Simulations were conducted to investigate the migration of sand particles in different circumstances. In this simulation, the fluid flows through a porous media from one end, which causes sand production at the other end through a small outlet. As it is shown in Figure 6, the sand particles were generated in the simulation region. The total length of the region was 60 mm and the length of the porous media was 20 mm. Both the width and height of the computational domain were 20 mm. The outlet of the simulation region is a circle in the middle of the right plane and its diameter equals 4 mm.

Figure 6. Cross-section of the simulation region.

Boundaries in both the y-direction and z-direction for the fluid were set as periodic boundaries. The fluid was assumed to flow into the computational domain with a specific velocity and a zero-pressure boundary condition was employed at the right boundary of the computational domain. Particles were allowed to leave the simulation domain from the small outlet. Body force was ignored in the simulation. The gravitational force on the particle was set according to the density difference between the particle and the fluid. The fluid density was 1140 kg/m^3, and the dynamic viscosity was 1.14 mPa·s. The simulation parameters of sand particles can be seen in Table 1.

Table 1. Parameters for the sand particles in the simulation.

Parameter	Value
Particle density ρ_s (kg/m^3)	2700
Friction coefficient f_r	0.45
Elastic modulus E (Pa)	5×10^6
Poisson's ratio η	0.2

First, we conduct a simulation with fluid flow through the porous media to study the change of force during waterflooding. Then, absolute and relative permeability before and after the sand production process are calculated and the changes in the properties of porous media are studied. Finally, the effect of injection flow rate, cementation strength, and confining pressure are investigated.

3. Results and Discussion

3.1. Force Chain Network Analysis

Figure 7 shows the force chain of the porous media at different time steps after a period of waterflooding. From the figure, we can see the evolution of the force chain with the fluid injection. Comparing the force chain circled by the blue line in Figures 7a and 7b, we can spot strong force chain rupture during this process and the sand particles detach from the matrix and migrate through the porous media. Besides this, weak force chains circled by the red line in Figure 7a transform into strong force chains, which can be clearly recognized in Figure 7b. During sand production, the force chain network is reorganized continuously until it reaches a steady state.

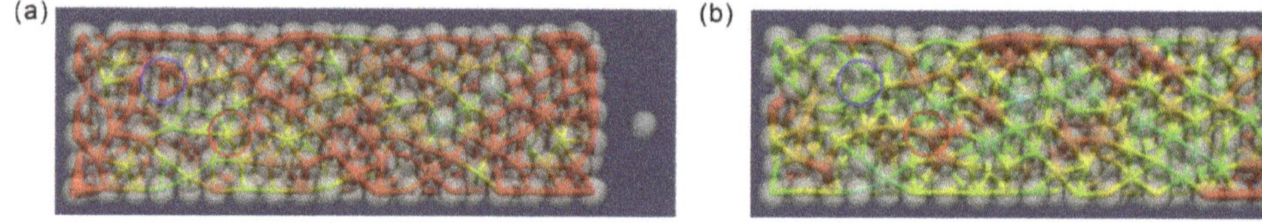

Figure 7. Force chain of the porous media. ((**a**) t = 0.05 s; (**b**) t = 0.1 s).

The force chain network can be analyzed by using complex network theory. The complex network theory is one of the approaches to describe a granular system. Many researchers used it to study granular packing in the past few decades [48,49]. Since the unconsolidated rock can be regarded as sand packing, it is appropriate to use a contact network to clarify the intrinsic mechanisms in the sand production process.

In our view, the network structure is full of randomicity and we can hardly see any principles in it. However, the certainty of the network structure is often hidden in the randomicity and we can reveal it by studying the statistical properties. Therefore, the description and study of the statistical properties of complex networks are of great significance for the development and application of complex network theory. In order to effectively characterize the mathematical and statistical characteristics of complex network structures, various scholars have proposed many methods and tried to use the network parameters to describe the statistical characteristics of complex networks. The three most basic and important parameters are the degree distribution, the mean shortest-path distance, and the clustering coefficient.

The adjacency matrix is often used in describing the information of a complex network. For the weighted and undirected network, the weighted adjacency matrix $\mathbf{W}$ can be expressed as [50]

$$\mathbf{W}_{ij} = \begin{cases} w_{ij}, & \text{if there is an edge between node } i \text{ and node } j \\ 0, & \text{otherwise} \end{cases} \tag{28}$$

where w_{ij} is the corresponding weight between node i and node j. For the rock system, w_{ij} is the force between two particles and the adjacency matrix should be symmetric.

In the weighted network, the connectivity of nodes is also worth noticing regardless of how strongly two particles interact with each other. Therefore, the binary adjacency matrix $\mathbf{A}$ which only focuses on the link between nodes can be expressed as [50]

$$\mathbf{A}_{ij} = \begin{cases} 1, & \text{if } w_{ij} \neq 0 \\ 0, & \text{otherwise} \end{cases} \tag{29}$$

The degree of a node can be described as the number of edges that are attached to it in an undirected network. Therefore, the degree of the node i can be described as [50]

$$de_i = \sum_{j=1}^{N} \mathbf{A}_{ij} \tag{30}$$

where N is the number of nodes in the network. Therefore, the mean degree of the network can be computed as [50]

$$\langle de \rangle = \frac{1}{N} \sum_i de_i \tag{31}$$

For the porous media system, it is easy to find that the node degree is exactly the coordination number of the particle from its definition. Since the degree of nodes in the network follows a certain probability distribution, the degree distribution is an important topological factor to describe and analyze complex networks. The degree distribution function $p(k)$ represents the ratio of the number of nodes with degree equals to k to the number of all nodes in the network, that is, the proportion of nodes with a degree of k in the network.

The mean shortest-path distance L can be defined as the average of the distance between any two nodes and can be calculated by [50]

$$L = \frac{2}{N(N-1)} \sum_{i,j,i \neq j} d_{ij} \tag{32}$$

where d_{ij} denotes the shortest distance between node i and node j. It should be noticed that the distance should be calculated along edges in a path. The mean shortest-path distance, also known as the characteristic path distance of a network, can be used to measure the efficiency of information transfer between network nodes.

The local clustering coefficient C_i can be defined as the number of triangles containing i divided by the number of triples centered at node i [51,52]. It can be expressed as [50]

$$C_i = \begin{cases} \frac{\sum_{hj} A_{hj} A_{ih} A_{ij}}{k_i(k_i-1)}, & k_i \geq 2 \\ 0, & k_i = 0,1 \end{cases} \tag{33}$$

Therefore, the global clustering coefficient C can be calculated as [50]

$$C = \frac{1}{N} \sum_i^{N} C_i \tag{34}$$

The clustering coefficient of a network describes the degree of the collectivity of a complex network. The larger the clustering coefficient of the network, the higher collectivity of the network, which means that the relationship between nodes is closer. The clustering coefficient can also reflect the density of triangles from the definition. The triangles, also known as three-cycles, are related to the stability of the network.

The mean degree change during the simulation and the comparison of degree distribution between the initial and final time steps can be seen in Figure 8. The mean degree of the porous media continues to increase in the simulation but the slope of the curve is constantly decreasing and finally tends to be 0 at the end of the simulation. The larger mean degree means more contact between particles and the porous media tends to reach a steady state. This conclusion can also be revealed from the degree distribution since the number of particles with a larger degree has increased after the simulation.

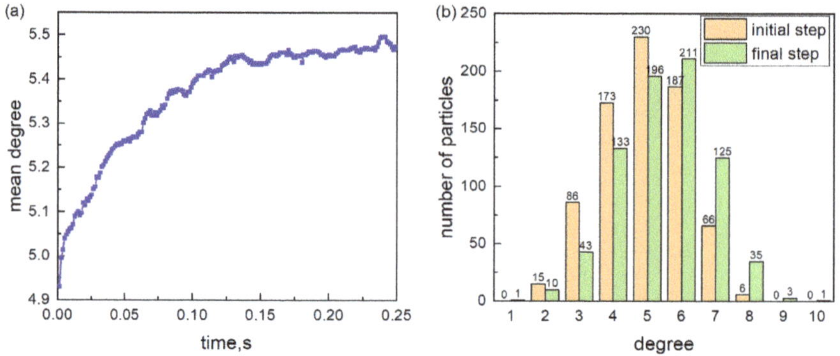

Figure 8. (**a**) Mean degree of porous media over time. (**b**) Comparison of degree distribution between the initial and final time steps.

The mean shortest-path distance and the clustering coefficient during the simulation process are presented in Figure 9. The mean shortest-path distance reduces gradually after a sharp initial decrease. At the end of the simulation, the curve of the simulation tends to be a horizontal line. The mean shortest-path distance indicates the efficiency of information transmission within the network. In the contact force network, it represents the efficiency of force transmission. Shorter mean shortest-path distance means higher force transmission efficiency between particles, which results in a more stable porous media. The tendency of the clustering coefficient is similar to that of the mean degree in the simulation. The clustering coefficient shows the cohesive tendency of the sand particles and the measure of the number of three-cycles in the network. A larger clustering coefficient means more three-cycles in the network, which reflects the network is more stable. All the network parameters indicate the porous media continues to evolve as simulations and gradually becomes stable.

3.2. Absolute Permeability and Relative Permeability

In this section, we present the analysis of absolute permeability and relative permeability of the porous media. For the absolute permeability, fluid flow was simulated by employing the single-phase Lattice Boltzmann method. First, we extract the structure of porous media at the different time steps of the sand production simulation for absolute permeability calculation. Then the fluid with a lattice density of 1 and zero velocity can be initialized throughout the simulation domain. A constant external body force $F = 1 \times 10^{-5}$ is applied along the x-direction to drive the flows. The periodic boundary condition is employed in all directions of the simulation domain, while the boundary condition for the solid particle is the bounce-back boundary condition. The simulation keeps calculating until reaching a steady state, which is defined by the relative error of permeability between n and n-1 time steps and can be expressed as

$$\left| \frac{k^n - k^{n-1}}{k^{n-1}} \right| < 1 \times 10^{-8} \tag{35}$$

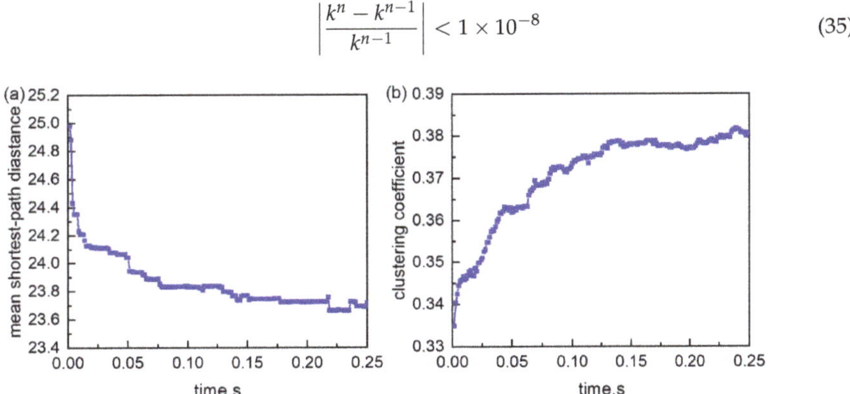

Figure 9. The network parameters change over time for porous media. (**a**) mean shortest-path distance. (**b**) clustering coefficient.

For the relative permeability, fluid flow was simulated by employing the two-phase Shen–Chen Lattice Boltzmann model. Only the initial and the final structure of the porous media in the sand production simulation was employed in the relative permeability calculation. Different water saturation was assigned in the simulation domain by initializing the nodes with wetting fluid of different fractions. A constant external body force $F = 1 \times 10^{-5}$ is assigned to drive the fluid flow along the x-direction. The periodic boundary condition is employed in the x-direction, while the bounce-back boundary condition is adopted for the y and z directions. As with the absolute permeability calculation, the simulation should reach a steady state to obtain reliable results. In relative permeability calculation, the relative error of permeability for both fluids should be less than the convergence limit.

Figure 10a shows the porosity and the absolute permeability as the sand production simulation. We can see both the porosity and permeability increase during this process and finally becomes almost unchanged in the later stage of simulation. This can be explained by the gradually steady state after a period of sand production. The relative permeability of the porous media at the initial and final steps can be seen in Figure 10b. It can be spotted that the connate water saturation and residual oil saturation become smaller at the end of the simulation, which means wider saturation for two-phase flow than that at the initial time step. Besides this, the figure also shows higher relative permeability at the endpoints. This indicates that permeability and connectivity have improved after the sand production process. Moreover, the figure also reveals that the points at which $k_{rw} = 0$ and $k_{ro} = k_{rw}$ shift to the right side of the relative permeability curve. This can reflect that the degree of preferential wettability for water has increased after the sand production process.

3.3. The Effect of Injection Flow Rate

Sand production simulations with different injection rates at different confining pressures are conducted in this section. The injection rates employed in the simulations are 2×10^{-6} m^3/s, 4×10^{-6} m^3/s, 2×10^{-5} m^3/s, and 4×10^{-5} m^3/s, respectively. The confining pressures in the simulation are 0.8 MPa and 3 MPa. The cementation strength employed in the simulation is 1×10^6 Pa. The cumulative weight of produced sand with different injection rates at different confining pressures can be seen in Figure 11.

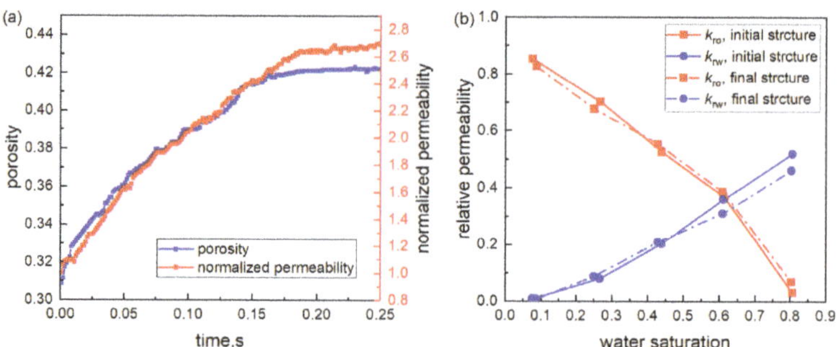

Figure 10. (**a**) porosity and normalized permeability change over time for porous media. (**b**) comparison of relative permeability for porous media at the initial and final steps.

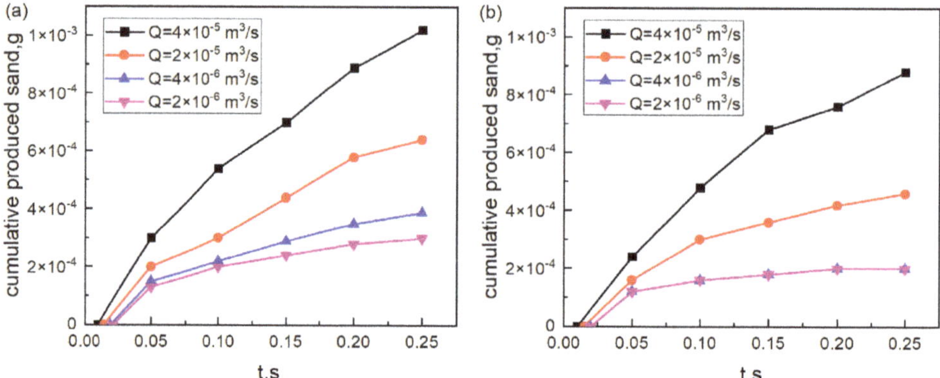

Figure 11. Cumulative weight of produced sand at different injection rates. (**a**) confining pressure equals 0.8 MPa (**b**) confining pressure equals 3 MPa.

At lower confining pressure, we can clearly see that the curves of cumulative produced sand weight with larger flow rate are steeper than that with smaller flow rate from Figure 11a. This is a direct reflection of more produced sand for the rock in larger injection rate conditions. Due to the larger injection rate, the larger hydrodynamic force will exert on the particles and lead to more bond failure. Furthermore, it can be spotted from the figure that the rock under a larger injection rate produces sand earlier than the smaller ones. This reveals that the bond in the rock is more easily broken with the larger injection rate. Although the overall trend of the cumulative weight of produced sand keeps increasing for all simulations, the increase becomes smaller in the later stage for the simulation with a smaller injection flow rate. This indicates that the porous media will gradually tend to reach a steady state. From the figure, we can find that the increment of cumulative produced sand weight between different flow rates becomes larger as the flow rate increases. Therefore, we can conclude that productivity can be improved by allowing a limited amount of produced sand in the oil reservoirs, which corresponds with previous research [53,54].

However, the curves of cumulative produced sand weight reach a steady state after the initial slow increase for smaller flow rate conditions at a higher confining pressure, which can be clearly seen in Figure 11b. For injection rates equal to 2×10^{-5} m^3/s and 4×10^{-5} m^3/s, the cumulative weight of produced sand continuously increases as the process goes by for simulations due to their larger injection rate. Compared with curves at the lower confining pressure, their overall slopes are smaller at the higher confining pressure. For injection rates equal to 2×10^{-6} m^3/s and 4×10^{-6} m^3/s, the curves of cumulative produced sand weight tend to overlap each other as seen in the figure. This phenomenon can be explained by the existence of a critical flow rate. For the conditions with a flow rate smaller than the critical value at a certain confining pressure, the total amount of produced sand is a definite value no matter what the flow rate is. It is important to determine the critical flow rate to avoid producing too much sand when the critical flow rate is exceeded. Therefore, many researchers proposed methods for predicting the critical flow rate [55–57]. In our simulation, the critical flow rate can be defined as the flow rate below which a steady state can be reached. From the figure, the weight of produced sand rapidly increases for flow rate larger than 4×10^{-6} m^3/s, which reveals the flow rate exceeds the critical flow rate at this confining pressure.

3.4. The Effect of Cementation Strength

We conduct sand production simulations with different cementation strengths at different confining pressures in this section. The cementation strengths are 5×10^5 Pa, 1×10^6 Pa, and 5×10^6 Pa, respectively. The confining pressures in the simulation are 0.8 MPa and 3 MPa. The injection rate employed in the simulation is 2×10^{-5} m^3/s. The cumulative weight of produced sand with different cementation strengths at different confining pressures can be seen in Figure 12.

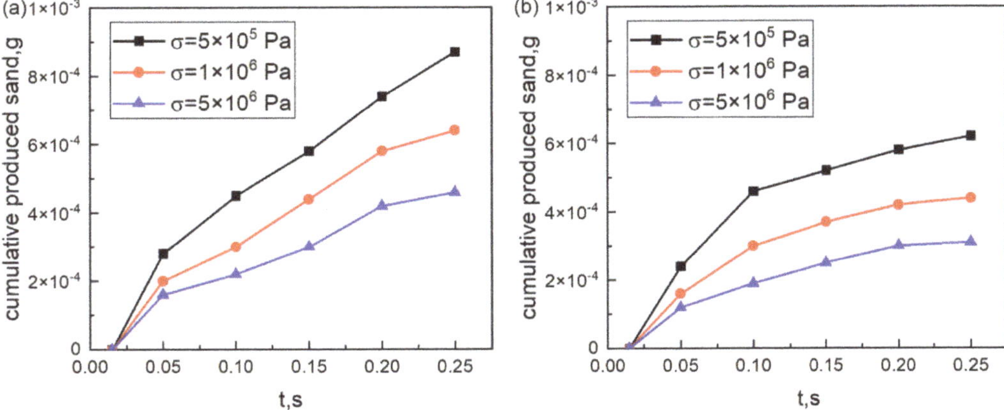

Figure 12. Cumulative weight of produced sand at different cementation strength (**a**) confining pres-sure equals 0.8 MPa (**b**) confining pressure equals 3 MPa.

For both higher and lower confining pressure, the overall trend of the cumulative weight of produced sand keeps increasing for all simulations. The rock with smaller cementation strength will produce more sand because weaker bonds are more easily broken by the hydrodynamic force at the same injection rate. The curves of cumulative produced sand weight at lower confining pressure are much steeper than that at higher confining pressure. This is due to higher confining pressure which keeps the porous media more stable. Besides this, the increase of cumulative weight of the produced sand tends to be smaller in the later stage of simulation than that at the beginning at higher confining pressure. This indicates that the porous media gradually evolve to steady conditions. In this stage, the rock with larger cementation strength changes less because fewer bonds can be broken due to the hydrodynamic force in this flow rate.

3.5. The Effect of Confining Pressure

Simulations with different confining pressure are conducted in this section. The confining pressures employed in the simulations are 0.1 MPa, 1 Mpa, and 3 Mpa, respectively. The injection rate employed in the simulations is 4×10^{-6} m^3/s. The cementation strength employed in the simulations is 1×10^6 Pa. The cumulative weight of the produced sand at different confining pressure can be seen in Figure 13.

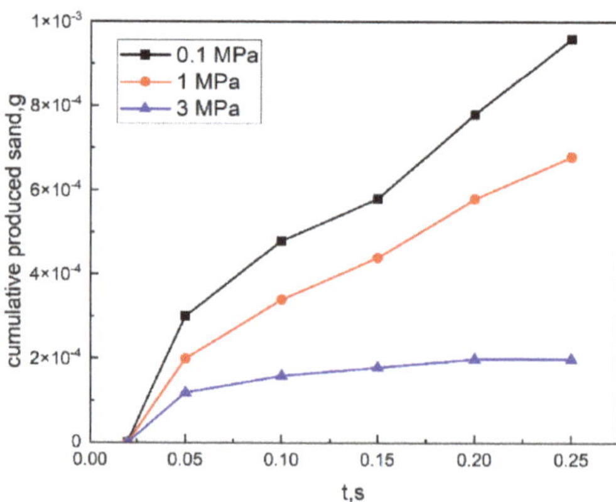

Figure 13. Cumulative weight of produced sand at different confining pressure.

It can be spotted that the cumulative weight of produced sand continuously increases as the process goes by for simulations with confining pressure equal to 1 MPa and 0.1 MPa. For the simulation with injection rate equals to 3 MPa, the cumulative weight of produced sand first slowly increases and then reaches a steady state in the simulation. Moreover, the rock under higher confining pressure will produce less sand. This is due to the better stability caused by the high confining pressure since it will increase the shear strength and the friction resistance between sand particles. Therefore, there will be enough resistance in the simulation with a smaller injection rate to avoid sand production at higher confining pressure conditions.

4. Conclusions

This paper presents a coupled lattice Boltzmann method and discrete element method (LBM-DEM) to study the sand production process of the formation. The method has been validated by simulating the sedimentation of a single particle. Simulation of the sand production process is conducted and the force chain network evolvement is analyzed. Absolute and relative permeability before and after the sand production process are calculated and the changes in the properties of porous media are studied. The effect of injection flow rate, cementation strength, and confining pressure are investigated and conclusions are listed in the following.

Strong force chain rupture can be recognized and force chain reorganization can be identified. Sand particles will detach from the reservoir and the formation will reach a steady state after a period. The total amount of produced sand first increases and remains constant after the force chain reorganization. Meanwhile, the mean degree and the clustering coefficient of the porous media continue to increase in the simulation but the slope of the curve is constantly decreasing and finally tends to be 0 at the end of the simulation. However, the mean shortest-path distance reduces gradually after a sharply initial decrease. This reveals that the state of the porous media becomes more stable in the sand production process.

By analyzing the change of relative permeability, wider saturation for two-phase flow and higher relative permeability at endpoints can be spotted at the end of the process. This indicates that permeability and connectivity have improved after the sand production process. The degree of preferential wettability for water has increased after the sand production process since the points at which $k_{rw} = 0$ and $k_{ro} = k_{rw}$ shift to the right side of the relative permeability curve.

The flow rate, cementation strength, and confining pressure have an important effect on the sand production process. A higher injection rate leads to more sand production than a lower injection rate. Besides this, a critical flow rate below which porous media can reach a steady state exists. Moreover, productivity can be improved by allowing a limited amount of produced sand in the oil reservoirs. A reservoir with lower cementation strength produces more sand than the higher one due to weak bonds. Results show that porous media under higher confining pressure will be more stable due to the higher friction resistance between particles to prevent sand production.

Author Contributions: Conceptualization, Q.F. and T.X; methodology, S.W.; simulation, T.X. and J.Z.; validation, W.Z.; formal analysis, T.X. and X.Z.; investigation, S.W.; writing—original draft preparation, T.X. and W.Z.; writing—review and editing, J.Z. and S.W.; supervision, Q.F. All authors have read and agreed to the published version of the manuscript.

Funding: This research was funded by the National Natural Science Foundation of China (Grant No. 51904319) and the National Science and Technology Major Project of China (Grant No. 2016ZX05011-001).

Institutional Review Board Statement: Not applicable.

Informed Consent Statement: Not applicable.

Data Availability Statement: Not applicable.

Conflicts of Interest: The authors declare that they have no known competing financial interest or personal relationship that could have appeared to influence the work reported in this paper.

Nomenclature

Symbols

m	particle mass, kg
v	particle velocity, m/s
t	time, s
$\mathbf{F}_b$	body force, N
$\mathbf{F}_c$	contact force, N
$\mathbf{F}_h$	hydrodynamic force, N
I	particle inertia, kg·m^2
$\mathbf{T}_c$	torque generated by the contact force, N·m
$\mathbf{T}_h$	hydrodynamic torque, N·m
k_n	normal stiffness, kg/m^2
F_{max}	critical tensile force, N
$\mathbf{x}$	position in the lattice space, lu (lattice unit)
$\mathbf{e}_a$	directional lattice velocity, lu/ts
f	particle distribution function, dimensionless
f^{eq}	equilibrium distribution function, dimensionless
a	discretized direction, dimensionless
c	basic speed on the lattice, lu/ts
u	macroscopic velocity, m/s
c_s	lattice sound velocity, lu/ts
p	pressure, Pa
f_a^m	bounce-back of the non-equilibrium part of the distribution function, dimensionless
$\mathbf{F}_f$	fluid-fluid interaction force, N
$\mathbf{F}_{ads}$	fluid-solid interaction force, N
r	particle radius, m
l	distance between the particle center and the wall, m
Re	Reynold number, dimensionless
E	elastic modulus, Pa
f_r	coefficient of surface friction, dimensionless
F_{sim}	force calculated by the simulaiton method, N
F_a	force calculated by the analytical method, N
de	degree of the node, dimensionless

L	mean shortest-path distance, dimensionless
C	global clustering coefficient, dimensionless
k	permeability, μm^2
Q	flow rate, m^3/s

Greek Letters

ω	particle angular velocity, rad/s
η	Poisson's ratio of the particle, dimensionless
δ_n	overlapping length, m
τ	collision relaxation time, dimensionless
Δt	time interval, ts (time step)
Δx	lattice spacing, lu
v	kinetic viscosity, m^2/s
ρ	fluid density, kg/m^3
ρ_s	particle density, kg/m^3
γ	solid fraction of a cell, dimensionless
β	weighting function, dimensionless
ε	cumulative error, dimensionless
σ	cementation strength, Pa

Subscripts

b	body
c	contact
h	hydrodynamic
a	the ath
σ	the σth
i	the ith
j	the jth

Superscripts

eq	the equilibrium state
σ	the σth
n	the n step

References

1. Walton, I.C.; Atwood, D.C.; Halleck, P.M.; Bianco, L.C. Perforating unconsolidated sands: An experimental and theoretical investigation. In Proceedings of the SPE Annual Technical Conference and Exhibition, New Orleans, LA, USA, 30 September–3 October 2001.
2. Bianco, L.; Halleck, P. Mechanisms of arch instability and sand production in two-phase saturated poorly consolidated sandstones. In Proceedings of the SPE European Formation Damage Conference, The Hague, The Netherlands, 21–22 May 2001.
3. Xia, T.; Feng, Q.; Wang, S.; Zhao, Y.; Xing, X. A Bonded 3D LBM-DEM Approach for Sand Production Process. In Proceedings of the SPE Asia Pacific Oil & Gas Conference and Exhibition, Virtual, 17–19 November 2020.
4. Terzaghi, K. Stress distribution in dry and in saturated sand above a yielding trap-door. In Proceedings of the International Conference on Soil Mechanics, Cambridge, MA, USA, 22–26 June 1936.
5. Hall, C.; Harrisberger, W. Stability of sand arches: A key to sand control. *J. Pet. Technol.* **1970**, *22*, 821–829. [CrossRef]
6. Clearly, M.P.; Melvan, J.J.; Kohlhaas, C.A. The effect of confining stress and fluid properties on arch stability in unconsolidated sands. In Proceedings of the SPE Annual Technical Conference and Exhibition, Las Vegas, NV, USA, 23–26 September 1979.
7. Cook, J.; Bradford, I.; Plumb, R. A study of the physical mechanisms of sanding and application to sand production prediction. In Proceedings of the European Petroleum Conference, London, UK, 25–27 October 1994.
8. Yim, K.; Dusseault, M.; Zhang, L. Experimental study of sand production processes near an orifice. In Proceedings of the Rock Mechanics in Petroleum Engineering, Delft, The Netherlands, 29–31 August 1994.
9. Geilikman, M.; Dusseault, M.; Dullien, F. Sand production as a viscoplastic granular flow. In Proceedings of the SPE formation damage control symposium, Lafayette, LA, USA, 7–10 February 1994.
10. Bratli, R.K.; Risnes, R. Stability and failure of sand arches. *Soc. Pet. Eng. J.* **1981**, *21*, 236–248. [CrossRef]
11. Vardoulakis, I.; Stavropoulou, M.; Papanastasiou, P. Hydro-mechanical aspects of the sand production problem. *Transp. Porous Media* **1996**, *22*, 225–244. [CrossRef]
12. Papamichos, E.; Vardoulakis, I.; Tronvoll, J.; Skjaerstein, A. Volumetric sand production model and experiment. *Int. J. Numer. Anal. Methods Geomech.* **2001**, *25*, 789–808. [CrossRef]
13. Isehunwa, O.; Farotade, A. Sand failure mechanism and sanding parameters in Niger Delta oil reservoirs. *Int. J. Eng. Sci. Tecnol.* **2010**, *2*, 777–782.

14. Detournay, C.; Wu, B.; Cotthem, A.; Charlier, R.; Thimus, J. Semi-analytical models for predicting the amount and rate of sand production. In *EUROCK 2006: Multiphysics Coupling and Long Term Behaviour in Rock Mechanics*; Taylor & Francis Group: London, UK, 2006; pp. 373–380.

15. OConnor, R.i.M.; Torczynski, J.R.; Preece, D.S.; Klosek, J.T.; Williams, J.R. Discrete element modeling of sand production. *Int. J. Rock Mech. Min. Sci. Geomech. Abstr.* **1997**, *34*, 231.e1–231.e15. [CrossRef]

16. Jensen, R.P.; Preece, D.S. *Modeling sand Production with Darcy-Flow Coupled with Discrete Elements*; Sandia National Labs.: Albuquerque, NM, USA, 2000.

17. Li, L.; Papamichos, E.; Cerasi, P. Investigation of sand production mechanisms using DEM with fluid flow. *Multiphysics Coupling Long Term Behav. Rock Mech.* **2006**, *1*, 241–247.

18. Han, Y.; Cundall, P.A. LBM–DEM modeling of fluid–solid interaction in porous media. *Int. J. Numer. Anal. Methods Geomech.* **2013**, *37*, 1391–1407. [CrossRef]

19. Kefayati, G. Lattice Boltzmann simulation of double-diffusive natural convection of viscoplastic fluids in a porous cavity. *Phys. Fluids* **2019**, *31*, 013105. [CrossRef]

20. Kefayati, G.; Tolooiyan, A.; Bassom, A.P.; Vafai, K. A mesoscopic model for thermal–solutal problems of power-law fluids through porous media. *Phys. Fluids* **2021**, *33*, 033114. [CrossRef]

21. Kefayati, G.R. Thermosolutal natural convection of viscoplastic fluids in an open porous cavity. *Int. J. Heat Mass Transf.* **2019**, *138*, 401–419. [CrossRef]

22. Kefayati, G.R.; Tang, H.; Chan, A.; Wang, X. A lattice Boltzmann model for thermal non-Newtonian fluid flows through porous media. *Comput. Fluids* **2018**, *176*, 226–244. [CrossRef]

23. Cook, B.K.; Noble, D.R.; Williams, J.R. A direct simulation method for particle-fluid systems. *Eng. Comput.* **2004**, *21*, 151–168. [CrossRef]

24. Boutt, D.F.; Cook, B.K.; McPherson, B.J.; Williams, J. Direct simulation of fluid-solid mechanics in porous media using the discrete element and lattice-Boltzmann methods. *J. Geophys. Res. Solid Earth* **2007**, *112*, 1978–2012. [CrossRef]

25. Feng, Y.T.; Han, K.; Owen, D.R.J. Coupled lattice Boltzmann method and discrete element modelling of particle transport in turbulent fluid flows: Computational issues. *Int. J. Numer. Methods Eng.* **2007**, *72*, 1111–1134. [CrossRef]

26. Han, K.; Feng, Y.T.; Owen, D.R.J. Coupled lattice Boltzmann and discrete element modelling of fluid–particle interaction problems. *Comput. Struct.* **2007**, *85*, 1080–1088. [CrossRef]

27. Han, K.; Feng, Y.T.; Owen, D.R.J. Numerical simulations of irregular particle transport in turbulent flows using coupled LBM-DEM. *Cmes-Comp. Model. Eng.* **2007**, *18*, 87–100.

28. Cundall, P.A.; Strack, O.D. A discrete numerical model for granular assemblies. *Geotechnique* **1979**, *29*, 47–65. [CrossRef]

29. Potyondy, D.O.; Cundall, P. A bonded-particle model for rock. *Int. J. Rock Mech. Min. Sci.* **2004**, *41*, 1329–1364. [CrossRef]

30. Cundall, P.; Strack, O. *Particle Flow Code in 2 Dimensions*; Itasca Consulting Group, Inc.: Minneapolis, MN, USA, 1999.

31. Wang, M.; Feng, Y.; Wang, C. Coupled bonded particle and lattice Boltzmann method for modelling fluid–solid interaction. *Int. J. Numer. Anal. Methods Geomech.* **2016**, *40*, 1383–1401. [CrossRef]

32. Utili, S.; Nova, R. DEM analysis of bonded granular geomaterials. *Int. J. Numer. Anal. Methods Geomech.* **2008**, *32*, 1997–2031. [CrossRef]

33. Wang, Y.; Leung, S. Characterization of cemented sand by experimental and numerical investigations. *J. Geotech. Geoenviron. Eng.* **2008**, *134*, 992–1004. [CrossRef]

34. Obermayr, M.; Dressler, K.; Vrettos, C.; Eberhard, P. A bonded-particle model for cemented sand. *Comput. Geotech.* **2013**, *49*, 299–313. [CrossRef]

35. Eshghinejadfard, A.; Daróczy, L.; Janiga, G.; Thévenin, D. Calculation of the permeability in porous media using the lattice Boltzmann method. *Int. J. Heat Fluid Flow* **2016**, *62*, 93–103. [CrossRef]

36. Wang, M.; Feng, Y.; Wang, C. Numerical investigation of initiation and propagation of hydraulic fracture using the coupled Bonded Particle–Lattice Boltzmann Method. *Comput. Struct.* **2017**, *181*, 32–40. [CrossRef]

37. Timm, K.; Kusumaatmaja, H.; Kuzmin, A.; Shardt, O.; Silva, G.; Viggen, E. *The Lattice Boltzmann Method: Principles and Practice*; Springer International Publishing: Cham, Switzerland, 2016.

38. Shan, X.; Chen, H. Lattice Boltzmann model for simulating flows with multiple phases and components. *Phys. Rev. E* **1993**, *47*, 1815. [CrossRef]

39. Ladd, A.; Verberg, R. Lattice-Boltzmann simulations of particle-fluid suspensions. *J. Stat. Phys.* **2001**, *104*, 1191–1251. [CrossRef]

40. Buick, J.; Greated, C. Gravity in a lattice Boltzmann model. *Phys. Rev. E* **2000**, *61*, 5307. [CrossRef]

41. Guo, Z.; Zheng, C.; Shi, B. Discrete lattice effects on the forcing term in the lattice Boltzmann method. *Phys. Rev. E* **2002**, *65*, 046308. [CrossRef]

42. Kupershtokh, A.; Medvedev, D.; Karpov, D. On equations of state in a lattice Boltzmann method. *Comput. Math. Appl.* **2009**, *58*, 965–974. [CrossRef]

43. Sukop, M.; Thorne, D. *Lattice Boltzmann Modeling: An Introduction for Geoscientists and Engineers*; Springer: Berlin, Germany, 2005.

44. Landry, C.; Karpyn, Z.; Ayala, O. Relative permeability of homogenous-wet and mixed-wet porous media as determined by pore-scale lattice Boltzmann modeling. *Water Resour. Res.* **2014**, *50*, 3672–3689. [CrossRef]

45. Sega, M.; Sbragaglia, M.; Kantorovich, S.S.; Ivanov, A.O. Mesoscale structures at complex fluid–fluid interfaces: A novel lattice Boltzmann/molecular dynamics coupling. *Soft Matter* **2013**, *9*, 10092–10107. [CrossRef]

46. Shan, X.; Doolen, G. Multicomponent lattice-Boltzmann model with interparticle interaction. *J. Stat. Phys.* **1995**, *81*, 379–393. [CrossRef]
47. Noble, D.; Torczynski, J. A lattice-Boltzmann method for partially saturated computational cells. *Int. J. Mod. Phys. C* **1998**, *9*, 1189–1201. [CrossRef]
48. Luding, S. Stress distribution in static two dimensional granular model media in the absence of friction. *Phys. Rev. E Statal Phys. Plasmas Fluids Relat. Interdiplinary Top.* **1997**, *55*, 4720–4729. [CrossRef]
49. Silbert, L.E.; Grest, G.S.; Landry, J.W. Statistics of the contact network in frictional and frictionless granular packings. *Phys. Rev. E* **2002**, *66*, 61303. [CrossRef] [PubMed]
50. Papadopoulos, L.; Porter, M.A.; Daniels, K.E.; Bassett, D.S. Network analysis of particles and grains. *J. Complex Netw.* **2018**, *6*, 485–565. [CrossRef]
51. Watts, D.J.; Strogatz, S.H. Collective dynamics of 'small-world' networks. *Nature* **1998**, *393*, 440–442. [CrossRef]
52. Newman, M.E. The structure and function of complex networks. *SIAM Rev.* **2003**, *45*, 167–256. [CrossRef]
53. Vaziri, H.; Lemoine, E.; Palmer, I.; McLennan, J.; Islam, R. How can sand production yield a several-fold increase in productivity. In Proceedings of the Paper SPE 63235 presented at the Society of Petroleum Engineers Annual Technical Conference, Dallas, TX, USA, 1–4 October 2000.
54. Ghassemi, A.; Pak, A. Numerical simulation of sand production experiment using a coupled Lattice Boltzmann–Discrete Element Method. *J. Pet. Sci. Eng.* **2015**, *135*, 218–231. [CrossRef]
55. Al-Awad, M.N.; El-Sayed, A.-A.H.; Desouky, S.E.-D.M. Factors affecting sand production from unconsolidated sandstone Saudi oil and gas reservoir. *J. King Saud Univ. Eng. Sci.* **1999**, *11*, 151–172. [CrossRef]
56. Risnes, R.; Bratli, R.K.; Horsrud, P. Sand stresses around a wellbore. *Soc. Pet. Eng. J.* **1982**, *22*, 883–898. [CrossRef]
57. Shirazi, A.K.; Mowla, D.; Esmaeilzadeh, F. Mathematical Modeling to Estimate the Critical Oil Flow Rate for Sand Production Onset in the Mansouri Oil Field with a Stabilized Finite Element Method. *J. Porous Media* **2009**, *12*, 379–386. [CrossRef]

Article

Seismic-Geological Integrated Study on Sedimentary Evolution and Peat Accumulation Regularity of the Shanxi Formation in Xinjing Mining Area, Qinshui Basin

Bo Liu [1,2], Suoliang Chang [1,2], Sheng Zhang [1,2,3,*], Yanrong Li [1], Zhihua Yang [4], Zuiliang Liu [4] and Qiang Chen [1,2]

[1] Department of Earth Science and Engineering, Taiyuan University of Technology, Taiyuan 030024, China; liu_deeper@163.com (B.L.); changsuoliang@tyut.edu.cn (S.C.); li.dennis@hotmail.com (Y.L.); chenqiang@tyut.edu.cn (Q.C.)
[2] Shanxi Key Laboratory of Coal & Coal-Measure Gas Geology, Taiyuan 030024, China
[3] Shanxi Institute of Geological Survey, Taiyuan 030006, China
[4] Department of Geology and Survey, Huayang New Material Technology Group Co., Ltd., Yangquan 045000, China; ymdc2009@163.com (Z.Y.); liuzuiliang@163.com (Z.L.)
* Correspondence: zhangsheng@tyut.edu.cn

Citation: Liu, B.; Chang, S.; Zhang, S.; Li, Y.; Yang, Z.; Liu, Z.; Chen, Q. Seismic-Geological Integrated Study on Sedimentary Evolution and Peat Accumulation Regularity of the Shanxi Formation in Xinjing Mining Area, Qinshui Basin. *Energies* **2022**, *15*, 1851. https://doi.org/10.3390/en15051851

Academic Editor: Nikolaos Koukouzas

Received: 31 January 2022
Accepted: 28 February 2022
Published: 2 March 2022

Publisher's Note: MDPI stays neutral with regard to jurisdictional claims in published maps and institutional affiliations.

Abstract: Accurate identification of the lithofacies and sedimentary facies of coal-bearing series is significant in the study of peat accumulation, coal thickness variation and coal-measured unconventional gas. This research integrated core, logging and 3D seismic data to conduct a comprehensive seismic–geological study on the sedimentary evolution characteristics and peat accumulation regularity of the Shanxi Formation in the Xinjing mining area of the Qinshui Basin. Firstly, the high-resolution sequence interface was identified, and the isochronous stratigraphic framework of the coal-bearing series was constructed. Then, the temporal and spatial evolution of sedimentary filling and sedimentary facies was dynamically analyzed using waveform clustering, phase rotation, stratal slice and frequency–division amplitude fusion methods. The results show that the Shanxi Formation in the study area can be divided into one third-order sequence and two fourth-order sequences. It developed a river-dominated deltaic system, mainly with delta plain deposits, and underwent a constructive–abandoned–constructive development stage. The locally distributed No. 6 coal seam was formed in a backswamp environment with distribution constrained by the distributary channels. The delta was abandoned at the later stage of the SS1 sequence, and the peat accumulation rate was balanced with the growth rate of the accommodation, forming a large-area distributed No. 3 thick coal seam. During the formation of the SS2 sequence, the No. 3 coal seam was locally thinned by epigenetic erosion of the river, and the thin coal belt caused by erosion is controlled by the location of the distributary channels and their extension direction. This study can provide a reference for the research on the distribution of thin sand bodies, sedimentary evolution and peat accumulation regularity in the coal-bearing series under the marine–continental transitional environment.

Keywords: seismic sedimentology; sedimentary facies evolution; peat accumulation regularity; frequency–division amplitude fusion; thin sand bodies

1. Introduction

The study of sedimentary facies is important for the in-depth understanding of the peat-forming environment, peat accumulation regularity and unconventional gas in coal measures during the formation and evolution of a coal-bearing basin [1–6]. Previous studies have shown that the Carboniferous–Permian coal-bearing series in the Yangquan mining area of the Qinshui Basin was formed in the marine–continental transitional environment, and the main coal-bearing strata, the Taiyuan Formation and Shanxi Formation, underwent sedimentary evolution from barrier coast systems to deltaic systems [7–9]. Generally, the distribution characteristics of sand bodies and sedimentary facies near coal seams usually have significant effects on peat accumulation, coal thickness variation, coalbed methane enrichment and resource recoverability [8,10–12]. Therefore, it is significant to carry out fine characterization of the sedimentary facies of coal-bearing series under small-scale conditions and their controlling effect on coal seams.

Traditionally, the sequence classification, lithology and sedimentary facies distribution characteristics of coal-bearing strata can be studied based on field outcrops, well logging curves and core data [2,13]. However, there is a problem of low resolution in identifying the lateral changes in sedimentary bodies due to the limitations of the research data. Meanwhile, seismic data have the characteristics of dense spatial sampling, which can effectively make up for this deficiency. The study of a stratigraphic sequence and sedimentary system using seismic techniques can be traced back to the seismic stratigraphy proposed by Vail et al., in the 1970s [14]. The internal structure and external morphology of seismic reflections are studied mainly based on the seismic facies, and the temporal and spatial distribution of sedimentary systems is restored through the identification and combination of seismic facies [15–17]. Seismic stratigraphy is more suitable for basin analysis, thick layer sequence classification and early- to middle-term resource exploration and evaluation, but the control accuracy of small-scale sedimentary units and thin sand bodies is still insufficient because of the limitation of seismic resolution.

With the deepening of resource exploration and development, researchers have higher requirements for the accuracy of seismic data interpretation in small-scale geological bodies. Within this context, benefiting from the advancement of 3D seismic technology and the improvement of the quality of seismic data, a new discipline—seismic sedimentology—was developed [18,19]. Seismic sedimentology is based on the disciplines of geophysics, seismic stratigraphy and sequence stratigraphy, and the technical means mainly include 90° phase rotation, stratal slice and frequency division calibration [20,21]. Seismic sedimentology studies the stratigraphic lithology, sedimentary genesis, basin filling history and sedimentary system through the integrated analysis of seismic lithology and seismic geomorphology [18,22]. It has played an important role in delineating thin sand bodies, predicting favorable reservoirs and comprehensively evaluating potential blocks in the hydrocarbon-bearing sedimentary basin [20,23–25]. However, there are few studies on the sedimentary facies of coal-bearing series by seismic sedimentology, and previous studies have made preliminary attempts to investigate the sedimentary seal ability and the gas content of coal seams by means of seismic sedimentology [11,26]. In fact, the lithological distribution of coal-bearing basins in China is complex and thin layers are commonly developed [2,12], which makes the relationship between lithology and wave impedance more complicated. How to use 3D seismic technology, including but not limited to the seismic sedimentology method, to predict the distribution of thin sand bodies and to finely delineate sedimentary facies in coal-bearing series deserves further study.

In view of the characteristics of thin layers development and rapid lithological changes in the coal-bearing series, this study comprehensively utilized core, well logging and high-quality broadband 3D seismic data, and employed the high-resolution sequence stratigraphy and various seismic interpretation methods, such as waveform clustering, stratal slice and frequency–division amplitude fusion, to finely delineate the distribution of sand bodies and sedimentary facies in each sequence of the Shanxi Formation of the Xinjing mining area. Besides, the peat accumulation regularity of the main coal seams and the applicability of seismic sedimentology in the prediction of sedimentary facies in coal-bearing series were discussed.

2. Geological Background

2.1. Tectonic Setting

Qinshui Basin is a Mesozoic basin between the Taihang Uplift and the Lvliang Uplift, presenting a large-scale NNE strike complex syncline [10]. The research area is located in the northwest of the Yangquan mining area and the northeastern margin of the Qinshui

Basin. The main body of the stratigraphy is a monoclinic structure with NW strike and SW dipping tendency, and the sublevel NNE-NE oriented spreading of the syncline and anticline are developed alternately in en echelon pattern, and the geological structures are relatively simple (Figure 1).

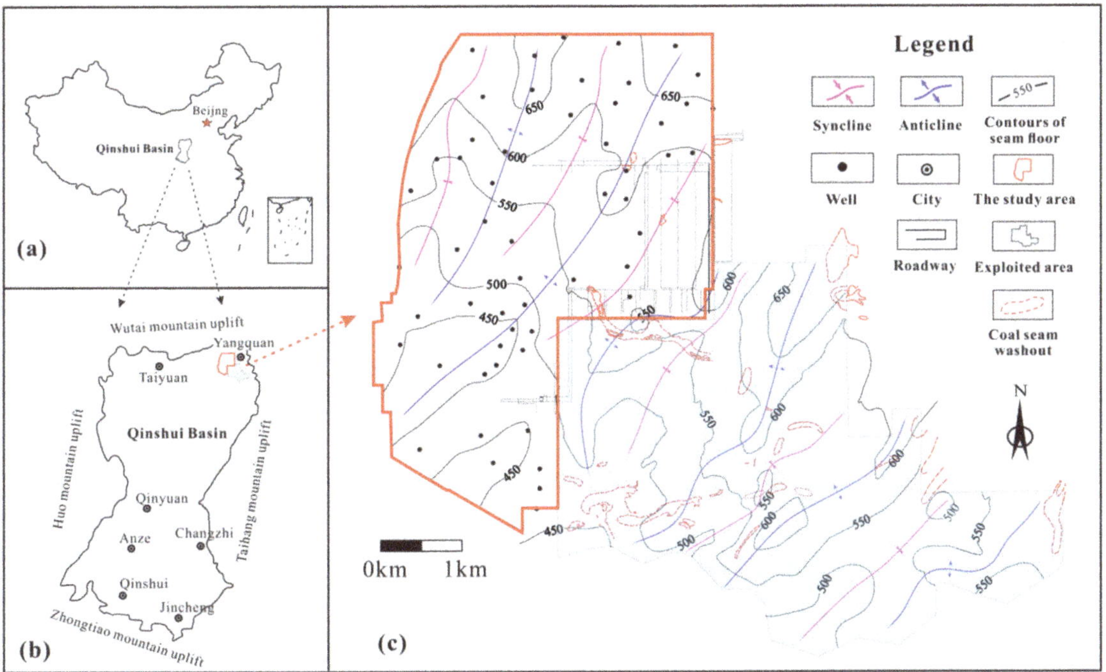

Figure 1. Map of the research area showing the geological structures and wells. (**a**) Location map of the Qinshui Basin. (**b**) Location of the research area in Qinshui Basin. (**c**) The major structures and wells in research area. Notably, the exploited area shows that coal seam was significantly affected by erosion.

2.2. Coal-Bearing Formations and Coal Seams

In the research area, the strata preserved include Ordovician, Pennsylvanian, Permian, Triassic to Quaternary units. The main coal-bearing strata are Pennsylvania Taiyuan Formation (C_3t) and Permian Shanxi Formation (P_1s). The thickness of the Shanxi Formation ranges from about 45 to 72 m, with an average of 56 m (Figure 2). The No. 3 coal seam in the Shanxi Formation has a thickness of 0.75 to 4.31 m (average 2.33 m), which is stable and mineable in the whole area, while the No. 6 coal seam is 0 to 3.10 m thick (average 1.38 m), which is unstable and locally recoverable. Other coal seams are unstable thin seams or coal lines. Meanwhile, the exploited region in the southeastern part of the study area shows that the No. 3 coal seam was obviously eroded by the river, and the sandstones on the roof of the coal seam are in scour contact with the No. 3 coal seam (Figure 1). The coal seam was thinned or even completely pinched out in the erosion zone, with coal loss of 10% to 60% and ash content increasing by about 2%. This study focuses on the accumulation regularity of the No. 3 and No. 6 coal seams.

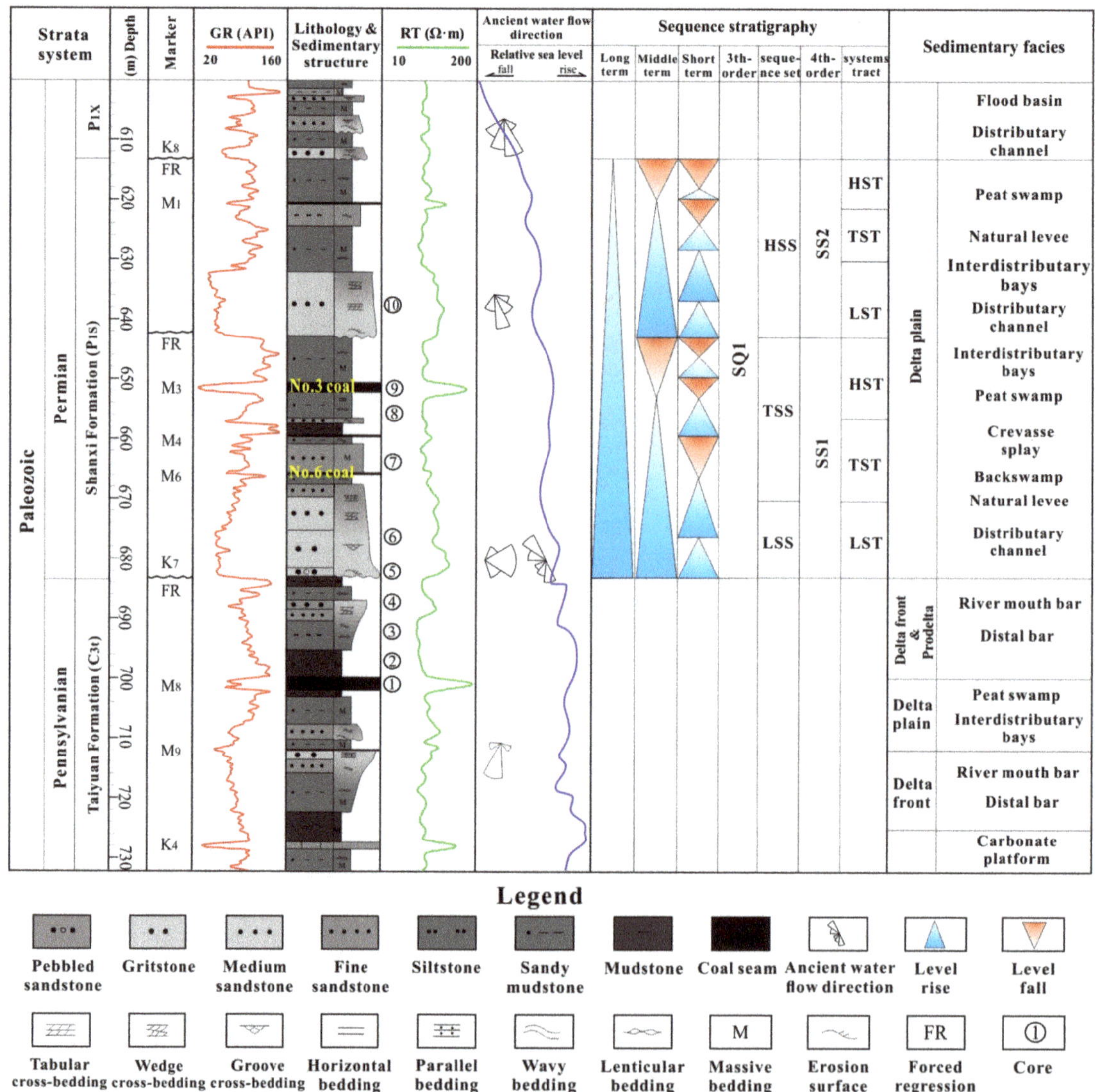

Figure 2. Main stratigraphy, logging curve, lithology, sedimentary facies, sedimentary structure and sequence stratigraphic framework of the coal-bearing strata of well 3-193 in the Qinshui Basin. HSS: high-stand sequence sets; TSS: transgressive sequence sets; LSS: low-stand sequence sets; HST: high-stand systems tract; TST: transgressive systems tract; LST: low-stand systems tract; P_{1X}: Xiashihezi Formation. Note that the ancient water flow was modified from Ge et al., 1985.

2.3. Lithofacies and Sedimentary Facies Types

Previous studies have shown that the Shanxi Formation in the Yangquan mining area of the Qinshui Basin developed a deltaic sedimentary system under the background of the epicontinental sea [7–9]. According to the outcrops, well logging and core data, the sedimentary facies of coal-bearing series of the Shanxi Formation in the study area can be further subdivided into delta front and delta plain sub-facies, which developed pebbled coarse sandstones, coarse- to fine-grained sandstones, siltstones, mudstones and coal seams (Figures 2 and 3 and Table 1). The sandstones mainly developed massive bedding, parallel bedding, wedge cross-bedding and tabular cross-bedding; the mudstones mostly developed massive bedding and horizontal bedding with fossil fragments of plant relics.

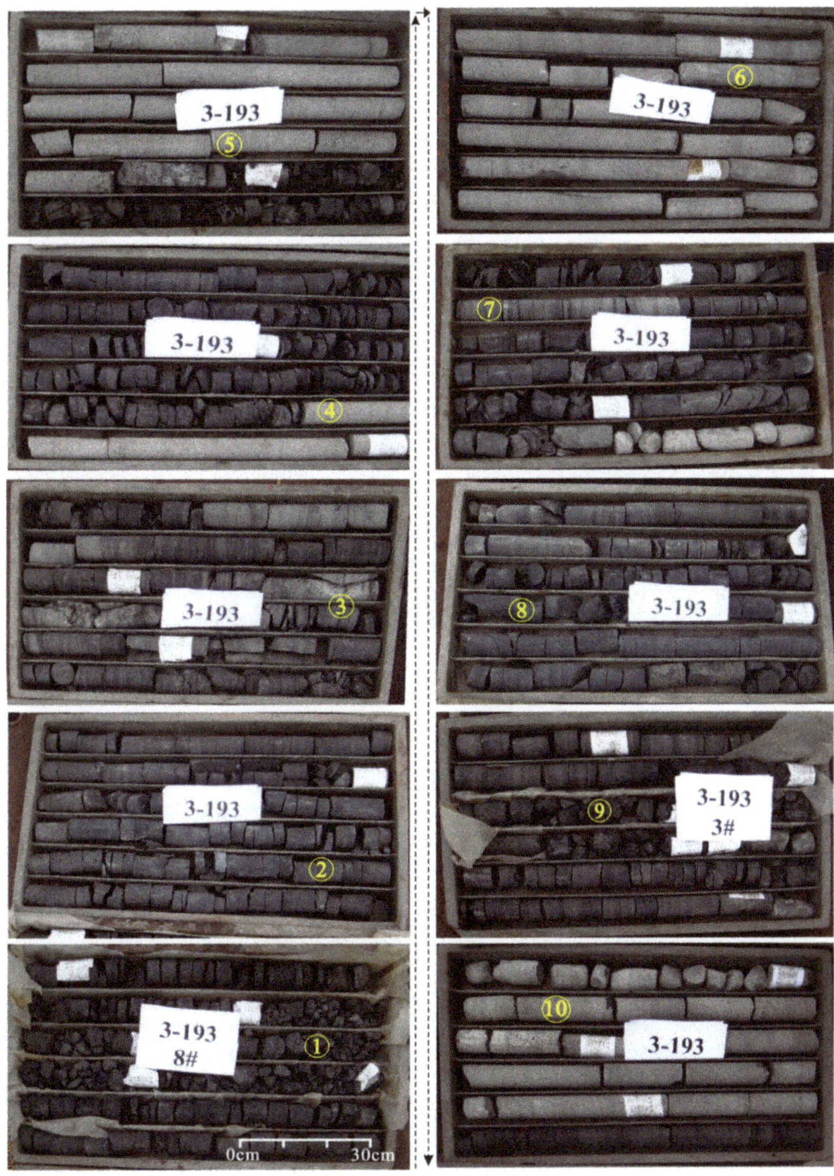

Figure 3. Core information of the coal-bearing series, showing a typical deltaic sedimentary sequence (well 3-193).

Table 1. Sedimentary facies types, lithofacies and sedimentary structure characteristics of the Shanxi Formation in the study area.

Sedimentary Facies	Sedimentary Sub-Facies	Sedimentary Microfacies	Lithofacies and Sedimentary Structure
River-dominated delta	Delta plain	Distributary channel	10 Greyish to dark grey medium sandstones with tabular cross-bedding and wedge cross-bedding; poor to moderate psephicity
		Peat swamp	1,9 Coal and carbonaceous mudstones
		Interdistributary bay	8 Gray–black mudstones and sandy mudstones with horizontal bedding
		Crevasse splay	7 Gray fine sandstones and siltstones with massive bedding
		Distributary channel	5,6 Greyish to dark grey pebbled gritstone, coarse- to medium-grained sandstones; tabular cross-bedding and groove cross-bedding; poor to moderate psephicity
	Delta front	River mouth bar	4 Gray–white medium sandstones and fine sandstones with wedge cross-bedding
		Distal bar	3 Gray–black siltstones with wavy bedding
	Prodelta		2 Dark gray–black mudstones and sandy mudstones with massive bedding and horizontal bedding

Note: the numbers 1–10 in the table correspond to Figures 2 and 3.

3. Data and Method

3.1. Data Base

3D seismic data with high signal-to-noise ratio, broad bandwidth and good amplitude fidelity are conducive to the research work of fine characterization of sedimentary facies. In the process of seismic data processing, on the one hand, well-controlled processing techniques, such as well-constrained wavelet deconvolution and phase consistency adjustment, were used to improve the resolution, wave group characteristics and well seismic matching degree. Furthermore, five-dimensional interpolation in offset vector tile (OVT) domain, pre-stack migration and anisotropy correction technologies were used to solve the amplitude distortion caused by the uneven change of offset and azimuth. The 3D seismic grid covers an area of about 20.04 km^2, with a common-midpoint (CMP) bin size of 5 m × 10 m. Effective bandwidth of the processed seismic data ranges from 20 to 100 Hz, and the dominant frequency is up to approximately 60 Hz.

3.2. Method

Based on high quality 3D seismic data, the corresponding workflow was developed with the integrated interpretation ideas of high-resolution sequence stratigraphy, seismic sedimentology and multiple seismic attribute techniques (Figure 4), mainly including: (1) identifying base level cycle interfaces and delineating high-resolution stratigraphic sequences using logging and core data, and interpreting single-well sedimentary microfacies; (2) seismic data phase analysis and 90° phase adjustment, establishing seismic–geological isochronous framework through frequency division calibration and tracking isochronous marker layers; (3) petrophysical analysis of the target sequence to determine the correspondence between different lithologies and seismic amplitudes/polarities; (4) making stratal slices on the full-band or frequency–division tuning seismic data, interpreting the seismic sedimentary facies under the constraints of core/logging facies and seismic facies and analyzing the dynamic evolution process of sedimentary facies based on seismic lithological information and seismic geomorphological features; (5) predicting the distribution of sand bodies and the evolution characteristics of sedimentary facies, and further analyzing the control of sedimentation on peat accumulation and coal thickness variation.

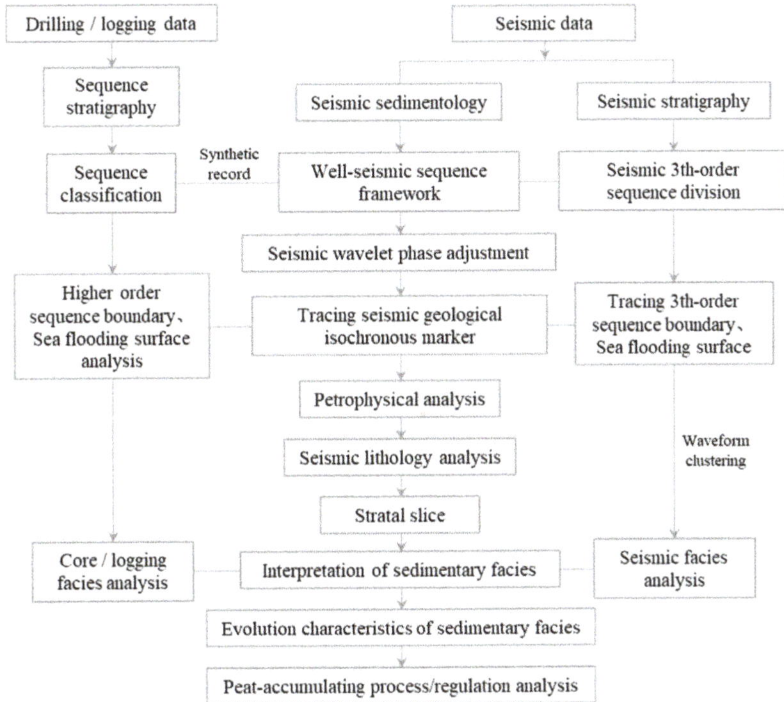

Figure 4. Workflow of integrated seismic–geological interpretation for sedimentary microfacies (referred to Zeng et al., 2012).

4. Results and Discussion

4.1. Isochronous Stratigraphic Frame

4.1.1. Sequence Classification

Identification of Sequence Boundary

The sequence strata can be classified by identifying key interfaces, such as the sequence boundary, transgressive surface and maximum flooding surface [13]. In general, regional unconformities, basal erosional surfaces of the fluvial incised valley, transgression direction switching surfaces and associated paleosols can be taken as the third-order sequence boundaries [12,13]. In terms of the Shanxi Formation in the study area, the large-scale distributed sand bodies, such as Beichagou sandstone (K_7 sandstone) and Luotuobozi sandstone (K_8 sandstone), are generally related to fluvial conditions. For instance, incised valley filling deposits in the low-stand stage and their bottoms are a kind of erosional unconformity, which can be taken as the third-order sequence boundaries. The bottom boundaries of the K_7 and K_8 sandstones were regional scour surfaces formed by the strong undercutting action of the river, and there are usually noticeable differences in the depositional environment, terrestrial clastic composition and trace element above and below them [9].

The fourth-order sequence boundaries are often marked by a fluvial erosion surface, coal seam bottom, paleosol layer, etc. [9,12]. In the river deltaic sedimentary system, the fourth-order sequence boundaries are often bounded by the fluvial erosion surface, for example, the sandstones bottom surface in the central part of the Shanxi Formation.

Sequence Classification

In this study, the Shanxi Formation can be divided into a third-order sequence (SQ1) from the bottom of the K_7 sandstone to the bottom of the K_8 sandstone, and this third-order composite sequence mainly consists of transgressive sequence sets (TSS) and high-stand sequence sets (HSS), with local development of low-stand sequence sets (see Figure 2). The SQ1 sequence was divided into two fourth-order sequences, namely SS1 and SS2. The fourth-order sequences include the low-stand systems tract (LST) represented by the thick channel-belt sands (e.g., K_7, K_8 sandstones) as the incised valley, the transgressive systems tract (TST) represented by the sand–mud interbed and the high-stand systems tract (HST) represented by the coal seam (Figure 2). The No. 3 coal seam, with stable distribution throughout the area, was formed in the HST of the fourth-order sequence, indicating a gradual increase in the increasing rate of the accommodation.

4.1.2. Seismic–Geological Isochronous Stratigraphic Framework

Tracing the reflection events on the seismic time section and establishing the isochronous stratigraphic framework are the prerequisites to ensure that the stratal slices are close to the isochronous sedimentary interface [18,20]. The sequence boundaries with isochronous significance can be demarcated into the seismic time section by making a synthetic record so that the seismic reflected waves can have clear geological meanings.

The seismic model observations, however, show that the seismic events do not simply reflect the isochronous interface nor the lithologic interface but are controlled by the frequency of the seismic data [22,27]. Using the wavelet transform method, the original seismic data were decomposed into a series of data volumes with different dominant frequencies, and, subsequently, the stable events K_8, M_3 and K_7 that do not change with the frequency were finally determined through frequency–division calibration. Geologically, incised valley ravinement surface, thin limestone, coal seam, etc., usually have isochronous significance, and their reflection characteristics can be used as good reference isochronous events [21,27]. These isochronous horizons were systematically tracked in the seismic time section, and a seismic–geological isochronous framework was constructed to ensure the reliability of stratal slices, as can be seen in Figure 5.

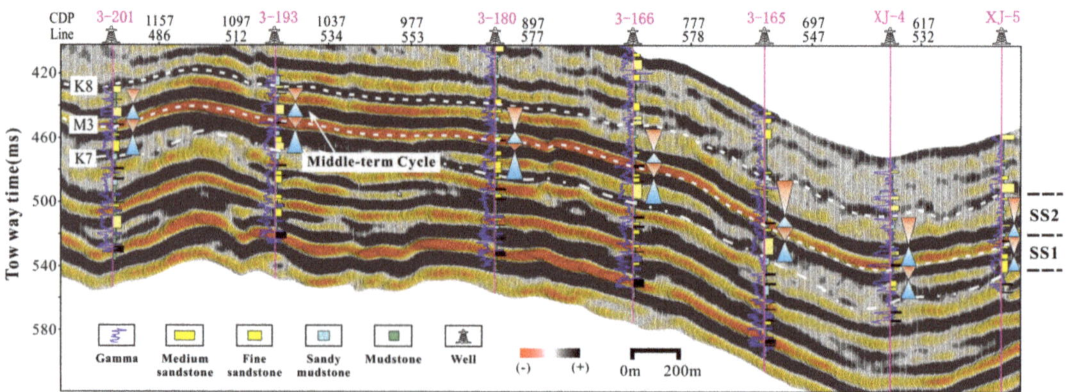

Figure 5. Seismic–geological isochronous stratigraphic frame. The Shanxi Formation was divided into two 4th-order sequences, SS1 and SS2.

4.2. Petrophysics and Seismic Lithology

4.2.1. Petrophysical Analysis

Petrophysical analysis can effectively obtain the differences in physical properties of rocks with different lithologies in terms of velocity, density, gamma, resistivity, etc., which is a prerequisite for whether seismic data can effectively identify different lithologies [28–30]. The petrophysical analysis results of the logging data (Figure 6) in the study area show that: (1) P-wave velocity of the coal seam is the lowest, about 2200 to 2400 m/s; (2) the fine-grained siltstones and sandy mudstones have part of the overlapping interval in velocity, but the overall velocity of the sandstones is larger than that of the mudstones, and the peak velocity statistic of the sandstones is about 3600 m/s and that of the mudstones is about 3200 m/s; (3) impedance difference between sandstones and mudstones is about 1000 g/cm^3·m/s (density is estimated by 2.5 g/cm^3), which provides a petrophysical basis for seismic lithology interpretation.

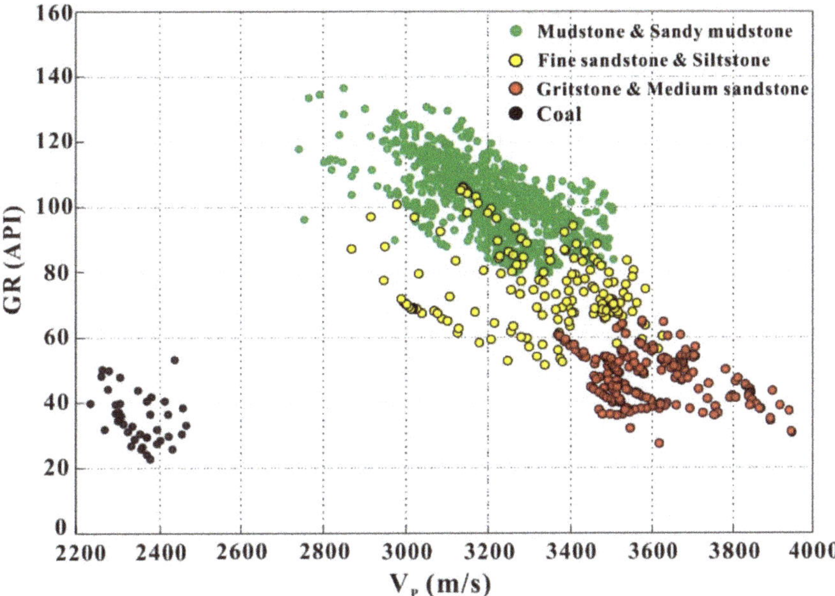

Figure 6. Crossplot of P-wave velocity and gamma of different lithologies in the Shanxi Formation.

4.2.2. Seismic Lithology Characteristics

The relationship between lithology and seismic parameters, such as seismic amplitude, polarity and phase, can be established through the analysis of seismic lithology, which provides a basis for subsequent interpretation of stratal slices [18,19]. In the seismic–geological isochronous framework (Figure 5), for the SS1 sequence of the Shanxi Formation, the strong positive amplitude (black event) reflects thick sandstones, while the weaker positive amplitude value represents thin sandstones or muddy sandstones and the negative amplitude (red event) corresponds to mudstone deposits. Most of the wells conform to the above relationships, but some cases do not follow, for example, wells 3-201 and 3-166, which indirectly indicate the complexity of the lithological distribution of coal-bearing series in the study area. Moreover, in view of the corresponding relationship between seismic waveform characteristics (e.g., amplitude and polarity) and lithology in the SS1 sequence, seismic waveform clustering was used to obtain the overall seismic facies characteristics of the sequence and subsequently converted to the sedimentary facies under the constraint of single well.

For the SS2 sequence, there is no clear indication relationship between lithology and amplitude or polarity in the seismic time section (Figure 5). Therefore, the conventional stratal slice technology cannot meet the lithology and sedimentary facies characterization requirements of this sequence. Similar to the studied coal basin, the lithological distribution of the coal-bearing basins in China is much more complex, and, in many cases, sandstones, mudstones, limestones and coal seams coexist, and the wave impedance would be multipolar [2,18,22], especially in areas where thin layers are commonly developed.

In general, coal-bearing strata can be simplified as a layered medium model for geophysical research work, and the thickness of the sedimentary bodies is a key parameter that needs to be considered [31,32]. When the thickness of the sedimentary body is thin, its top and bottom positions are difficult to determine due to the limited seismic resolution. It has been shown that it would be difficult and limited to resolve geological bodies smaller than 1/4 wavelength (1/4λ) based on the conventional idea of improving seismic resolution [33–35]. Spectrum analysis found that the dominant frequency of the target layer segment is about 60 Hz. Assuming that the formation velocity is 3600 m/s, the seismic resolution can be estimated by 1/4λ to be about 15 m. The statistical results of the well data show that the median thicknesses of the sandstones in the SS1 and SS2 sequences are about 16 m and 7 m, respectively, with a large difference between them (Figure 7). The sandstone thickness of the SS2 sequence is generally less than 1/4λ, and, thus, it may be difficult to meet the needs of fine characterization of the SS2 sequence's thin sand bodies and sedimentary facies by using the amplitude information of the original data.

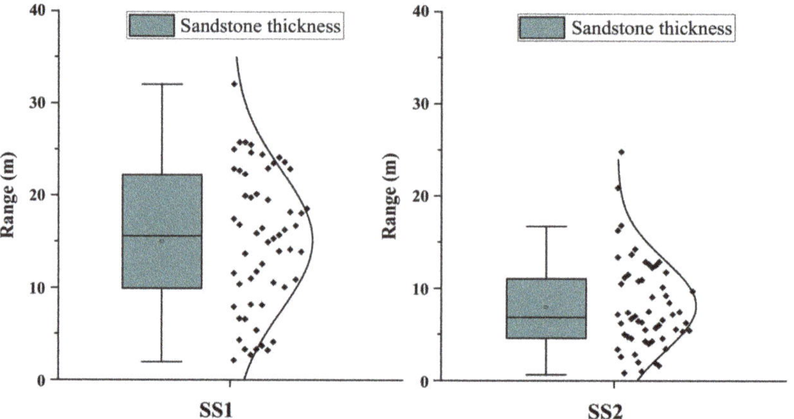

Figure 7. Thickness distribution characteristics of sandstones in the SS1 and SS2 sequences.

It is well known that the amplitude of seismic reflected waves in the layered medium model has a tuning effect, and the amplitude is affected by the thickness of the layered medium [32,36]. It has been reported in the literature that the interpretation of sedimentary facies and the prediction of thin layer thicknesses less than 1/4λ have been achieved by using the tuning amplitude [34,36,37]. Therefore, an attempt was made to use tuning amplitude to predict the thin sand bodies of the SS2 sequence in the study area, and thus to obtain higher resolution sedimentary facies interpretation results.

In summary, the waveform clustering method was adopted to study the overall characteristics of the sedimentary facies of the SS1 sequence, and the stratal slice technology was used to research the dynamic development process of the sedimentary facies. For the SS2 sequence in the study area, where the thickness of the sand bodies is generally less than 1/4λ, the method of frequency–division amplitude fusion was used to characterize the distribution characteristics of thin sand bodies and sedimentary facies.

4.3. Coal-Bearing Series Sedimentary Facies and Evolution Characteristics

4.3.1. Sedimentary Facies Characteristics of SS1 Sequence

Waveform clustering mainly uses the similarity of seismic waveform to acquire the spatial distribution of the same geological information and has promising effects in sedimentary boundary detection and sedimentary facies identification [38,39]. The waveform clustering was performed by the self-organizing neural network method, taking 5 ms downward from the T_7 reflected wave to 5 ms upward from the M_3 reflected wave as the time window (about 35 ms), to obtain the seismic facies characteristics of the SS1 sequence. Subsequently, the seismic facies were reasonably converted to the corresponding sedimentary facies based on the core and log facies, and the results are shown in Figure 8.

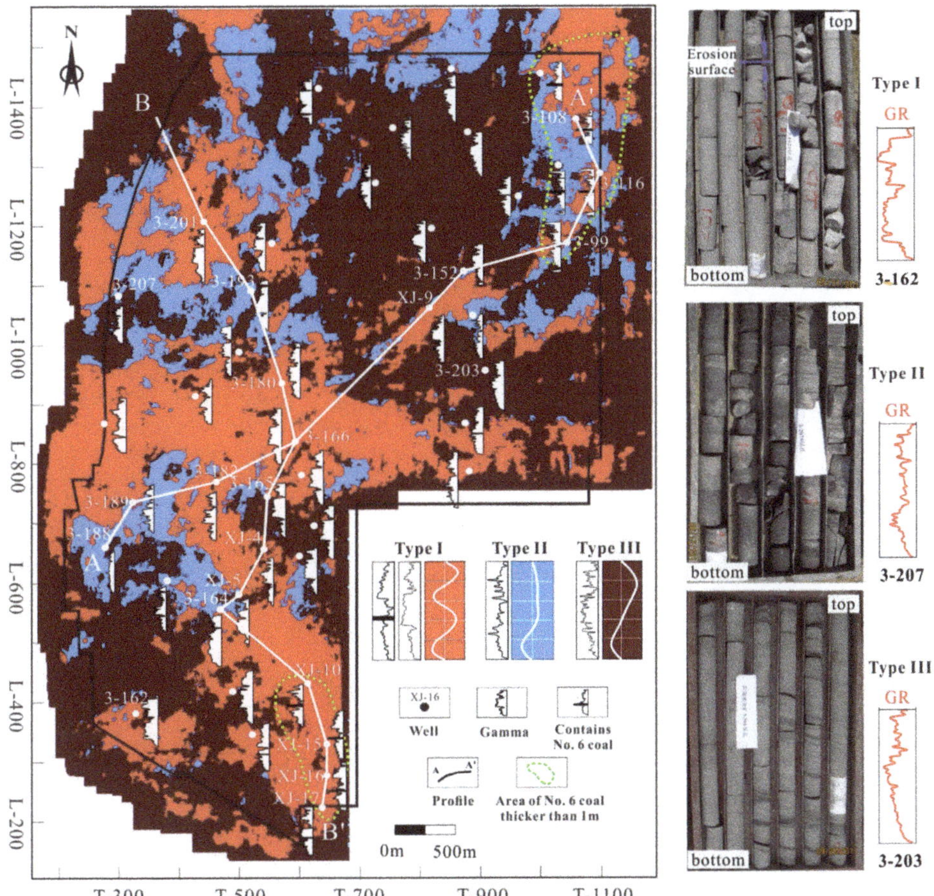

Figure 8. Waveform clustering results of the SS1 sequence, core and logging curve form characteristics. Type I represents distributary channel or backswamp facies. Type II refers to natural levee and crevasse splay facies. Type III reflects river mouth bar face.

Type I (red areas in Figure 8) waveform has the characteristics of higher amplitude with two positive and two negative peaks, corresponding to two kinds of sedimentary facies. One represents distributary channel facies, the gamma curves are box-shaped or bell-shaped and obvious erosion surface could be seen from the core, and sandstones are poorly sorted with low textural maturity. The other refers to backswamp facies, and the logging curves are finger-shaped, representing the local development of No. 6 coal seam (thickness > 1 m).

Type II (blue areas in Figure 8) waveform is a negative peak, the gamma curves are mostly straight and the lithology is fine-grained sandstones, siltstones and mudstones, representing the natural levees and crevasse splays. Type III (dark areas in Figure 8) waveform is characterized by one positive and one negative peak, reflecting the river mouth bar deposits of the delta front beneath the SS1 sequence. The gamma curves are funnel-shaped, and the cores show well-sorted sandstones with high textural maturity. The sand bodies are mostly elliptical in plane distribution, and the sand quality is relatively pure due to the stronger effect of longshore current or tidal reformation.

In total, a large amount of debris was transported here during the formation of the SS1 sequence, and the channel sand bodies as a whole bifurcate from west to east and southeast. The waveform clustering results are more indicative of the overall developmental characteristics of the sedimentary facies of the SS1 sequence.

4.3.2. Sedimentary Facies Characteristics of SS2 Sequence

The net-to-gross ratio, reflecting the proportion of sandstone content in a section of the stratum, is an important parameter for analyzing sand bodies distribution and sedimentary microfacies [9,40]. An attempt was made to create stratal slices on the raw seismic data to predict the net-to-gross ratio of the SS2 sequence, but the correlation coefficient was about 0.5 when compared with the well data (Figure 9a–c), and the prediction results were unsatisfactory. This also indirectly indicates that the resolution of the original seismic data may not be able to meet the needs of finely delineating thin sand bodies of the SS2 sequence. Thus, the frequency–division amplitude fusion method was used in order to predict the thin sand bodies. Firstly, the original seismic data were spectrally decomposed to generate low, medium and high frequency data volumes; Then, stratal slices were made on seismic data volumes with different dominant frequencies, and the amplitude values at the well points were cross-analyzed with the net-to-gross ratio to screen out the stratal slices with high correlation. Finally, the selected stratal slices were weighted and fused under the constraint of well data to predict the plane distribution characteristics of the net-to-gross ratio of the SS2 sequence (Table 2).

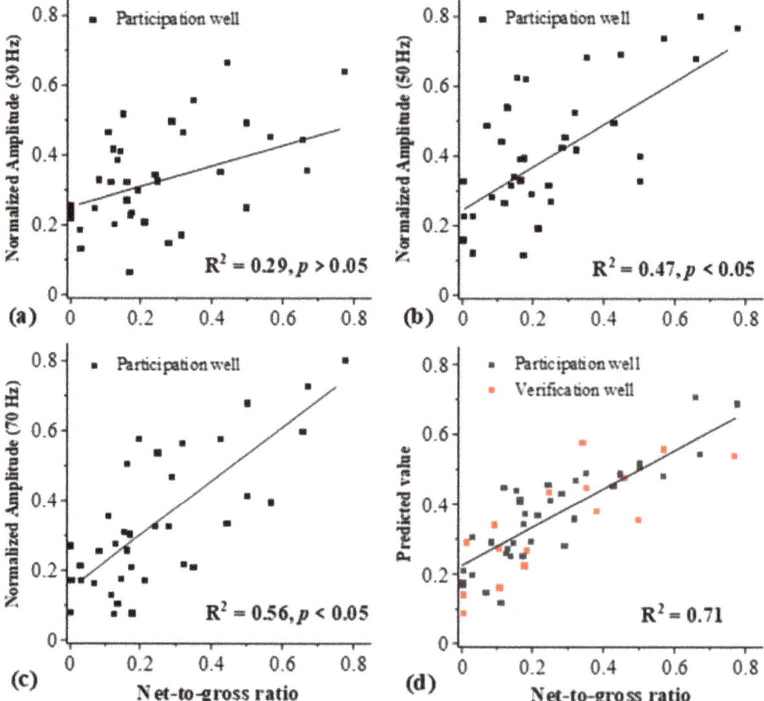

Figure 9. (a–c) are the crossplot results of 30 Hz, 50 Hz and 70 Hz normalized amplitude and the net-to-gross ratio of SS2 sequence, respectively. (d) Crossplot of the predicted net-to-gross ratio of the SS2 sequence by frequency–division amplitude fusion method vs. the actual value. Note: (1) net-to-gross ratio represents the proportion of sandstones content in a section of the stratum; (2) R^2 was calculated only by participation wells; (3) the frequency–division amplitudes were normalized to eliminate the effect of order of magnitude.

Table 2. Results of regression analysis of the net-to-gross ratio and frequency division amplitude in SS2 sequence.

	Coefficients	Standard Error	*p*-Value	Lower 95%	Upper 95%	VIF	R Square	F
Intercept	−0.089	0.046	0.002	−0.254	−0.064	-		
$NA_{50\,Hz}$	0.480	0.102	0.001	0.273	0.687	1.179	0.72	47.436
$NA_{70\,Hz}$	0.592	0.099	0.000	0.392	0.795	1.179		

Note: NA is the normalized amplitude; VIF is variance inflation factor.

Since there is no correlation between the 30 Hz amplitude and the target sequence net-to-gross ratio ($p > 0.05$, as shown in Figure 9a), it was not involved in the regression analysis. Meanwhile, the 50 Hz and 70 Hz amplitudes were used to build the SS2 sequence net-to-gross ratio prediction model, as shown in Equation (1). It can be seen from Table 2 that the model passed the F test (F = 47.436), and the R^2 of the model was 0.72, which means that $NA_{50\,Hz}$ and $NA_{70\,Hz}$ can explain 72% of the variation in the net-to-gross ratio of the SS2 sequence. In addition, the multicollinearity test of the model found that the VIF values in the model were all less than 5, suggesting that there was no collinearity problem and the model was good.

$$Y_{SS2} = -0.089 + 0.480_{x1} + 0.592_{x2} \tag{1}$$

where Y_{SS2} represents the net-to-gross ratio of SS2 sequence; x_1 is the normalized amplitude of 50 Hz; x_2 is the normalized amplitude of 70 Hz.

The correlation coefficient between the net-to-gross ratio predicted by the frequency–division amplitude fusion method and the actual values reaches 0.71, and most of the verification wells fall within the predicted trend (Figure 9d), suggesting that the prediction results are credible. The method of frequency–division amplitude fusion has improved the ability of seismic data to identify thin sand bodies in coal-bearing strata. However, the stratigraphic structure of the study area is complex, and some factors were not considered in the seismic lithology analysis, such as the influence of the surrounding rock on the reflected signal [37], and there is still room for improvement in prediction accuracy.

The statistical results of well data show that the net-to-gross ratio of distributary channels, natural levees and interdistributary bays in the Shanxi Formation are 0.5–1, 0.3–0.5, and 0–0.3, respectively. Taking the net-to-gross ratio as a sensitive parameter to reflect sedimentary facies, the development characteristics of the sedimentary facies of the target sequence were comprehensively interpreted under the constraints of the regional depositional environment and single-well sedimentary facies (Figure 10a,b). The delta plain sub-facies developed in the study area during the formation of the SS2 sequence. The distribution of sand bodies is dominated by north–south trend, and the sand bodies in the north are thicker (Figure 10a), indicating that the sediments continuously prograde toward the center of the basin from north to south. Distributary channels are developed in the eastern, northwestern and central-southern parts, and the sediments are mainly medium sandstones and fine sandstones. Meanwhile, crevasse-splay deposits are developed locally. There are natural levees on both sides of the distributary channels, and the sediments are fine sandstones and siltstones. Other parts of the study area are mostly developed interdistributary bays, which connected with the sea and had weak hydrodynamics. The sediments in these parts are mainly mudstones, with a small amount of siltstones.

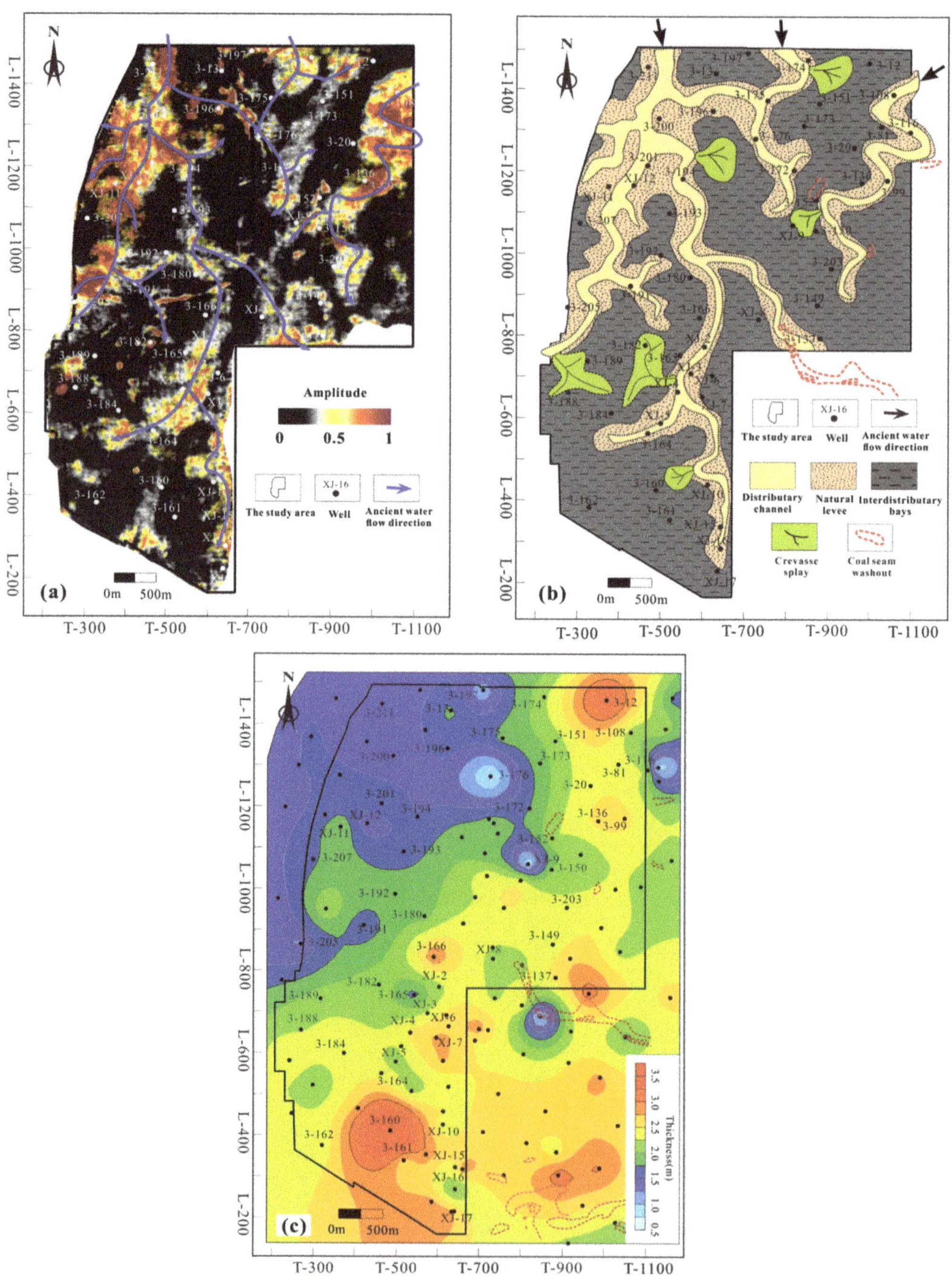

Figure 10. (**a**) SS2 sequence frequency–division amplitude fusion results; (**b**) sedimentary microfacies of the SS2 sequence; (**c**) thickness of the No. 3 coal seam obtained by Kriging interpolation.

Notably, this study found that the thickness of the No. 3 coal seam has a good correspondence on the plane with the distribution of the distributary channels of the SS2 sequence. Comparing the thickness of the No. 3 coal seam and sedimentary facies maps (Figure 10b,c), it can be seen that the coal thickness in the northwestern part of the study area is significantly thinner (average less than 1.5 m), which coincides well with the distribution position and extension direction of the sand bodies of the distributary channels. The cores data show that the direct roofs of the No. 3 coal seams in most wells in the northwest regions are coarse- to medium-grained sandstones, which are generally thicker than those of other areas. This is clear evidence that the coal seam had been thinned by river erosion. In addition, there are thin coal seams with banded or spotted distribution in the northeast and central parts (such as the regions where wells 3-116, 3-152 and XJ-9 are located), which is presumed to be caused by the lateral erosion of the distributary channels on the concave bank.

4.3.3. Evolution Characteristics of Sedimentary Facies

On the basis of the seismic–geological isochronous stratigraphic framework, continuous stratal slices of the Shanxi Formation in the study area were made to obtain the lithological distribution characteristics of the coal-bearing series and the dynamic evolution process of sedimentary facies. A total of thirty-six stratal slices were made between the K_7 and K_8 isochronous layers, and four typical slices were selected for description and interpretation, namely slice 1 to slice 4 from bottom to top. It should be noted that slice 1 to slice 3 were acquired on the phase-rotated data volume of the original seismic data, mainly to interpret the lithological distribution characteristics of the SS1 sequence in different periods, while slice 4 was acquired on the frequency–division seismic data volume by attribute fusion technique to better characterize the thin sand bodies in the SS2 sequence.

Stratal Slices Description and Interpretation

The stratal slice 1 reflects the sedimentary characteristics of the lower part of the SS1 sequence. The eastern part of the slice shows that the distributary channels (red areas in Figure 11a, slice 1) began to bifurcate to form two major branches, and the sand bodies distributed in an east–west direction (Figure 11). The blue areas are relatively concave bay areas, and there are sporadic deposits of river mouth bar and distal bar. This slice mainly shows the depositional environment of the delta plain and delta front developed in the deltaic system.

Stratal slice 2 reflects the sedimentary characteristics of the middle part of the SS1 sequence, and the sand bodies are characterized by widening from west to east. It can be seen that the distributary channels (red areas in Figure 11a, slice 2) further advance to the southeast and northeast. At the same time, the bifurcation and swing at the end of the river were obviously enhanced, and the river mouth bars were developed (Figure 11). This progradation feature is most obvious between well 3-189 and 3-152 in profile A–A' (Figure 12), and the sand bodies in different periods were arranged to the east in imbricate arrangement. Profile B–B' shows that the sand bodies were vertically incised and superimposed, indicating that these areas have strong hydrodynamics, and some river mouth bars or early channel sand bodies may be washed away by downward erosion (Figure 13). Abandoned channels were identified in the logging data of XJ-4 and XJ-5, indicating that the channels migrated frequently. This slice mainly shows the depositional environment of the delta plain.

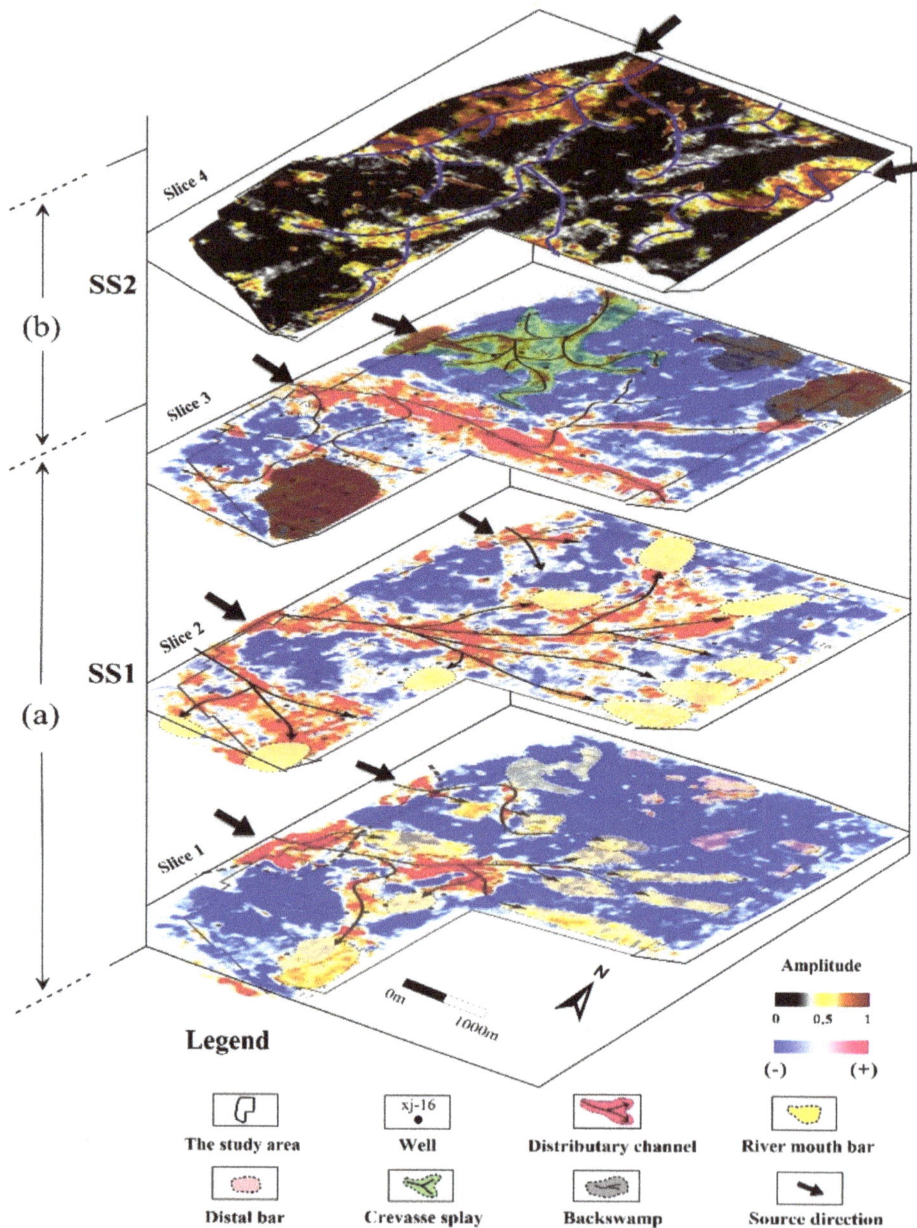

Figure 11. Interpretations of typical stratal slices of the SS1 sequence (**a**) and SS2 sequence (**b**) in the Shanxi Formation. Note that slices 1–3 reflect the sedimentary characteristics of the SS1 sequence and were made on the basis of the phase rotation of the original seismic data. Slice 4 reflects the SS2 sequence, obtained by the frequency–division amplitude fusion method.

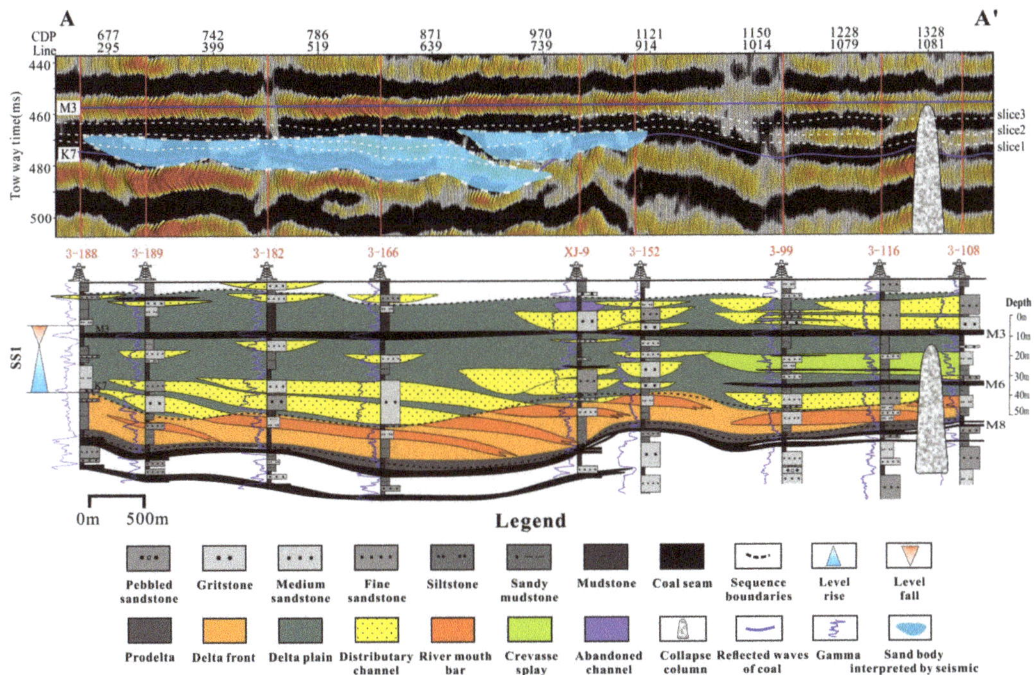

Figure 12. Seismic–geological comprehensive interpretation of sand bodies distribution and sedimentary facies ((**A**–**A′**) profile).

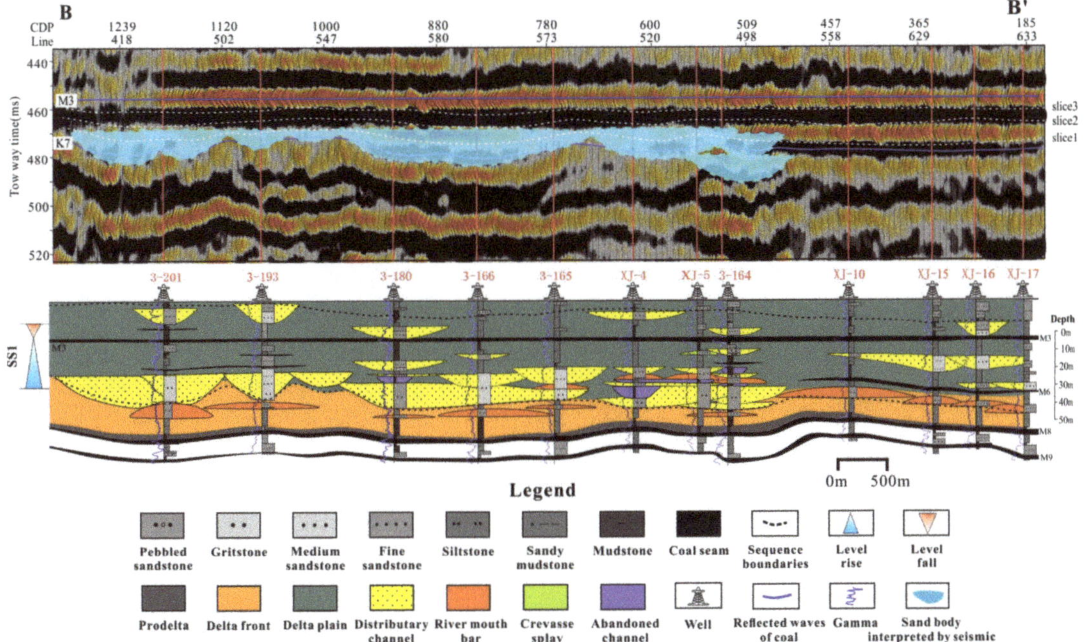

Figure 13. Seismic–geological comprehensive interpretation of sand bodies distribution and sedimentary facies ((**B**–**B′**) profile).

Stratal slice 3 mainly reflects the depositional characteristics of the late formation stage of the SS1 sequence. In the northeast and southeast, there are three flaky backswamp depositional areas, which are characterized by the deposition of the No. 6 coal seam, and the coal thickness is generally greater than 1 m, and even more than 3 m in the northeast part of the area (Figure 11a, slice 3). In addition, the logging curve shape and geomorphological features suggest that crevasse splays are developed in the northwest part of the area with a relatively complete morphology. It is worth noting that, due to the complex lithological distribution of coal-bearing strata, sandstones, mudstones, limestones and coal seams coexist in most cases, and the large differences in impedance between coal seams and surrounding rocks may cause variations in reflected wave amplitudes. Therefore, the local areas that do not conform to the geological laws need to be corrected in combination with well data during the interpretation of stratal slices.

Stratal slice 4 comprehensively reflects the depositional characteristics of the SS2 sequence. This slice shows that the sand bodies are distributed in a north–south direction as a whole and are thick in the north and thin in the south, and the sand bodies distribution direction is obviously different from the east–west oriented sand bodies of the SS1 sequence (Figure 11b, slice 4). This phenomenon was supposed to be mainly caused by the diversion of rivers back to the study area. Combined with the sandstone percentage of SS2 sequence and the logging curve characteristics, stratal slice 4 was interpreted as sedimentary microfacies, such as distributary channels, natural levees and crevasse splays.

Sedimentary Evolution Characteristics of Shanxi Formation

The comprehensive analysis results of cores, log curves characteristics and seismic interpretation results indicate that the Shanxi Formation in the study area developed a deltaic sedimentary system, mainly with delta plain sub-facies (Figure 14). In the early stage of the formation of the SS1 sequence, the base level rose slowly and the sediments supply was sufficient. The sediments show the characteristics of high energy and high construction during the process of progradation from west to southeast and northeast. The mouth bars in the delta front are relatively developed, and the upper distributary channel sand bodies are closely superimposed, with fewer muddy interlayers. During the early delta construction stage, the palaeomires of No. 6 coal seams were developed in the backswamp environment.

In the later stage of the SS1 sequence, it was speculated that the sediments supply was significantly reduced due to the diversion of the river, resulting in a smaller distribution scale of sand bodies. Eventually, the delta was abandoned and the interdistributary bays were slowly filled and silted up to form a vast platform, and then palaeomires were developed, where the precursor peats of the No. 3 coal seam were accumulated in a large area.

The sand bodies distribution direction of the SS2 sequence is obviously different from that of the SS1 sequence (Figure 14), which is presumed to be caused by the diversion of rivers and returning to the research area. This finding is consistent with previous studies on the change of paleocurrent direction in the middle stage of the formation of the Shanxi Formation in the entire Yangquan mining area [7,41]. After the accumulation of the No. 3 coal seam, the delta began to enter the constructive stage again. The SS2 sequence is composed of several cycles, with a finer grain size upward, and developed sedimentary microfacies, such as distributary channels, interdistributary bays, natural levees and crevasse splays. To sum up, the Shanxi Formation in the study area developed a deltaic sedimentary system and underwent the constructive–abandoned–constructive development stage.

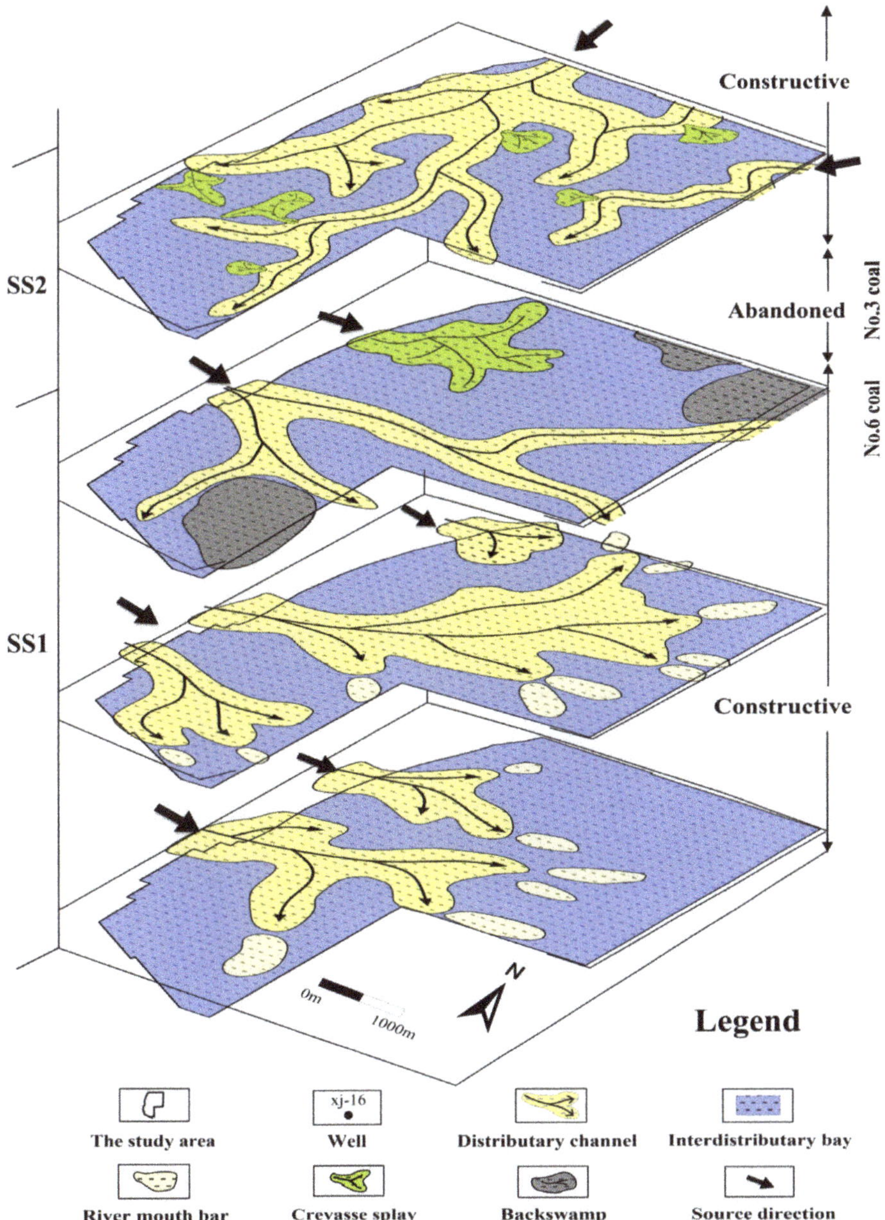

Figure 14. Sedimentary evolution process of the Shanxi Formation. Note that the Shanxi Formation was divided into SS1 and SS2 sequences.

4.4. Peat-Accumulating Regularity of Main Coal Seams

Peat accumulation often occurs in a certain evolution stage and specific part of the sedimentary basin and shows a certain regularity in time and space [9,12,42]. Usually, factors such as paleotectonic, paleogeography, paleoclimate and the peat-forming plants all have certain influences on peat accumulation [1,2,5,43,44]. Nevertheless, in this study, we only focused on the control of sedimentation on the accumulation of the No. 3 coal seam and the No. 6 coal seam in the Shanxi Formation.

The palaeomires of No. 6 coal seam were accumulated in the delta construction stage and developed in the backswamp of the delta plain, which is conducive to the reproduction of peat-forming plants and the accumulation of peat. The distribution of the No. 6 coal seam is restricted by the distributary channels, and the extension direction is usually consistent with the distributary channels. At the same time, the river diversion would destroy the continuous accumulation of peat, resulting in a large variation in the thickness of the No. 6 coal seam in this environment, with coal thickness ranging from 0 to 3.1 m, average 1.4 m and only locally mineable (see Figure 14).

There are many factors that affect the thickness variation of the No. 3 coal seam, mainly due to the depositional environment during the peat accumulation period and the epigenetic erosion of the river. The No. 3 coal seam is mainly distributed on the land side of the delta plain and was formed near the maximum flooding surface. The higher rate of peat accumulation rate was balanced with the higher growth rate of the accommodation, which makes the No. 3 coal seam thicker, with a thickness of 0.5 to 4.3 m, an average of 2.3 m.

The abandoned stage of the delta during the formation of the SS1 sequence was very favorable for peat accumulation. On the abandoned delta lobes, because of differential compaction and imbalanced crustal subsidence, the terrain became low-lying to form an interdistributary bay environment that connected with the sea. The filling and silting of the interdistributary bays connected the peat-forming swamp on the abandoned delta lobes, providing an ideal place for the accumulation of precursor peat of the No. 3 coal seam. Since the early formation of the SS2 sequence, the delta entered the constructive stage again, and the distributary channels caused local erosion of the deposited No. 3 coal seam (Figure 10). The thin and unminable coal zones caused by river erosion are controlled by the location of the channels and their extension direction.

5. Conclusions

The Shanxi Formation in the study area mainly developed delta plain sub-facies, including sedimentary microfacies, such as distributary channels, natural levees, interdistributary bays and crevasse splays, and it can be divided into a third-order sequence, SQ1, and two fourth-order sequences, SS1 and SS2. The sedimentary filling direction changed from the west–east direction of the SS1 sequence to the north–south direction of the SS2 sequence, and the delta development underwent a constructive–abandoned–constructive stage.

The palaeomires of the No. 6 coal seam in the Shanxi Formation developed in the natural levee and backswamp depositional environment, and their distribution was controlled by the distributary channels. The delta in the later stage of the SS1 sequence was abandoned, and the filling and silting of the interdistributary bays provided an ideal place for the precursor peat accumulation of the No. 3 coal seam. The higher peat accumulation rate was balanced with the growth rate of the accommodation, resulting in the formation of a thicker No. 3 coal seam over a wide area. Moreover, the No. 3 coal seam was locally thinned by the epigenetic erosion of the river, and the thin coal zone caused by river erosion is controlled by the location of the distributary channels.

By combining static and dynamic analysis methods, such as waveform clustering, stratal slice and frequency–division amplitude fusion, the planar distribution of the sedimentary facies and sedimentary evolution process of coal-bearing series in the study area can be finely delineated with the comprehensive use of multi-source data, such as cores, logs and seismic data. On the basis of tuning amplitude, the stratal slice technique has been improved to enhance the prediction accuracy of thin sand bodies in the fourth-order sequence, extending the application of seismic sedimentology in coal-bearing strata. Meanwhile, this research can provide a reference for the study of sedimentary facies before and after peat accumulation in coal-bearing strata under a marine–continental transitional environment.

Author Contributions: Conceptualization, S.C. and Q.C.; methodology, S.C.; software, B.L.; validation, B.L. and Q.C.; formal analysis, B.L.; investigation, S.C.; resources, Y.L., Z.Y. and Z.L.; data curation, Z.Y. and Z.L.; writing—original draft preparation, B.L.; writing—review and editing, S.Z.; supervision, S.Z. and S.C.; project administration, S.C.; funding acquisition, S.Z. and S.C. All authors have read and agreed to the published version of the manuscript.

Funding: This research was funded by the National Natural Science Foundation of China (grant no. 42102212), the Natural Science Foundation of Shanxi Province (grant no. 201901D211005), the unveiling projects of the Department of Science and Technology of Shanxi Province (20191101018), Shanxi Province Science and Technology Major Project (20191102001) and the Research and Development Project of Yangquan Coal Industry (Group) Co., Ltd. (GY18027).

Data Availability Statement: Data available on request due to restrictions of privacy or ethical.

Conflicts of Interest: The authors declare no conflict of interest.

References

1. Dai, S.; Bechtel, A.; Eble, C.F.; Flores, R.M.; French, D.; Graham, I.T.; Hood, M.M.; Hower, J.C.; Korasidis, V.A.; Moore, T.A.; et al. Recognition of peat depositional environments in coal: A review. *Int. J. Coal Geol.* **2020**, *219*, 103383. [CrossRef]
2. Shao, L.; Wang, X.; Lu, J.; Wang, D.; Hou, H. A Reappraisal on Development and Prospect of Coal Sedimentology in China. *Acta Sedimentol. Sin.* **2017**, *35*, 1016–1031. [CrossRef]
3. Wang, Y.; Zhu, Y.; Liu, Y.; Chen, S. Reservoir characteristics of coal–shale sedimentary sequence in coal-bearing strata and their implications for the accumulation of unconventional gas. *J. Geophys. Eng.* **2018**, *15*, 411–420. [CrossRef]
4. Cecil, C.B. An overview and interpretation of autocyclic and allocyclic processes and the accumulation of strata during the Pennsylvanian–Permian transition in the central Appalachian Basin, USA. *Int. J. Coal Geol.* **2013**, *119*, 21–31. [CrossRef]
5. Çelik, Y.; Karayigit, A.I.; Oskay, R.G.; Kayseri-Özer, M.S.; Christanis, K.; Hower, J.C.; Querol, X. A multidisciplinary study and palaeoenvironmental interpretation of middle Miocene Keles lignite (Harmancık Basin, NW Turkey), with emphasis on syngenetic zeolite formation. *Int. J. Coal Geol.* **2021**, *237*, 103691. [CrossRef]
6. Opluštil, S.; Lojka, R.; Pšenička, J.; Yilmaz, Ç.; Yilmaz, M. Sedimentology and stratigraphy of the Amasra coalfield (Pennsylvanian), NW Turkey–New insight from a 1 km thick section. *Int. J. Coal Geol.* **2018**, *195*, 317–346. [CrossRef]
7. Ge, B.; Yin, G.; Li, C. A preliminary study on sedimentary environments and law of coal-bearing formation in Yangquan, Shanxi. *Acta Sedimentol. Sin.* **1985**, *3*, 33–44. [CrossRef]
8. Jiao, X.; Wang, Y. Depositional environments of the coal-bearing strata and their controls on coal seams in the Yangquan mining district, Shanxi. *Sediment. Facies Palaeogeogr.* **1999**, *19*, 30–39.
9. Shao, L.; Dong, D.; Li, M.; Wang, H.; Wang, D.; Lu, J.; Zheng, M.; Cheng, A. Sequence-paleogeography and coal accumulation of the Carboniferous-Permian in the North China Basin. *J. China Coal Soc.* **2014**, *39*, 1725–1734. [CrossRef]
10. Cai, Y.; Liu, D.; Yao, Y.; Li, J.; Qiu, Y. Geological controls on prediction of coalbed methane of No. 3 coal seam in Southern Qinshui Basin, North China. *Int. J. Coal Geol.* **2011**, *88*, 101–112. [CrossRef]
11. Chang, S.; Chen, Q.; Liu, D.; Pan, Y.; Gui, W.; Liu, Z.; Cao, L.; Feng, A. Preliminary discussion on coal bed methane storage unit and its seismic geology comprehensive identification method. *J. China Coal Soc.* **2016**, *41*, 57–66. [CrossRef]
12. Ren, J.; Wang, H.; Sun, M.; Gan, H.; Song, G.; Sun, Z. Sequence stratigraphy and sedimentary facies of Lower Oligocene Yacheng Formation in deepwater area of Qiongdongnan Basin, Northern South China Sea: Implications for coal-bearing source rocks. *J. Earth Sci.* **2014**, *25*, 871–883. [CrossRef]
13. Catuneanu, O.; Abreu, V.; Bhattacharya, J.; Blum, M.; Dalrymple, R.; Eriksson, P.; Fielding, C.; Fisher, W.; Galloway, W.; Gibling, M.; et al. Towards the standardization of sequence stratigraphy. *Earth Sci. Rev.* **2009**, *92*, 1–33. [CrossRef]
14. Vail, P. Seismic stratigraphy and global changes of sea level. *Bull. Am. Assoc. Petrol. Geol. Mem.* **1977**, *26*, 49–212.
15. Ferreira, D.; Lupinacci, W.M.; Neves, I.D.A.; Zambrini, J.P.R.; Ferrari, A.L.; Gamboa, L.A.P.; Azul, M.O. Unsupervised seismic facies classification applied to a presalt carbonate reservoir, Santos Basin, offshore Brazil. *AAPG Bull.* **2019**, *103*, 997–1012. [CrossRef]
16. De Ruig, M.J.; Hubbard, S.M. Seismic facies and reservoir characteristics of a deep-marine channel belt in the Molasse foreland basin, Puchkirchen Formation, Austria. *AAPG Bull.* **2006**, *90*, 735–752. [CrossRef]
17. Du, H.-K.; Cao, J.-X.; Xue, Y.-J.; Wang, X.-J. Seismic facies analysis based on self-organizing map and empirical mode decomposition. *J. Appl. Geophys.* **2015**, *112*, 52–61. [CrossRef]
18. Zeng, H. What is seismic sedimentology? A tutorial. *Interpretation* **2018**, *6*, SD1–SD12. [CrossRef]
19. Zhu, X.; Dong, Y.; Zeng, H.; Huang, H.; Liu, Q.; Qin, Y.; Ye, L. New development trend of sedimentary geology: Seismic sedimentology. *J. Palaeogeogr.* **2019**, *21*, 189–201. [CrossRef]
20. Liu, H.; Zhang, M.; Chi, X. Application of seismic sedimentology in a fluvial reservoir: A case study of the Guantao Formation in Dagang Oilfield, Bohai Bay Basin, China. *Geol. J.* **2021**, *56*, 5125–5139. [CrossRef]
21. Zeng, H.; Backus, M.M. Interpretive advantages of 90°-phase wavelets: Part 1–Modeling. *Geophys.* **2005**, *70*, C7–C15. [CrossRef]

22. Zeng, H.; Zhu, X.; Zhu, R.; Zhang, Q. Guidelines for seismic sedimentologic study in non-marine postrift basins. *Pet. Explor. Dev.* **2012**, *39*, 295–304. [CrossRef]

23. Abbas, A.; Zhu, H.; Zeng, Z.; Zhou, X. Sedimentary facies analysis using sequence stratigraphy and seismic sedimentology in the Paleogene Pinghu Formation, Xihu Depression, East China Sea Shelf Basin. *Mar. Pet. Geol.* **2018**, *93*, 287–297. [CrossRef]

24. Luo, Y.; Huang, H.; Yang, Y.; Li, Q.; Zhang, S.; Zhang, J. Deepwater reservoir prediction using broadband seismic-driven impedance inversion and seismic sedimentology in the South China Sea. *Interpretation* **2018**, *6*, SO17–SO29. [CrossRef]

25. Zeng, H.; Zhao, W.; Xu, Z.; Fu, Q.; Hu, S.; Wang, Z.; Li, B. Carbonate seismic sedimentology: A case study of Cambrian Longwangmiao Formation, Gaoshiti-Moxi area, Sichuan Basin, China. *Pet. Explor. Dev.* **2018**, *45*, 830–839. [CrossRef]

26. Cao, L.; Chang, S.; Yao, Y. Application of seismic sedimentology in predicating sedimentary microfacies and coalbed methane gas content. *J. Nat. Gas Sci. Eng.* **2019**, *69*, 102944. [CrossRef]

27. Zeng, H.; Backus, M.M.; Barrow, K.T.; Tyler, N. Stratal slicing, Part I: Realistic 3-D seismic model. *Geophysics* **1998**, *63*, 502–513. [CrossRef]

28. Grana, D.; Della Rossa, E. Probabilistic petrophysical-properties estimation integrating statistical rock physics with seismic inversion. *Geophysics* **2010**, *75*, O21–O37. [CrossRef]

29. Verma, S.; Bhattacharya, S.; Lujan, B.; Agrawal, D.; Mallick, S. Delineation of early Jurassic aged sand dunes and paleo-wind direction in southwestern Wyoming using seismic attributes, inversion, and petrophysical modeling. *J. Nat. Gas Sci. Eng.* **2018**, *60*, 1–10. [CrossRef]

30. Zhang, S.; Huang, H.; Dong, Y.; Yang, X.; Wang, C.; Luo, Y. Direct estimation of the fluid properties and brittleness via elastic impedance inversion for predicting sweet spots and the fracturing area in the unconventional reservoir. *J. Nat. Gas Sci. Eng.* **2017**, *45*, 415–427. [CrossRef]

31. Meng, Z.-P.; Guo, Y.-S.; Wang, Y.; Pan, J.-N.; Lu, J. Predicting Models of Coal Thickness Based on Seismic Attributions and Their Applications. *Chin. J. Geophys.* **2006**, *49*, 450–457. [CrossRef]

32. Puryear, C.I.; Castagna, J.P. Layer-thickness determination and stratigraphic interpretation using spectral inversion: Theory and application. *Geophysics* **2008**, *73*, R37–R48. [CrossRef]

33. Ling, Y.; Gao, J.; Sun, D.; Lin, J. Spatially relative resolution based on geological concept and its application in seismic exploration. *Geophys. Prospect. Pet.* **2007**, *46*, 19.

34. Li, W.; Yue, D.; Wang, W.; Wang, W.; Wu, S.; Li, J.; Chen, D. Fusing multiple frequency-decomposed seismic attributes with machine learning for thickness prediction and sedimentary facies interpretation in fluvial reservoirs. *J. Pet. Sci. Eng.* **2019**, *177*, 1087–1102. [CrossRef]

35. Manzi, M.S.D.; Hein, K.; Durrheim, R.; King, N. Seismic attribute analysis to enhance detection of thin gold-bearing reefs: South Deep gold mine, Witwatersrand basin, South Africa. *J. Appl. Geophys.* **2013**, *98*, 212–228. [CrossRef]

36. Lyu, B.; Qi, J.; Li, F.; Hu, Y.; Zhao, T.; Verma, S.; Marfurt, K.J. Multispectral coherence: Which decomposition should we use? *Interpretation* **2020**, *8*, T115–T129. [CrossRef]

37. Zeng, H.; Marfurt, K.J. Recent progress in analysis of seismically thin beds. *Interpretation* **2015**, *3*, SS15–SS22. [CrossRef]

38. Hong, Z.; Li, K.-H.; Su, M.-J.; Hu, G.-M.; Yang, J.; Gao, G.; Hao, B. A DTW distance-based seismic waveform clustering method for layers of varying thickness. *Appl. Geophys.* **2020**, *17*, 171–181. [CrossRef]

39. Song, C.; Liu, Z.; Wang, Y.; Li, X.; Hu, G. Multi-waveform classification for seismic facies analysis. *Comput. Geosci.* **2017**, *101*, 1–9. [CrossRef]

40. Van Toorenenburg, K.A.; Donselaar, M.E.; Noordijk, N.; Weltje, G. On the origin of crevasse-splay amalgamation in the Huesca fluvial fan (Ebro Basin, Spain): Implications for connectivity in low net-to-gross fluvial deposits. *Sediment. Geol.* **2016**, *343*, 156–164. [CrossRef]

41. Li, C.; Ge, B. The sedimentary characteristic and sedimentary environments of the Shanxi Formation in the Yangquan coal district, Province Shanxi. *J. Jiaozuo Min. Inst.* **1985**, *02*, 33–51.

42. Omodeo-Salé, S.; Deschamps, R.; Michel, P.; Chauveau, B.; Suárez-Ruiz, I. The coal-bearing strata of the Lower Cretaceous Mannville Group (Western Canadian Sedimentary Basin, South Central Alberta), PART 2: Factors controlling the composition of organic matter accumulations. *Int. J. Coal Geol.* **2017**, *179*, 219–241. [CrossRef]

43. Liu, J.; Dai, S.; Hower, J.C.; Moore, T.A.; Moroeng, O.M.; Nechaev, V.P.; Petrenko, T.I.; French, D.; Graham, I.T.; Song, X. Stable isotopes of organic carbon, palynology, and petrography of a thick low-rank Miocene coal within the Mile Basin, Yunnan Province, China: Implications for palaeoclimate and sedimentary conditions. *Org. Geochem.* **2020**, *149*, 104103. [CrossRef]

44. Prinz, L.; Zieger, L.; Littke, R.; McCann, T.; Lokay, P.; Asmus, S. Syn- and post-depositional sand bodies in lignite—The role of coal analysis in their recognition. A study from the Frimmersdorf Seam, Garzweiler open-cast mine, western Germany. *Int. J. Coal Geol.* **2017**, *179*, 173–186. [CrossRef]

Article

Stable Crack Propagation Model of Rock Based on Crack Strain

Xiao Huang [1,2,*], **Chong Shi** [1,2], **Huaining Ruan** [1,2], **Yiping Zhang** [1,2] and **Wei Zhao** [3]

[1]   Key Laboratory of Ministry of Education for Geomechanics and Embankment Engineering, Hohai University, Nanjing 210024, China; scvictory@hhu.edu.cn (C.S.); hnruan@hhu.edu.cn (H.R.); 190404010015@hhu.edu.cn (Y.Z.)

[2]   Research Institute of Geotechnical Engineering, Hohai University, Nanjing 210024, China

[3]   School of Earth Sciences and Engineering, Hohai University, Nanjing 211100, China; 211609010047@hhu.edu.cn

*   Correspondence: 160204010006@hhu.edu.cn

Abstract: The establishment of a rock constitutive model considering microcrack propagation characteristics is an important channel to reflect the progressive damage and failure of rocks. The prepeak crack strain evolution curve of rock is divided into three stages based on the triaxial compression test results of granite and the definition of crack strain. According to the nonlinear variation characteristics of crack strain in the stage of rock crack stable propagation, rock deformation is expressed as the sum of matrix strain and crack strain. Then, the exponential constitutive relationship of rock crack stable propagation is deduced. The axial crack strains of the rock sample and its longitudinal section are equal. Thus, the longitudinal symmetry plane of the rock sample is abstracted as a model containing sliding crack structure in an elastic body, and the evolution equation of crack geometric parameters in the process of stable crack propagation is obtained. Compared with the experimental data, results show that the rock crack stable propagation model based on crack strain can adequately describe the evolution law of crack strain and wing crack length. In addition, the wing crack propagates easily when the elastic body with small width contains an initial crack with a large length and an axial dip angle of 45° under compressive load. This study provides a new idea for the analysis of the stable propagation characteristics and laws of rock cracks under compressive load.

Keywords: stable crack propagation; crack strain; stress–strain curve; model; wing crack

1. Introduction

The damage and failure of rock mass are closely related to the evolution behavior of cracks. For many rock engineering disasters, such as rock burst in deep tunnels, splitting of high sidewalls of large caverns, and coal/water outburst in mining engineering, the key to exploring their mesoforming mechanism is to understand the propagation and evolution process of microcracks [1–3]. However, the fracture combination of natural rock mass is complex and changeable, and the expansion and penetration path under load cannot be determined easily; thus, the study of mesoscopic damage and fracture of rock mass has always been a popular and difficult issue in rock mechanics [4]. In recent years, acoustic emission (AE) and other methods have effectively identified the evolution characteristics of cracks in loaded rock samples [5,6]. The corresponding relationship between the evolution process of cracks in rocks and the stress–strain curve of rocks has been also clarified. Therefore, the establishment of a rock stress–strain constitutive model considering microcrack propagation is an important channel to reflect progressive damage and rock failure. This approach is essential to understand the crack evolution law and construct rock engineering disaster assessment and warning systems.

The deformation and failure of rocks result from closure, initiation, propagation, and coalescence of the internal cracks under the action of external load. Theoretical studies related to cracking processes can be broadly classified into three categories [7–10],

Citation: Huang, X.; Shi, C.; Ruan, H.; Zhang, Y.; Zhao, W. Stable Crack Propagation Model of Rock Based on Crack Strain. *Energies* **2022**, *15*, 1885. https://doi.org/10.3390/en15051885

Academic Editor: Kamel Hooman

Received: 25 January 2022
Accepted: 2 March 2022
Published: 3 March 2022

namely, development and verification of crack initiation criteria, studies based on analytical methods, and studies based on numerical methods. The effects of loading conditions, sample boundary size, crack geometric characteristics, crack angle, crack number, and crack surface mechanical parameters on crack phenomena and crack patterns on rock samples were observed and studied through a large number of laboratory and numerical tests [11–16]. The effects of off-notch location on the crack initiation and propagation behaviors were examined by in situ observation, which indicated that the strength, failure patterns, and deformation properties of the rocks containing the pre-existing flaws evidently depended on the crack evolution [17]. Lee et al. [18] carried out a numerical simulation for the coalescence characteristics in Hwangdeung granite containing two unparallel fissures by using PFC2D, and the simulated peak strength, crack initiation stress and ultimate failure mode of Hwangdeung granite were compared with the experimental results. Martin proposed the definition of crack strain [19,20] to describe the crack size quantitatively and provided a method to determine the damage stress based on the inflection point of volume crack strain. Cai et al. [21] proposed a generalized crack initiation stress and crack damage stress threshold of rock mass according to crack strain.

The existence of initial cracks in hard and brittle rocks and the evolution of crack propagation during loading have an important influence on the stress–strain relationship of rocks. Liu et al. [22] and Zhao et al. [23] proposed the stress–strain relationship of porous rock under different stress states by using the double-strain Hooke law. Li et al. [24] theoretically analyzed the macro stress–strain relationship of rocks caused by crack propagation based on the airfoil crack model assuming crack angle. Zhang et al. [25] abstracted brittle rock as fracture material and skeleton material. They also established a constitutive model of brittle rock by considering initial void closure and its influence. Zuo et al. [26] regarded rock matrix and porous materials as "hard part" and "soft part" and proposed axial crack closure model and axial crack propagation model.

Mesoscopic damage and macroscopic deformation failure caused by crack evolution in rocks have been widely studied, mainly because of the randomness of crack distribution and uncertainty of evolution. However, the crack strain ignores the complex distribution shape of the crack and only considers the overall macrodeformation of the crack, effectively avoiding the difficulties caused by this uncertainty [27,28]. Based on the axial crack strain, the prepeak crack evolution behavior and its propagation model of rock have been preliminarily studied. However, the crack propagation in the model starts from the yield damage of rocks. The crack has been generated and propagated slowly and stably before the differential stress reaches the yield damage stress. When AE/microseismic technology is used to monitor the damage degree of rock mass in the actual rock engineering site to evaluate the stability of surrounding rock [29], the crack characteristics in the stable crack propagation stage before the yield damage (long-term strength) of surrounding rock must be mastered. Therefore, studying the stable propagation characteristics and laws of rock cracks under compressive load is necessary.

Given the above knowledge, the evolution characteristics of crack strain in the process of rock deformation and failure are analyzed in this study based on the stress–strain curve data of typical granite peaks. A rock crack stable propagation constitutive model was also established based on axial crack strain. This model was verified by comparing the results with the experimental data. Based on the principle that the axial crack strains of a rock sample and its longitudinal plane are equal, the equation describing the variation of crack geometric parameters in the process of stable crack propagation of rock sample is deduced, and the law of stable crack propagation is revealed.

2. Constitutive Model of Rock Crack Stable Propagation Based on Axial Crack Strain

Spatial AE for real-time positioning combined with stress–strain relationship is commonly used to describe the evolution process of crack damage during rock deformation [30,31]. Figure 1 shows the stage characteristics and stress levels at key points of the prepeak deformation process of granite. In general, the prepeak stress–strain curve of rocks can be divided into four stages, namely, crack closure stage, linear elastic deformation stage, stable crack propagation stage, and unstable crack propagation stage. The characteristic stresses corresponding to different stages are crack closure stress σ_{cc}, crack initiation stress σ_{ci}, crack damage stress σ_{cd}, and peak stress σ_c. The crack initiation stress level is the beginning of crack initiation and stable propagation, accompanied by AE events. The crack damage stress level indicates the beginning of unstable crack propagation, which is usually determined by the inflection point of volume strain. In addition, the crack in the rock may close under hydrostatic pressure when the confining pressure is sufficiently large. Thus, the differential stress–strain curve has no crack closure stage.

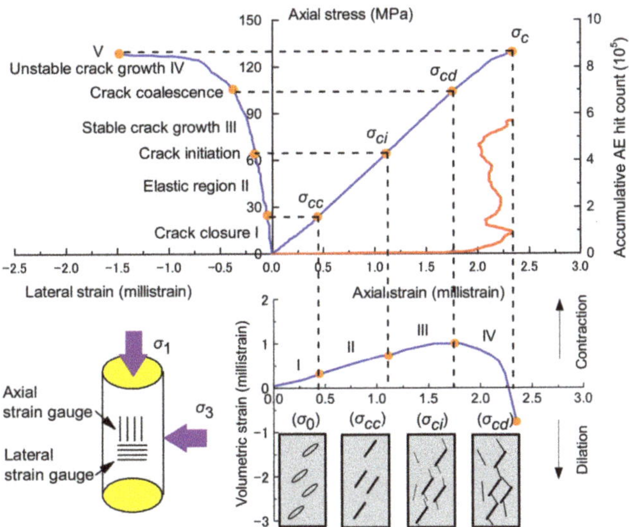

Figure 1. Stage division of the prepeak deformation and failure process of brittle rock. Reproduced from [19], Elsevier Science Ltd.: 1994.

2.1. Evolution Characteristics of Prepeak Crack in the Rock

The closure, initiation, propagation, and coalescence of internal cracks in rocks under compressive load cause rock deformation. The crack strain represents the rock deformation caused by the crack or the deformation of the cracked body, and its value is equal to the total measured strain minus the elastic strain of the rock. The crack strain ignores the complicated distribution patterns of the crack in the rock deformation process and directly represents the total macrodeformation of cracks, which can quantitatively describe the characteristics of crack evolution. Under conventional triaxial compression, the crack strain [19] can be expressed as follows:

$$\varepsilon_1^c = \varepsilon_1 - \frac{1}{E}(\sigma_1 - 2\mu\sigma_3) \tag{1}$$

$$\varepsilon_v^c = \varepsilon_v - \frac{1-2\mu}{E}(\sigma_1 + 2\sigma_3) \tag{2}$$

where ε_1^c and ε_v^c represent axial crack strain and volumetric crack strain, respectively; ε_1 and ε_v are axial strain and volumetric strain, respectively; σ_1 and σ_3 represent axial stress and lateral stress, respectively; E and μ are the elastic modulus and Poisson's ratio at the linear elastic stage, respectively.

This section analyzes the variation law of crack strain based on the triaxial compression test data of typical rock to describe the prepeak crack evolution characteristics of the rock. The triaxial test rock samples were taken from fine-grained granite with a buried depth of

450–460 m, mainly composed of plagioclase, quartz, alkaline feldspar, and biotite. Figure 2 shows the prepeak stress–strain curve of granite under different confining pressures. The increase in confining pressures increases the peak strength, elastic modulus, and peak strain of granite, and the initial compaction stage almost does not appear. According to the definition of crack strain in Equations (1) and (2), the variation trend of crack strain before granite peak under 10 MPa confining pressure is calculated and plotted, as shown in Figure 3. Under confining pressure, the prepeak differential stress–axial crack strain curve can be roughly divided into three stages, namely, linear elastic stage, stable crack growth stage, and unstable crack growth stage. First, the increase in the differential stress makes the axial crack strain nearly equal to 0, indicating that the cracks in the rock are closed. Second, the axial crack strain increases when the differential stress increases to the effective crack initiation stress (σ_{ci}–σ_3). However, the rate of increase is slow, indicating that the crack initiation and propagation in the rock are stable. The axial crack strain increases at a fast rate when the differential stress reaches the effective crack damage stress (σ_{cd}–σ_3), indicating that the crack is in a state of unstable propagation. The variation trend of crack volumetric strain can also reflect the evolution characteristics of cracks in the rock. The variation law of crack volume strain is similar to that of axial crack strain, which is not explained further.

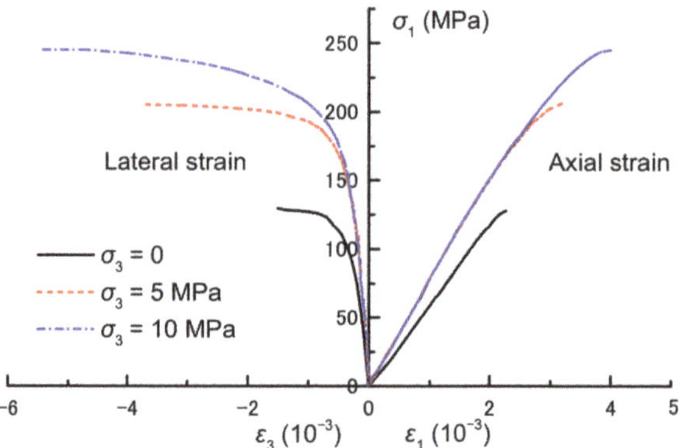

Figure 2. Prepeak stress–strain curves of typical granites under different confining pressures.

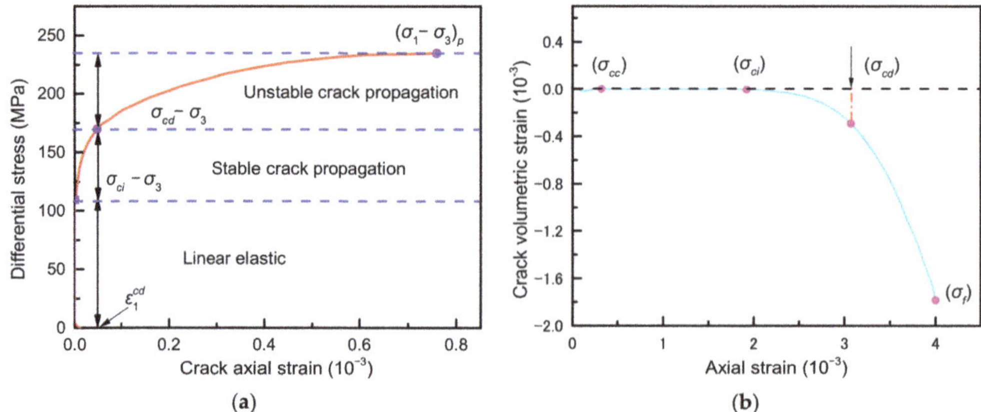

Figure 3. Prepeak crack strain trend of typical granites under confining pressure (10 MPa): (**a**) the differential stress–axial crack strain curve; (**b**) the axial strain–volume crack strain curve.

2.2. Constitutive Model of Stable Crack Propagation

Rock is a natural heterogeneous material. Its physical composition includes hard materials, such as rock grain skeleton, and soft materials, such as a large number of cracks or pores. Therefore, the macroscopic deformation of rock can be expressed as the sum of matrix strain and crack strain [22], as shown in Figure 4. The crack body is assumed as a unique material with its own physical and mechanical properties to reflect reasonably the deformation in the mechanical properties of soft and hard materials, and its deformation can be described by natural strain [32]. In comparison, the rock matrix is an elastic material, and its deformation is described by engineering strain. On the basis of this idea, Zuo studied the stress–strain relationship model of the prepeak crack closure stage and the unstable crack propagation stage [26] and regarded the stable crack propagation stage as the linear elastic stage. However, the crack strain changes nonlinearly during the stable crack propagation in rock, as shown in the above section. On the basis of the above ideas, a model of stable axial crack propagation in front of rock peak is attempted to be established in this section.

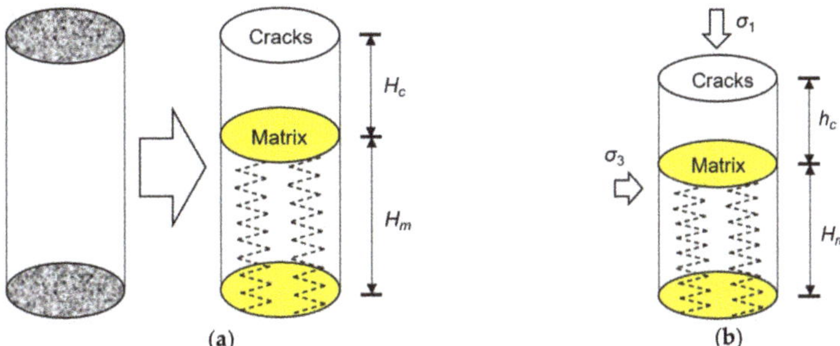

Figure 4. Schematic of rock deformation analysis: (**a**) mesoscopic treatment of rock; (**b**) rock deformation after loading.

The natural strain is the ratio of the absolute deformation to the existing size of the sample, which is suitable for describing the deformation of soft matter. If the compression direction is positive and the tensile direction is negative, then the strain of the crack body during stable propagation is as follows:

$$d\varepsilon_1^c = \frac{dh_c}{h_c} \tag{3}$$

where h_c is the equivalent height of crack in the process of stable crack propagation.

The crack propagation in the rock is stable when the differential stress is between the effective crack initiation stress (σ_{ci}–σ_3) and effective crack damage stress (σ_{cd}–σ_3). In this study, the effective differential stress (σ_e) is defined, that is, the differential stress minus the effective crack initiation stress, as expressed in Equation (4). A uniformly distributed force is imposed on the surface of the rock specimen (including cracks and the matrix). Therefore, the stress on the crack body in the process of stable crack propagation is the effective differential stress σ_e.

$$\sigma_e = (\sigma_1 - \sigma_3) - (\sigma_{ci} - \sigma_3) \tag{4}$$

$$d\sigma_e = E^c \, d\varepsilon_1^c \tag{5}$$

where E^c is the equivalent elastic modulus of the crack body.

The substitution of Equation (3) into Equation (5) and their integration obtain σ_e.

$$\sigma_e = E^c \ln h_c + C \tag{6}$$

where C is the integration constant.

Cracks in the rock begin to crack when the differential stress reaches the effective crack initiation stress. Therefore, the crack initiation conditions can be obtained as follows: $\sigma_e = 0$, $h_c = H_{ci} \approx 0$. H_{ci} is the equivalent crack height at the moment of crack initiation in rock. However, this condition makes Equation (6) mathematically meaningless. Assuming that the equivalent height of the crack is H_{cd}, the crack damage conditions can be obtained as follows: $\sigma_e = \sigma_{cd} - \sigma_{ci}$, $h_c = H_{cd}$. The integral constant C can be obtained by substituting this condition into Equation (6).

$$C = \sigma_{cd} - \sigma_{ci} - E^c \ln H_{cd} \tag{7}$$

h_c can be obtained by substituting Equation (7) into Equation (6).

$$h_c = H_{cd} \exp\left[\frac{\sigma_e - (\sigma_{cd} - \sigma_{ci})}{E^c}\right] \tag{8}$$

Based on Equation (8), the relationship between axial crack strain and effective differential stress of the crack body at the stage of stable crack propagation can be obtained.

$$\varepsilon_1^c = \frac{H_{cd}}{H} \exp\left[\frac{\sigma_e - (\sigma_{cd} - \sigma_{ci})}{E^c}\right] \tag{9}$$

where H is the height of the rock sample.

The stress–strain relationship model at the stage of steady crack growth can be obtained by replacing H_{cd}/H with ε_1^{cd} and substituting Equation (4) into Equation (9).

$$\varepsilon_1^c = \varepsilon_1^{cd} \exp\left(\frac{\sigma_1 - \sigma_{cd}}{E^c}\right) \tag{10}$$

where ε_1^{cd} is the axial crack strain of rock under yield damage.

The relation between stress and axial strain can be obtained by substituting Equation (10) into Equation (1) in the stable crack propagation stage ($\sigma_{ci} < \sigma_1 < \sigma_{cd}$).

$$\varepsilon_1 = \varepsilon_1^{cd} \exp\left(\frac{\sigma_1 - \sigma_{cd}}{E^c}\right) + \frac{1}{E}(\sigma_1 - 2\mu\sigma_3) \tag{11}$$

2.3. Model Validation

The stable crack propagation growth model is verified by the test data, as shown in Figure 5. The test data were taken from the stable crack propagation stage of the typical granite stress–strain curve (Figure 2). By using the definition of crack strain, the test data of differential stress–axial crack strain and axial stress–axial strain of granite under different confining pressures can be obtained. From Equations (10) and (11), combined with relevant fitting parameters, the corresponding theoretical curve can be drawn. Figure 5a shows that the stable crack growth model is highly consistent with the test data of differential stress–axial crack strain, and the model can well describe the nonlinear evolution characteristics of crack strain in the stable crack growth stage before the rock peak. The stress–strain constitutive relation curve obtained based on this model at the stage of stable crack growth also coincides well with the test data, verifying the correctness of the model again, as shown in Figure 5b. The relevant fitting parameters are shown in Table 1.

The evolution starting point of axial crack strain shown in Figure 5a is not zero because a certain number of wing cracks are generated at the moment of crack initiation under low confining pressure. The axial crack strain at the moment of crack initiation in rock is denoted as ε_1^{ci}, and its theoretical value can be obtained by substituting $\sigma_1 = \sigma_{ci}$ into Equation (10). ε_1^{ci} is related to the confining pressure, and the model parameters ε_1^{cd} and E^c are affected by the confining pressure.

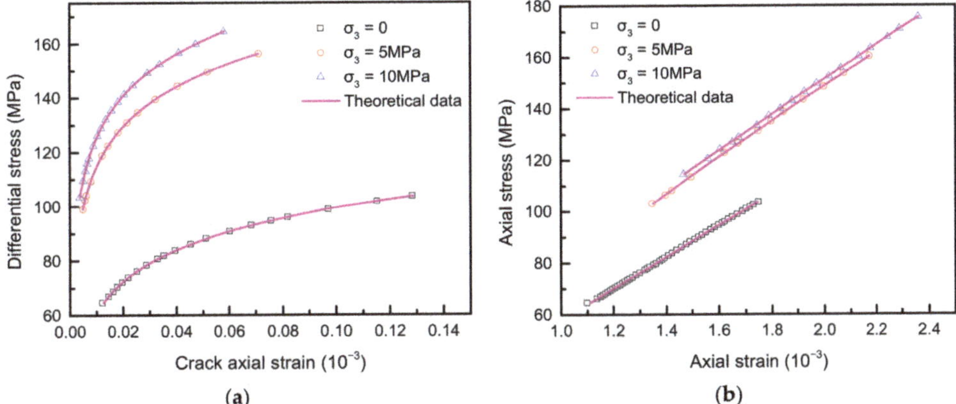

Figure 5. Stress–strain test values and theoretical curves at the stage of stable crack propagation in rock: (**a**) differential stress–axial crack strain; (**b**) axial stress–axial strain.

Table 1. Theoretical parameters of granite crack stable propagation model.

σ_3/MPa	E/GPa	μ	σ_{cd}/MPa	$\varepsilon_1^{cd}/10^{-3}$ Test Value	$\varepsilon_1^{cd}/10^{-3}$ Theoretical Value	E^c/MPa
0	64.05	0.15	103.70	0.1281	0.1284	16.81
5	75.69	0.17	161.27	0.0709	0.0710	21.14
10	77.98	0.19	174.47	0.0580	0.0583	22.15

The effects of low confining pressure σ_3 (0–10 MPa) on the damage axial crack strain ε_1^{cd}, crack equivalent elastic modulus E^c, and instantaneous crack initiation axial crack strain ε_1^{ci} are shown in Figure 6. The equivalent elastic modulus E^c of crack increases with the increase in confining pressure, whereas the damage axial crack strain ε_1^{cd} and the instantaneous crack initiation axial crack strain ε_1^{ci} show a decreasing trend. Compared with no confining pressure, the confining pressure of 10 MPa increased E^c by 31.8%, whereas ε_1^{cd} and ε_1^{ci} decreased by 54.6% and 71.7%, respectively. These results indicate that confining pressure has an obvious inhibition effect on crack initiation and stable propagation.

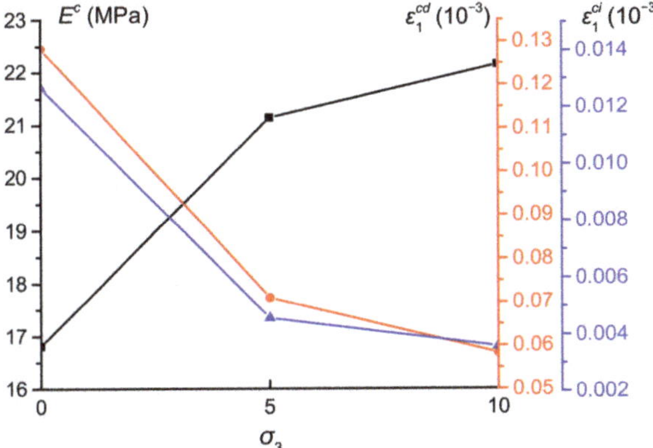

Figure 6. Effect of confining pressure on model parameters ε_1^{cd}, E^c, and ε_1^{ci}

3. Evolution Model of Crack Geometric Parameters in the Process of Stable Crack Propagation in Rock

3.1. Mechanical Model of Structures with Cracks

New cracks begin to sprout in the rock when the axial stress reaches the crack initiation stress σ_{ci}. A large number of microscopic observation test results [33,34] show that the new cracks generated at the end of the initial crack are tensile cracks. Most of them are consistent with the direction of the maximum compressive stress, and the new cracks only expand to some extent under a given stress increment. On the basis of this finding, some scholars proposed a two-dimensional plane strain model with crack structure and performed mechanical analysis. If the rock sample is considered an assembly of extremely thin longitudinal sections, the axial crack strain of the rock sample is equal to the axial crack strain of each longitudinal section. Therefore, the longitudinal plane of symmetry of the rock sample is abstracted as a model containing a sliding crack structure in an elastic body. An equation that can characterize the change of crack geometric parameters (wing crack length) during the stable crack propagation process of the rock sample can be obtained by calculating the axial crack strain and combining this calculation with the formula of the axial crack strain of the rock sample proposed above.

No interaction exists between microcracks at the stage of stable crack propagation. Each microcrack extends along different propagation paths until the axial stress reaches the crack damage stress σ_{cd}, and the intersection and penetration occur between them. Therefore, the deformation caused by each microcrack in the longitudinal symmetry plane of the rock sample in the process of stable expansion is superimposed as equivalent to the impact of a single crack structure on the rock, as shown in Figure 7a. Some scholars proposed an actual sliding crack model with a curved wing crack, but the stress intensity factor solution of this structure has no analytical form [35]. Therefore, the sliding crack model with straight axial wing crack is often adopted, as shown in Figure 7b, and this structure is confirmed to be an effective approximation in the analysis [36].

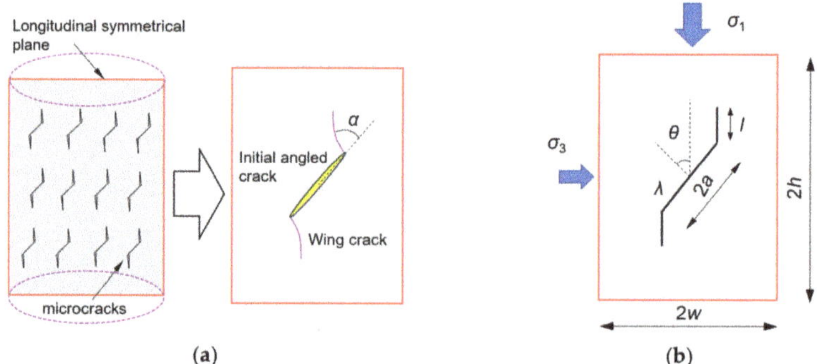

Figure 7. Rock model with crack structure: (**a**) superposition equivalence of microcracks; (**b**) sliding cracks with long wing crack limit.

3.2. Evolution Equation of Crack Geometric Parameters

In the two-dimensional plane strain mechanical model shown in Figure 7b, the length of the elastic body is $2w$, the height is $2h$, the sliding crack with half-length a is at an angle θ with the horizontal plane, and the initial length of the wing crack is l. They are subjected to axial and lateral compressive stresses σ_1 and σ_3 and resist sliding through friction. The stress intensity factor solution for this structure [21] is as follows:

$$K_{\mathrm{I}} = \frac{2a\tau\cos\theta}{\sqrt{\pi l}} - \sigma_3\sqrt{\pi l} \tag{12}$$

where K_I is the I-type stress intensity factor at the crack tip. When $K_I = K_{IC}$, the crack begins to grow stably; K_{IC} is the fracture toughness of the rock. τ is the driving shear stress, which is expressed as follows:

$$\tau = \frac{1}{2}\{(\sigma_1 - \sigma_3)\sin 2\theta - \lambda[\sigma_1 + \sigma_3 + (\sigma_1 - \sigma_3)\cos 2\theta]\} \tag{13}$$

where λ is the coefficient of friction.

The axial displacement of the model is the sum of the axial displacement caused by the applied stress in the noncracked body and the axial displacement caused by the sliding crack. The axial displacement caused by sliding cracks can be derived from elastic strain energy, and the derivation process can be referred to in Reference [37]. The effective axial strain of the crack under the applied load is finally obtained by dividing the axial displacement by the height of the model.

$$\varepsilon_1^c = \frac{2a \sin \theta \left(1 - \mu^2\right)}{Ewh} \left[\frac{\pi \tau a}{4 \cos \theta} + \frac{2a\tau \cos \theta}{\pi} \ln \frac{l}{a} - \sigma_3(l - a)\right] \tag{14}$$

where E is Young's modulus, and μ is Poisson's ratio.

The stress–strain curves of the crack body are obtained by abstracting the rock sample and its longitudinal symmetry plane into a matrix–crack composite model and an elastomer–slip crack structure model, respectively, i.e., Equations (10) and (14). The evolution equation with the geometric parameters of the crack in the stable crack propagation stage ($\sigma_{ci} < \sigma_1 < \sigma_{cd}$) under low confining pressure can be obtained by combining the two models.

$$\varepsilon_1^{cd} \exp\left(\frac{\sigma_1 - \sigma_{cd}}{E^c}\right) = \frac{2a \sin \theta \left(1 - \mu^2\right)}{Ewh} \left[\frac{\pi \tau a}{4 \cos \theta} + \frac{2a\tau \cos \theta}{\pi} \ln \frac{l}{a} - \sigma_3(l - a)\right] \tag{15}$$

3.3. Evolution of Wing Crack Length in the Stage of Stable Crack Propagation

The relevant parameters in the evolution equation of geometric parameters with cracks in the stable crack propagation stage ($\sigma_{ci} < \sigma_1 < \sigma_{cd}$) of rock are assigned, the variation law of wing crack length with axial stress is analyzed, and the influence of some parameters on it is discussed. The related parameters are assigned as follows: the height of the elastic body with sliding crack structure $2h = 100$ mm and its mechanical parameters (E, μ) are the same as those of the rock sample at the linear elastic stage, and $\lambda = 0.3$ is the friction coefficient between initial cracks. Other related parameters are shown in Table 1. The effects of the initial crack inclination angle θ and the elastic half-width w on the evolution of the wing crack length with the axial stress are analyzed and compared with the experimental results by taking the case without confining pressure as an example. The effect of the initial crack half-length a on the wing crack growth is also discussed.

Figures 8 and 9 show the growth of wing crack length with axial stress under different initial crack inclination angles θ and different model half-widths w, respectively. In the stable crack growth stage, the wing crack length increases slowly at first, and then the increasing rate increases continuously. When the initial crack inclination angle is 45°, the wing crack tends to expand. Cracks in the narrow elastic body spread easily. The results of the progressive compression test of a prefabricated single crack in PMMA plates carried out by Ashy [36] showed that the cracks grow much more easily in narrow samples and cracks which lie near 45° to the compression axis grow most easily, though all those in the range $30 < \theta < 60$ nucleated wings at nearly the same stress. These results are consistent with the results of the test carried out by Ashy, which verified the correctness of the geometric parameter equation with cracks in the stable crack propagation stage proposed in this study.

The initial crack length a also has an important influence on the wing crack growth, as shown in Figure 10. The increase in the initial crack length simplifies wing crack propagation. The longer the initial crack is, the larger the total length of the compacted cracks in the rock is, or the greater the number of microcracks is. The new cracks may be initiated under a low axial compression.

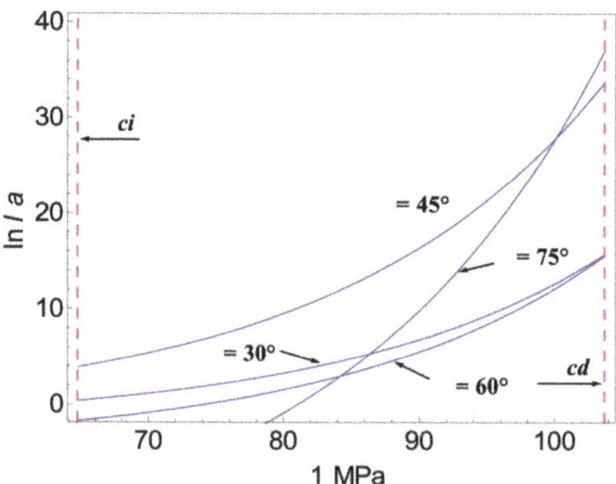

Figure 8. Variation curves of wing crack length during stable crack propagation under different initial crack inclination angles θ ($\sigma_3 = 0$, $w = 25$ mm, $a = 5$ mm).

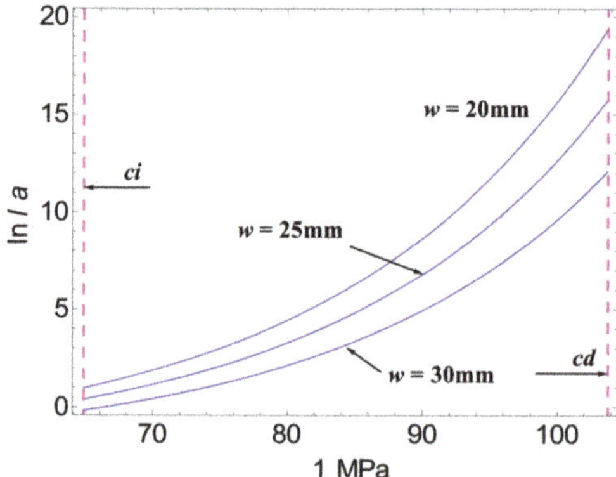

Figure 9. Variation curves of the wing crack length during stable crack propagation of rock under different model half-widths w ($\sigma_3 = 0$, $\theta = 30°$, $a = 5$ mm).

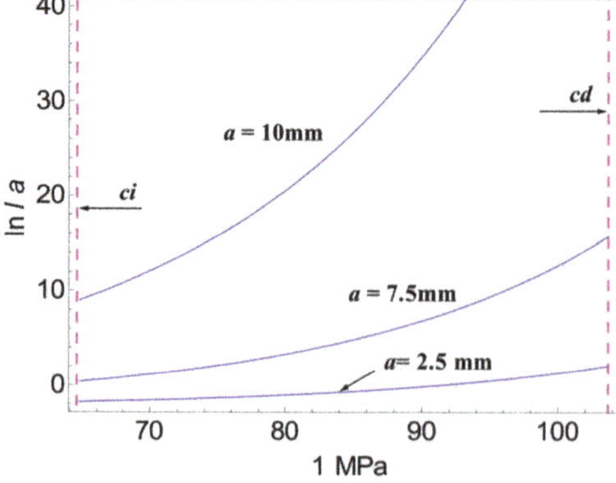

Figure 10. Variation curves of the wing crack length in rock crack stable propagation stage under different initial crack lengths a ($\sigma_3 = 0$, $\theta = 45°$, $w = 25$ mm).

4. Conclusions

The evolution characteristics of crack strain before peak are analyzed in this study based on the triaxial compression test results of typical granite and the definition of crack strain. On this basis, the constitutive relationship of stable crack propagation and the evolution equation of crack geometric parameters are derived by abstracting the rock sample and its longitudinal symmetry plane into a matrix–crack composite model and an elastomer–slip crack structure model, respectively. Moreover, the law of stable crack propagation is revealed. The main conclusions are listed below.

(1) The prepeak differential stress–axial/volume crack strain curve of granite can be roughly divided into three stages: linear elastic stage (crack strain is approximately 0), stable crack growth stage (nonlinear change with slow crack strain growth), and unstable crack growth stage (nonlinear change with fast crack strain growth).

(2) The exponential constitutive relation of rock crack stable propagation derived from the matrix–crack composite model can thoroughly describe the nonlinear evolution characteristics of crack strain in the stage of stable crack propagation.

(3) Based on the principle that the axial crack strain of the rock sample and its longitudinal section are equal, the equation of the change of crack geometric parameters in the process of rock crack stable propagation can well reflect the evolution law of wing crack length.

(4) The crack equivalent elastic modulus increases with the increase in confining pressure, whereas the damage axial crack strain and the instantaneous crack initiation axial crack strain show a decreasing trend. The initial crack inclination angle is 45°, the elastomer width is small, the initial crack length is large, and the wing crack is easy to expand.

(5) The stable crack propagation model of rock based on axial crack strain supplements the neglect of the stable crack growth stage in previous studies and can semiquantitatively describe the evolution law of crack strain and wing crack length with fewer parameters. By embedding the proposed model into numerical software, more extensive studies on crack stable propagation can be carried out in the future.

Author Contributions: Conceptualization, X.H. and H.R.; funding acquisition, C.S.; investigation, X.H.; software, X.H. and C.S.; supervision, H.R.; validation, X.H.; visualization, X.H.; writing—original draft, X.H.; writing—review and editing, X.H., Y.Z. and W.Z. All authors have read and agreed to the published version of the manuscript.

Funding: The work presented in this paper was financially supported by the National Natural Science Foundation of China (Grants Nos. 41831278, 51679071), the Fundamental Research Funds for the Central Universities (2017B642 × 14) and Postgraduate Research and Practice Innovation Program of Jiangsu Province (KYCX17_0463).

Institutional Review Board Statement: Not applicable.

Informed Consent Statement: Not applicable.

Data Availability Statement: Not applicable.

Conflicts of Interest: The authors declare no conflict of interest.

References

1. He, M.C.; Miao, J.L.; Feng, J.L. Rock burst process of limestone and its acoustic emission characteristics under true-triaxial unloading conditions. *Int. J. Rock Mech. Min. Sci.* **2010**, *47*, 286–298. [CrossRef]
2. Jiang, Q.; Feng, X.T.; Fan, Y.L.; Fan, Q.X.; Liu, G.F.; Pei, S.F.; Duan, S.Q. In situ experimental investigation of basalt spalling in a large underground powerhouse cavern. *Tunn. Undergr. Space Technol.* **2017**, *68*, 82–94. [CrossRef]
3. Shreedharan, S.; Kulatilake, P. Discontinuum-equivalent continuum analysis of the stability of tunnels in a deep coal mine using the distinct element method. *Rock Mech. Rock Eng.* **2016**, *49*, 1903–1922. [CrossRef]
4. Lai, Y.; Zhang, Y.X. Study on macro-and meso-damage composite model of rock and crack propagation rule. *Chin. J. Rock Mech. Eng.* **2008**, *27*, 534–542. [CrossRef]

5.	Zhang, H.J.; Zhang, X.H.; Zhou, H.B. Research on acoustic emission characteristics and constitutive model of rock damage evolution with different sizes. *Adv. Civ. Eng.* **2020**, *2020*, 6660595. [CrossRef]

6.	Dou, L.T.; Yang, K.; Chi, X.L. Fracture behavior and acoustic emission characteristics of sandstone samples with inclined precracks. *Int. J. Coal Sci. Technol.* **2021**, *8*, 77–87. [CrossRef]

7.	Li, H.Q.; Wong, L. Influence of flaw inclination angle and loading condition on crack initiation and propagation. *Int. J. Solids Struct.* **2012**, *49*, 2482–2499. [CrossRef]

8.	David, E.C.; Brantut, N.; Schubnel, A.; Zimmerman, R.W. Sliding crack model for nonlinearity and hysteresis in the uniaxial stress–strain curve of rock. *Int. J. Rock Mech. Min. Sci.* **2012**, *52*, 9–17. [CrossRef]

9.	Lajtai, E.Z.; Carter, B.J.; Ayari, M.L. Criteria for brittle fracture in compression. *Eng. Fract. Mech.* **1990**, *37*, 59–74. [CrossRef]

10.	Wang, E.Z.; Shrive, N.G. On the griffith criteria for brittle fracture in compression. *Eng. Fract. Mech.* **1993**, *46*, 15–26. [CrossRef]

11.	Zhao, C.; Zhou, Y.M.; Zhang, Q.Z.; Zhao, C.F.; Matsuda, H. Influence of inclination angles and confining pressures on mechanical behavior of rock materials containing a preexisting crack. *Int. J. Numer. Anal. Methods Geomech.* **2020**, *44*, 353–370. [CrossRef]

12.	Li, Y.P.; Chen, L.Z.; Wang, Y.H. Experimental research on pre-cracked marble under compression. *Int. J. Solids Struct.* **2005**, *42*, 2505–2516. [CrossRef]

13.	Yang, S.Q.; Huang, Y.H.; Jing, H.W.; Liu, X.R. Discrete element modeling on fracture coalescence behavior of red sandstone containing two unparallel fissures under uniaxial compression. *Eng. Geol.* **2014**, *178*, 28–48. [CrossRef]

14.	Dong, Q.Q.; Ma, G.W.; Wang, Q.S. Different damage parameters on failure properties with non-straight marble under uniaxial compression. *Electron. J. Geotech. Eng.* **2015**, *20*, 6151–6167.

15.	Maligno, A.R.; Rajaratnam, S.; Leen, S.B.; Williams, E.J. A three-dimensional (3D) numerical study of fatigue crack growth using remeshing techniques. *Eng. Fract. Mech.* **2010**, *77*, 94–111. [CrossRef]

16.	Tang, C.A.; Liu, H.; Lee, P.K.K.; Tsui, Y.; Tham, L. Numerical studies of the influence of microstructure on rock failure in uniaxial compression—Part I: Effect of heterogeneity. *Int. J. Rock Mech. Min. Sci.* **2000**, *37*, 555–569. [CrossRef]

17.	Zuo, J.P.; Wang, X.S.; Mao, D.Q. SEM in-situ study on the effect of offset-notch on basalt cracking behavior under three-point bending load. *Eng. Fract. Mech.* **2014**, *131*, 504–513. [CrossRef]

18.	Lee, H.; Jeon, S. An experimental and numerical study of fracture coalescence in pre-cracked specimens under uniaxial compression. *Int. J. Solids Struct.* **2011**, *48*, 979–999. [CrossRef]

19.	Martin, C.D.; Chandler, N.A. The progressive fracture of Lac du Bonnet granite. *Int. J. Rock Mech. Min. Sci.* **1994**, *31*, 643–659. [CrossRef]

20.	Martin, C.D. Seventeenth Canadian Geotechnical Colloquium: The effect of cohesion loss and stress path on brittle rock strength. *Can. Geotech. J.* **1997**, *34*, 698–725. [CrossRef]

21.	Cai, M.; Kaiser, P.K.; Tasaka, Y.; Maejima, T.; Morioka, H.; Minami, M. Generalized crack initiation and crack damage stress thresholds of brittle rock masses near underground excavations. *Int. J. Rock Mech. Min. Sci.* **2004**, *41*, 833–847. [CrossRef]

22.	Liu, H.H.; Rutqvist, J.; Berryman, J.G. On the relationship between stress and elastic strain for porous and fractured rock. *Int. J. Rock Mech. Min. Sci.* **2009**, *46*, 289–296. [CrossRef]

23.	Zhao, Y.; Liu, H.H. An elastic stress–strain relationship for porous rock under anisotropic stress conditions. *Rock Mech. Rock Eng.* **2012**, *45*, 389–399. [CrossRef]

24.	Li, X.Z.; Qi, C.Z.; Shao, Z.S. Study on strength weakening model induced by microcrack growth in rocks. *Chin. J. Undergr. Space Eng* **2020**, *16*, 30–38. Available online: https://kns.cnki.net/kcms/detail/detail.aspx?dbcode=CJFD&dbname=CJFDLAST2020&filename=BASE202001004&uniplatform=NZKPT&v=CGfGiiZyuN3E9HVQCoLLkWurJ5Ds26eroTznKZ33nHyUjzRD2-_LEXQCjrASHf4j (accessed on 8 October 2021).

25.	Zhang, C.; Chen, Q.N.; Yang, Q.J.; Lei, Y. Whole process simulation method of brittle rocks deformation and failure considering initial voids closure and its influence. *J. Chin. Coal Soc.* **2020**, *45*, 1044–1052. [CrossRef]

26.	Zuo, J.P.; Chen, Y.; Song, H.Q.; Xu, W. Evolution of pre-peak axial crack strain and nonlinear model for coal-rock combined body. *Chin. J. Geotech. Eng.* **2017**, *39*, 1609–1615. [CrossRef]

27.	Cai, M. Practical estimates of tensile strength and Hoek–Brown strength parameter m_i of brittle rocks. *Rock Mech. Rock Eng.* **2010**, *43*, 167–184. [CrossRef]

28.	Zuo, J.P.; Chen, Y.; Liu, X.L. Crack evolution behavior of rocks under confining pressures and its propagation model before peak stress. *J. Cent. South Univ.* **2019**, *26*, 3045–3056. [CrossRef]

29.	Liu, J.P.; Xu, S.D.; Li, Y.H. Analysis of rock mass stability according to power-law attenuation characteristics of acoustic emission and microseismic activities. *Tunn. Undergr. Space Technol.* **2019**, *83*, 303–312. [CrossRef]

30.	Zhao, X.G.; Cai, M.; Wang, J.; Ma, L.K. Damage stress and acoustic emission characteristics of the Beishan granite. *Int. J. Rock Mech. Min. Sci.* **2013**, *64*, 258–269. [CrossRef]

31.	Liang, D.X.; Zhang, N.; Rong, H.Y. An Experimental and Numerical Study on Acoustic Emission in the Process of Rock Damage at Different Stress Paths. *Geofluids* **2021**, *2021*, 1–13. [CrossRef]

32.	Hencky, H. The law of elasticity for isotropic and quasi-isotropic substances by finite deformations. *J. Rheol.* **1931**, *2*, 169–176. [CrossRef]

33.	Hoek, E.; Bieniawski, Z.T. Brittle fracture propagation in rock under compression. *Int. J. Fract.* **1965**, *1*, 137–155. [CrossRef]

34.	Wong, R.H.C.; Tang, C.A.; Chau, K.T.; Lin, P. Splitting failure in brittle rocks containing pre-existing flaws under uniaxial compression. *Eng. Fract. Mech.* **2002**, *69*, 1853–1871. [CrossRef]

35. Horii, H.; Nemat-Nasser, S. Compression-induced microcrack growth in brittle solids: Axial splitting and shear failure. *J. Geophys. Res.* **1985**, *90*, 3105–3125. [CrossRef]
36. Ashby, M.F.; Hallam, S.D. The failure of brittle solids containing small cracks under compressive stress states. *Acta Mater.* **1986**, *34*, 497–510. [CrossRef]
37. Kemeny, J.M.; Cook, N.G.W. Crack models for the failure of rocks in compression. *Fract. Mech.* **1986**, *7*, 1–13. Available online: https://escholarship.org/uc/item/13r6q6vs (accessed on 13 October 2020).

Article

Analysis of Outburst Coal Structure Characteristics in Sanjia Coal Mine Based on FTIR and XRD

Anjun Jiao [1], Shixiang Tian [1,2,*] and Huaying Lin [1]

[1]   College of Mining Engineering, Guizhou University, Guiyang 550025, China; jj182204759@gmail.com (A.J.); eternity37326@gmail.com (H.L.)

[2]   The National Joint Engineering Laboratory for the Utilization of Dominant Mineral Resources in Karst Mountain Area, Guizhou University, Guiyang 550025, China

*   Correspondence: husttsx@163.com

Abstract: In order to reveal the distribution characteristics of functional groups and the difference of microcrystalline structure parameters between outburst coal and primary coal, the coal samples inside and outside the outburst holes of the Sanjia coal mine were examined. The functional groups and microcrystalline structure parameters of outburst coal and primary coal in the Sanjia coal mine were studied by infrared spectroscopy, X-ray diffraction (XRD) experiment and peak-splitting fitting method. The results showed that the substitution mode of the benzene ring in an aromatic structure was mainly benzene ring tri-substituted, with primary coal of 32.71% and outburst coal of 31.6%. The primary coal contained more functional groups, from which hydrogen bonds can easily be formed, meaning that gas is not easily adsorbed by coal. The aromatic hydrogen rate (f_{Ha}) of the outburst coal was 0.271, the aromatic carbon rate (f_C) was 0.986, the aromaticity I_1 was 0.477, I_2 was 0.373 and the length of the aliphatic branched chain (A_{CH2}/A_{CH3}) was 0.850. Compared with the primary coal, the aromatic hydrogen rate, aromatic carbon rate and the aromaticity of the outburst coal were higher, indicating that the hydrogen and carbon elements in the aromatic functional groups of outburst coal were higher and that the aliphatic functional group was higher than the aromatic structural functional group. A_{CH2}/A_{CH3} and maturity (Csd) were slightly lower than those of primary coal, indicating that the coal has more straight chains than side chains, while aliphatic hydrocarbons are mostly short chains and have high branched degree. There were obvious 002 and 100 peaks in the XRD pattern. The d_{002} and d_{100} of outburst coal were 3.570 and 2.114, respectively, while the number of effective stacking aromatics was 3.089, which was lower than that of primary coal, indicating that the structure of the dense ring in the coal saw certain changes.

Keywords: outburst coal; primary coal; structural parameters; infrared spectrum; XRD

Citation: Jiao, A.; Tian, S.; Lin, H. Analysis of Outburst Coal Structure Characteristics in Sanjia Coal Mine Based on FTIR and XRD. *Energies* **2022**, *15*, 1956. https://doi.org/10.3390/en15061956

Academic Editor: Nikolaos Koukouzas

Received: 30 January 2022
Accepted: 3 March 2022
Published: 8 March 2022

Publisher's Note: MDPI stays neutral with regard to jurisdictional claims in published maps and institutional affiliations.

1. Introduction

Guizhou Province in south China has abundant coal resources, wide distribution and complete coal types; however, the natural occurrence conditions of coal seams in Guizhou are poor, the occurrence of gas is complex and its control is difficult at underground mines [1–3]. In its natural state, gas exists in coal or surrounding rock in a free state and suction state [4–8]. The functional group of coal determines the chemical adsorption characteristics of coal, which is mainly controlled by the maceral and mineralogical compositions of coal seams and could indirectly affect the adsorption state of gas by coal seams [9–18]. Therefore, it is of great significance to study the functional group distribution and microcrystalline structure parameters of coal in order to understand gas occurrence and predict gas disasters.

Due to the complex microstructure characteristics of coal, our understanding of coal chemical composition is still limited and under discussion. For a long time, the research on the chemical structure of coal and the characteristics of functional groups in coal has

---

been the mainly focus of several studies [19–23]. Infrared spectroscopy is a non-destructive structural characterization technique that is especially suitable for qualitative and semi-quantitative study of functional groups in coal [24–28]. It is very suitable for characterizing insoluble organic compounds in coal. It is the comprehensive absorbance estimation of organic components based on the main functional groups (such as alkyl CH, CH_2 and CH_3, aromatic C=C and C-H, carbonyl/carboxylic acid C=O, hydroxyl-OH) [29,30]. Based on this, several studies have been conducted during the last two decades. Ibarra et al. [31] assigned the absorption peak of the whole spectrum and divided the whole spectrum into four parts, namely, aromatic structure, oxygen-containing functional group, fatty functional group and hydroxyl functional group. Sobkowiak et al. [32] studied the distribution characteristics of aromatic and aliphatic functional groups in coal by infrared spectroscopy. Rosa et al. [33] and Alemany et al. [34], who focused on the characterization of Illinois No.6 coal samples, found that the aromaticity of the coal was 0.72, 0.7 and 0.68. Painter et al. [35] summarized the assignment of hydroxyl absorption peaks in infrared spectroscopy coal, finding that hydroxyl hydrogen bonds are mainly OH-π, OH-OH, OH-ether oxygen. Orrego-Ruiz et al. [36] used FTIR spectroscopy to determine five Colombian coals and obtained the contents of aliphatic hydrogen and aromatic hydrogen. Cao et al. [37] found that the substitution degree of high maturity coal decreases rapidly in the early stage of coalification. There are a large number of aromatic clusters in semi anthracite and anthracite, and the average aromatic ring size is above 3–4.

X-ray diffraction (XRD) is an important technique for determining the structural parameters of crystalline matter in coal, which is also widely used in studying the polynuclear aromatic structure and the composition of microcrystal aromatic ring lamellae of amorphous materials [38–42]. Saikia et al.'s [43] XRD and FTIR analysis of the coal samples from Makum and Assam states in India indicated that the intensities of carbonyl groups content in coal was related to the metamorphic degree of coal. Wu et al. [44] found that there was an exponential relationship between XRD structure parameters and C/H by analyzing the XRD spectra of nine coals. Jiang et al. [45], according to the XRD pattern of medium and high rank coal, concluded that there is a linear relationship between XRD structural parameters and vitrinite reflectance ($\%R_o$). Boral P et al. [46] studied five coal samples selected from different coal pools in India using X-ray diffraction. It was found that d_{002} and f_a values depend on coal rank, not only carbon content.

Even though the previous studies mainly focused on the functional groups and microstructure of primary coal and tectonic zone coal, the research on functional groups and the microstructure of coal samples in outburst sites is limited, and the above research is less related to the research and analysis of outburst coal in the Guizhou mining area. In order to explore the distribution of functional groups in outburst coal and the characteristics of microcrystalline structure parameters, this paper intends to use an infrared spectroscopy test and peak fitting method to study the infrared spectrum of coal samples and determine the structural parameters of its coal samples. Secondly, the microcrystalline structure parameters of coal samples were calculated by XRD analysis. Finally, the functional groups and microcrystalline structure parameters of outburst coal samples and primary structure coal samples were compared, and the differences between the two were analyzed to better explain the influence of outburst coal on gas occurrence.

2. Sample Preparation and Test Method

2.1. Sampling Cite

The Sanjia coal mine is located in the south-west section of the Guiyang complex tectonic deformation zone of Tailong in the north of the Yangtze platform in the south-east wing of Guanzhai syncline. In general, it is a monoclinic structure, with strata moving towards the northeast, tending to the northwest, and with inclination of about 10°. The faults found in the mining area are located in the southern part of the mining area and its edge. Affected by the faults, the strata have changed. On 25 November 2019, a prominent accident occurred at the heading face of the 41,601 transport roadway in Sipan District. The sampling point selected in the 41,601 transport roadway heading face was located near the northern boundary of the mining area. See Figure 1.

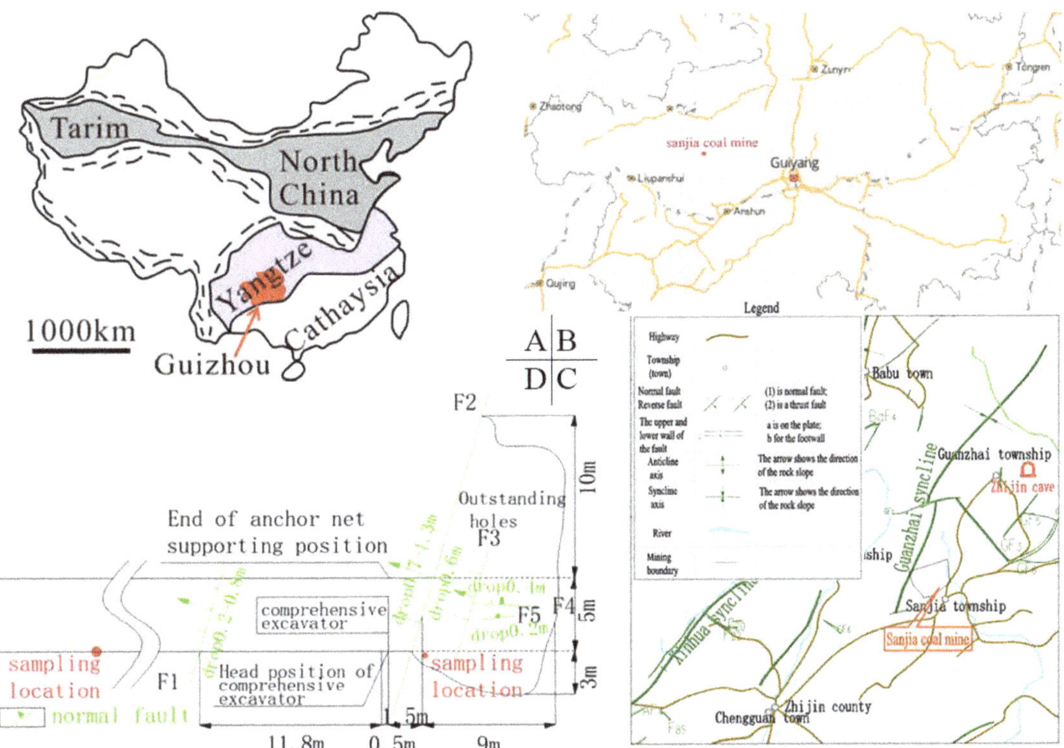

Figure 1. Prominent position diagram. Note: (**A,B**) are the geographical location of the mine; (**C**) is the geological structure near the mine; (**D**) is a prominent geological structure.

2.2. Proximate and Ultimate Analyses of Coal Samples

Samples were gathered soon after the outburst accident. The surface oxide layer of the sampling point was stripped and the outburst coal samples were taken in the outburst hole by the groove method, while the primary coal samples were taken outside the outburst hole. The collected coal samples were immediately sealed in the coal sample tank to reduce exposure to the external environment, the oxidation of the samples were slowed down during the period before the start of the experiment, and the sampling point information was recorded. The tank was fixed in the transport process to prevent secondary damage to coal. The samples were sent to the laboratory and ground into the specified particle size by agate bowl. According to GB/T 214-2007 and SN/T 4764-2017, coal ash, moisture, volatile content and C, H, N, O, S and other elements were determined by industrial analyzer and elemental analyzer.

2.2.1. Infrared Spectroscopy Experiments

The experiment was carried out by Nicolet 6700 Fourier transform infrared spectrometer. Before the experiment, the coal samples were placed in a constant temperature oven at 80 °C for 5 h for drying treatment. After drying, the coal samples were fully mixed with potassium bromide (KBr) at a ratio of 1:100, and then loaded into the grinding tool to make transparent sheets of 0.1–1 mm thickness on the tablet press. The equipment test conditions were: a wave number test range of 7800–375 cm^{-1}, a scanning rate of 0.158–6.33 cm/s, an infrared spectral resolution of 4 cm^{-1} and scanning times of 32.

2.2.2. X-ray Diffraction (XRD) Analysis

The X-ray diffractometer used the X-ray with a fixed wavelength to irradiate the coal. When the wavelength of the X-ray was consistent with the magnitude of the lattice in the coal, the diffraction phenomenon occured. Different diffraction light was formed at different positions. A series of X-rays irradiated at different angles were recorded to form. According to the map, the data provided about the crystalline structure of the coal was analyzed [47]. In the experiment, the X-ray diffractometer Shimadzu XRD-6100 was used for analysis and testing, and the coal samples that had been prepared in advance were loaded onto the support. Setting working conditions, the light source was an X-ray tube copper that targeted radiation (γ = 0.15405 nm) and that had a voltage of 40 kV, a current of 40 mA, a scanning speed of 10/mm and a scanning range of 0–90°.

3. Experimental Results and Analysis

3.1. Standard Coal Quality Parameters

It can be seen from the composition analysis of the coal samples that the outburst coal has higher volatile matter and ash content than the primary coal and that the moisture content of the primary coal is higher than that of the outburst coal. Except hydrogen, the elements of outburst coal are higher than that of the primary coal. The test results are shown in Table 1.

Table 1. Coal sample analysis table.

Component Coal Samples	Proximate Analysis			Ultimate Analysis				
	A_{ad}/%	V_{ad}/%	M_{ad}/%	C	H	N	S	O
outburst coal	6.1	5.4	4.6	82.9	2.87	1.51	0.42	4.05
primary coal	9.3	4.7	4.3	81.7	2.69	1.65	0.38	3.92

Note: M_{ad} = moisture content (wt%, dry basis), A_{ad} = ash (wt%, dry basis), V_{ad} = volatile (wt%, dry basis), FCad = fixed carbon (wt%, dry basis).

3.2. FTIR Fitting Spectral Characteristics Analysis

The infrared spectra of raw coal and outburst coal samples were obtained by an infrared spectrum test of coal samples, as shown in Figure 2 and Table 2. In the process of the infrared spectrum test, due to the complex composition of coal, the absorption bands of many functional groups in coal provided the infrared spectrum. In addition, the wave number distribution range of the spectrum is wide, meaning that it is easy to cause spectral superposition and difficult to determine the position and intensity of absorption peaks. Through the peak fitting of the original spectrum of the coal samples, the parameters and attribution of absorption peaks are determined, and the change characteristics of infrared spectrum of coal samples are thus more truly reflected. The infrared spectrum of coal samples can be divided into four absorption bands: aromatic structure absorption band, oxygen-containing functional group absorption band, fat structure absorption band and hydroxyl absorption band. The corresponding wave numbers are 700–900 cm^{-1}, 1000–1800 cm^{-1}, 2800–3000 cm^{-1}, 3000–3600 cm^{-1}, respectively, as shown in Figure 2.

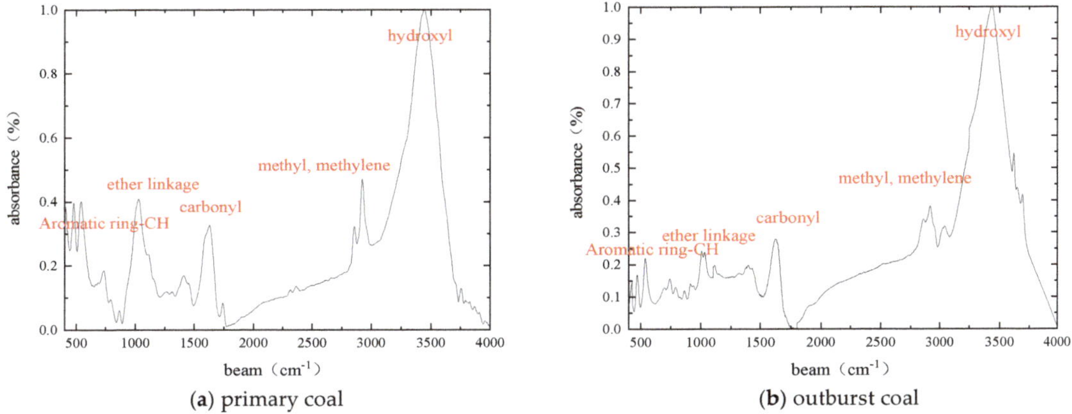

(a) primary coal   **(b)** outburst coal

Figure 2. Fourier infrared spectra of coal samples.

Table 2. Ratios of various functional groups.

Component Coal Samples	Functional Groups Ratios			
	AS/%	OC/%	FS/%	HY/%
outburst coal	2.19	16.10	7.38	43.34
primary coal	2.71	24.58	7.27	42.31

Note: AS = aromatic structure absorption band; OC = oxygen-containing functional group absorption band; FS = fat structure absorption band; HY = hydroxyl absorption band.

3.2.1. Absorption Band of Aromatic Structure

It can be seen from Figure 3 that the original spectra of aromatic structures in the two coals are different. There are three peaks in the primary structure coal and four peaks in the outburst coal. In order to make the fitted spectra closer to the experimental results, 11–14 peaks were fitted, and the correlation coefficients were more than 99.9%. The 700–900 cm^{-1} absorption band is the absorption vibration of the aromatic structure, which is an important characteristic peak for identifying the substitution mode of the aromatic rings in the coal. Among them, 730–750 cm^{-1} is the disubstituted absorption peak, 750–810 cm^{-1} is the trisubstituted absorption peak, 810–850 cm^{-1} is the tetrasubstituted absorption peak and 850–900 cm^{-1} is the pentasubstituted absorption peak.

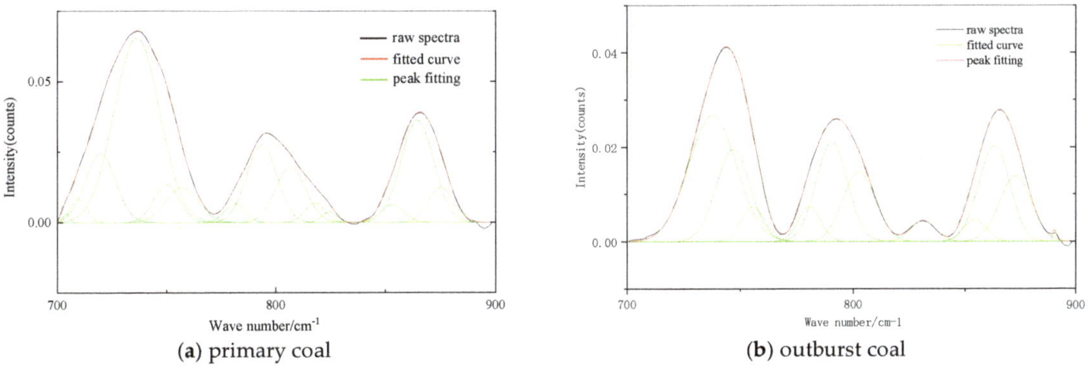

(a) primary coal   **(b)** outburst coal

Figure 3. FTIR Spectra of Aromatic Structure.

According to the above four kinds of wave number classification statistics, the different benzene ring substitution column diagram was obtained (see Figure 4). The substitution of the benzene ring in the aromatic structure was mainly the trisubstituted benzene ring, and the proportion of primary structure coal reached 32.71%. With the increase of heating impact, the outburst coal decreased to 31.6%, and the difference was small. The disubstituted benzene ring decreased from 19.56% to 12.98%, the tetrasubstituted benzene ring increased from 8.93% to 16.89% and the pentasubstituted benzene ring increased from 10.91% to 22.10%. Among them, the values of the tetrasubstituted benzene ring and the pentasubstituted benzene ring are 1.89 times and 2.02 times of primary coal, respectively. The factors that cause this phenomenon are diverse, but may include the substitution reaction of the cyclization of the aliphatic chain and the positioning group on the aromatic ring.

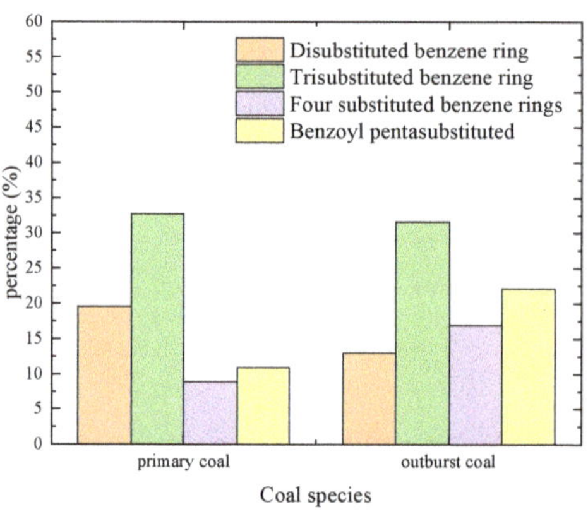

Figure 4. Statistical distribution of substituted benzene rings.

3.2.2. Absorption Band of Oxygen Functional Groups

It can be seen that the absorption band of 1000–1800 cm^{-1} is the stretching vibration of oxygen-containing functional groups, and the absorption peaks of 1000–1300 cm^{-1} are mainly attributable to the C-O stretching vibration of phenols, alcohols, ethers and esters (Figure 5). The peak positions at 1375–1460 cm^{-1} were symmetric and antisymmetric deformation vibrations of methyl, while the peak at 1800–1530 cm^{-1} had the stretching vibration of -COOH- and CO.

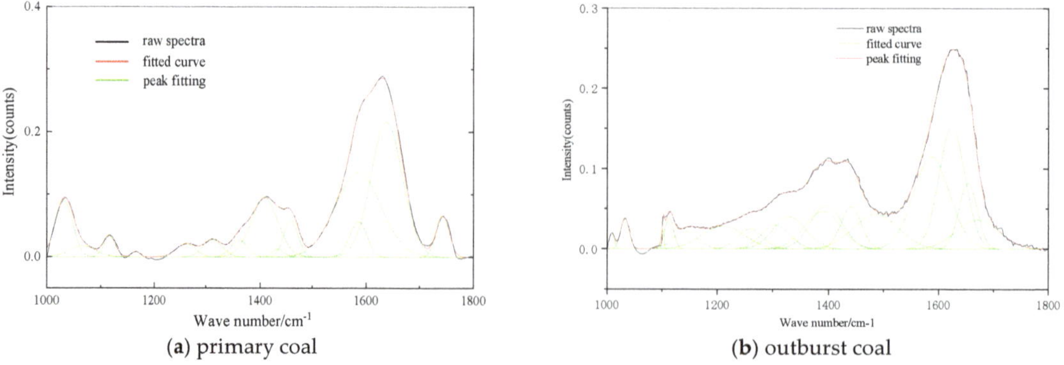

Figure 5. FTIR Spectra of Oxygen Functional Group Structure.

The percentage of C-O, C=O and C-C in these two coal samples were counted and the histogram was obtained, as shown in Figure 6. It can be concluded that with the increase of heating impact, the content of C-O increased from 14.97% of the primary structure coal to 19.86% of outburst coal, and the content of C-C increased from 17.33% of primary structure coal to 19.47% of outburst coal. Among them, the content of C=O changed most obviously, from 3.82% of primary structure coal to 6.84% of outburst coal, which increased by 44.15%, indicating that the content of carbon and oxygen elements in outburst coal was accumulated due to increased temperature during outburst.

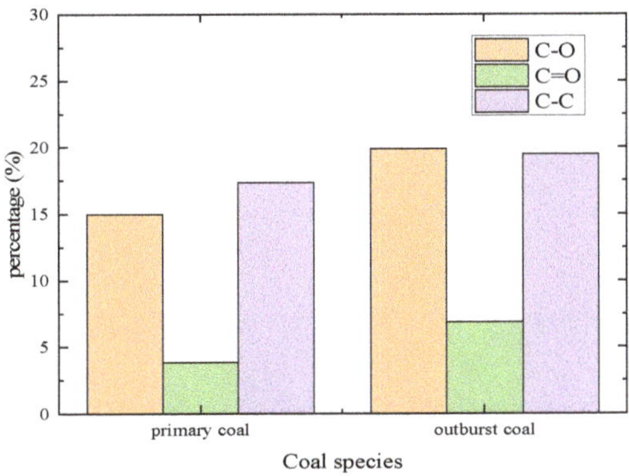

Figure 6. Statistical distribution histogram of oxygen-containing functional groups' structure content.

3.2.3. CHal Structure Absorption Band

It can be seen from Figure 7 that the absorption band of 2800–3000 cm^{-1} was the absorption vibration of CHal stretching structure. The spectrum was fitted with 6–7 peaks, and the fitting degree reached more than 99.9%. The coal contains methyl, methylene and methylene. The peak at 2853 cm^{-1} is the symmetric stretching vibration of methylene, the peak at 2871 cm^{-1} is the symmetric stretching vibration of methyl, the peak at 2895 cm^{-1} is the stretching vibration of methylene, the peak at 2923 cm^{-1} is the antisymmetric stretching vibration of methylene and the peak at 2953 cm^{-1} is the antisymmetric stretching vibration of methyl.

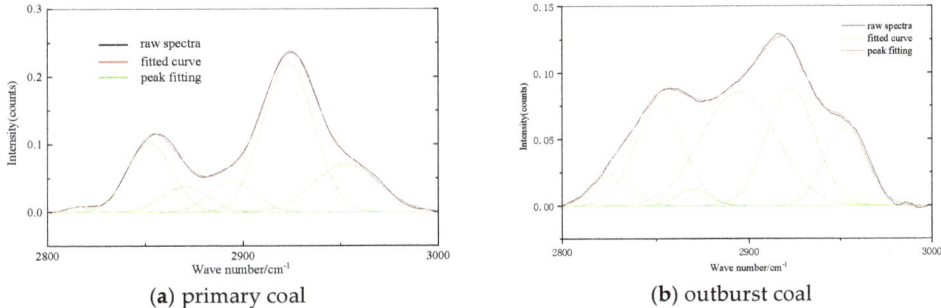

(a) primary coal   **(b)** outburst coal

Figure 7. FTIR spectra of CHal structure.

The aliphatic bands in the studied coal samples were statistically analyzed to get the histogram (see Figure 8). The content of methylene symmetric stretching vibration in raw coal and outburst coal was 18.73% and 20.10%, respectively. The content of symmetric stretching vibration of methyl was 7.14% and 2.31%, respectively. Compared with the primary coal, the outburst coal had a significant downward trend, which was reduced by 67.65%. The content of methylene stretching vibration was 8.47% and 35.11%, respectively.

The stretching vibration of outburst coal methylene increased rapidly, and the change value increased 4.15 times. The content of antisymmetric stretching vibration of methylene was 45.95% and 24.02%, respectively. The content of antisymmetric stretching vibration of methyl was 18.87% and 13.33%, respectively. It may be that the thermal effect of the outburst process leads to the volatilization of hydrogen or the fracture of the fat chain.

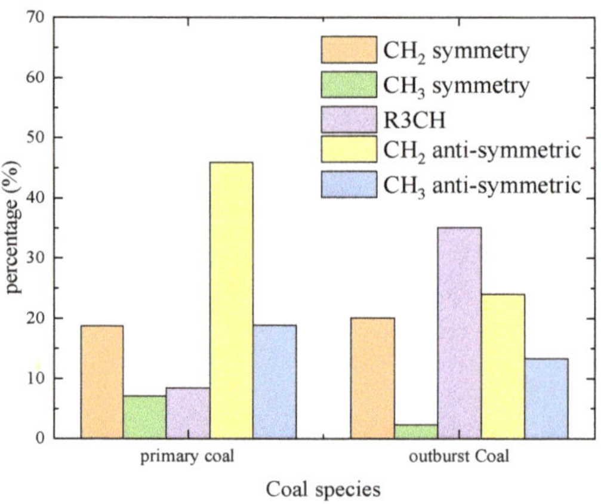

Figure 8. Statistical distribution histogram of CHal.

3.2.4. Hydroxyl Absorption Band

It can be seen from Figure 9 that the absorption band of 3000–3600 cm^{-1} was the absorption vibration of the hydroxyl structure. The spectrum was fitted with 4~5 peaks, and the fitting degree reached more than 99.9%, which was closer to the original spectrum. There are several O-H stretching vibration absorption peaks in 3200–3516 cm^{-1}, forming a wide and scattered absorption band. There is a sharp absorption peak at 3611 cm^{-1}, which is the absorption peak of free-OH.

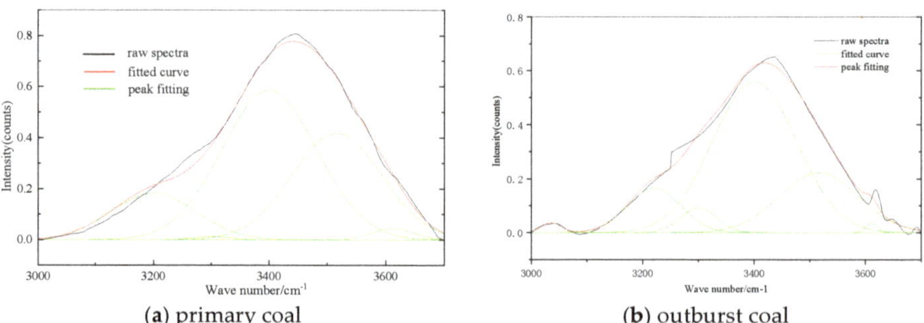

Figure 9. FTIR fitting spectra of hydroxyl structure.

The types of hydrogen bonds in the coal hydroxyl groups were analyzed, and the histogram was obtained, as shown in Figure 10. Hydrogen bonds exist in coal, and the content of cyclic associative hydrogen bonds in primary coal and outburst coal accounts for 13.97% and 11.24%, respectively. The content of self-associated hydroxyl accounted for 50.40% and 60.60%, respectively; the increase of self-association hydroxyl content is due to the tighter internal structure of the molecule, which increases the probability of self-association hydroxyl synthesis. When water molecules and methane compete on the surface of coal, water molecules are more likely to seize the adsorption site. Water molecules' adsorption replaces gas adsorption and indirectly reduces gas adsorption in coal seams.

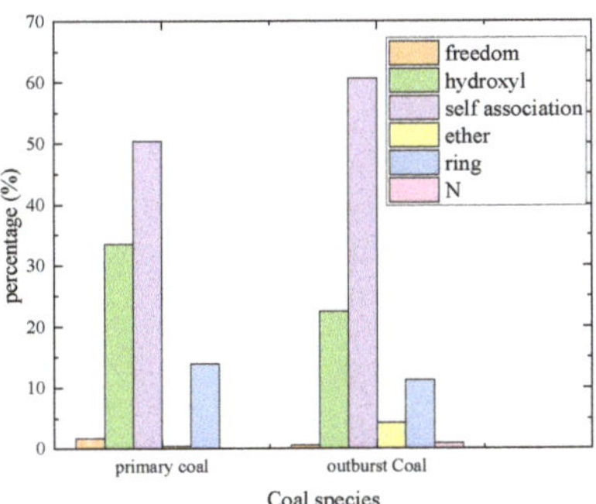

Figure 10. Column diagram of hydrogen bond type statistical distribution in hydroxyl group.

3.3. Analysis of FTIR Structural Parameters

In order to analyze the macromolecular structure of coal, the chemical structure parameters and significance of coal are generally quantitatively analyzed through structural parameters such as aromatic hydrogen rate, aromatic carbon rate, aromaticity, polycondensation degree of the aromatic ring, length of aliphatic chain, branching degree and the maturity of the coal.

3.3.1. Hydrogen Aromaticity f_{Ha}

Aromatic hydrogen rate [44] is the percentage of hydrogen atoms in aromatic compounds. It is assumed that only aromatic hydrogen and aliphatic hydrogen exist.

$$f_{Ha} = \frac{H_a}{H} \frac{A_{900\sim700\text{cm}^{-1}}}{A_{900\sim700\text{cm}^{-1}} + A_{3000\sim2800\text{cm}^{-1}}} \tag{1}$$

where H_a is the number of H atoms of aromatic compounds. H is the number of H atoms.

3.3.2. Aromatic Carbon Ratio f_C

Aromatization rate indicates the proportion of carbon atoms in aromatic hydrocarbons. Calculation of aryl carbon ratio supposes only aryl carbon and lipocarbon exist in carbon atom.

$$f_C = 1 - \frac{C_a}{C} \tag{2}$$

$$\frac{C_a}{C} = \frac{\left(\frac{H_a}{H} \cdot \frac{H}{C}\right)}{\frac{H_a}{C_a}} \tag{3}$$

$$\frac{H_a}{H} = \frac{A_{3000\sim2800}}{\left(A_{900\sim700} + A_{3000\sim2800}\right)} \tag{4}$$

where C_a/C is the proportion of fat carbon in the carbon element, H/C is the ratio of hydrogen to carbon and H_a/C_a is the proportion of hydrogen and carbon atoms in fat clusters (generally about 1.8. H_a/H is the percentage of aliphatic hydrogen in hydrogen atoms).

3.3.3. Aromaticity I

Aromaticity is an important parameter used to determine the aromaticity of substances. Aromaticity is defined by the ratio of aromaticity to aliphatic carbon.

$$I_1 = \frac{A_{3100\sim3000}}{A_{3000\sim2800}} \tag{5}$$

$$I_2 = \frac{A_{900\sim700}}{A_{3000\sim2800}} \tag{6}$$

where 3100–3000 cm^{-1} is mainly aromatic C-H stretching vibration, 3000–2800 cm^{-1} is aromatic CH$_x$ stretching vibration and 900–700 cm^{-1} absorption position of aromatic structure.

3.3.4. Degree of Aromatic Ring Condensation *DOC*

The degree of aromatic ring condensation can also be used to judge the degree of coal rank. The higher the coal metamorphism, the more the internal carbon content, and the more condensed aromatic substances.

$$DOC = \frac{A_{700\sim900}}{A_{1490\sim1600}} \tag{7}$$

where the ratio of 900–700 cm^{-1} to 1490–1600 cm^{-1} indicates the degree of aromatic ring condensation.

3.3.5. A_{CH2}/A_{CH3}

A_{CH2}/A_{CH3} is the fat structure parameter [48]. CH$_2$ mainly has a chain, ring and aromatic side hydrocarbon straight chain, while CH$_3$ mainly has a chain, ring side chain and aromatic side hydrocarbon branched chain.

$$\frac{A_{CH_2}}{A_{CH_3}} = \frac{A_{2920}}{A_{2950}} \tag{8}$$

3.3.6. Maturity *Csd*

The maturity of coal is indicated by the large change in the carbon–oxygen double bond relative to the carbon-carbon double bond.

$$Csd = \frac{A_{1800\sim1650}}{A_{1800\sim1650} + A_{1600}} \tag{9}$$

3.4. Summary of Structural Parameters of Infrared Spectroscopy

The outburst coal and primary coal f_{Ha}, f_C, I_1, I_2, DOC, A_{CH2}/A_{CH3} and *Csd* are calculated by Equations (1)–(9). The calculation results are shown in Table 3.

Table 3. Structure parameters summary table.

Coal Samples \ Component	f_{Ha}	f_C	I_1	I_2	DOC	A_{CH2}/A_{CH3}	Csd
outburst coal	0.271	0.986	0.477	0.373	1.560	0.850	0.969
primary coal	0.229	0.984	0.454	0.298	1.010	0.916	0.976

The analysis shows that the aromatic hydrogen rate (f_{Ha}) and aromatic carbon rate (f_C) of outburst coal are higher than those of primary coal, indicating that the hydrogen and carbon elements in aromatic functional groups are higher. The aromaticity I_1 and I_2 of outburst coal were higher than those of raw coal, indicating that the aliphatic functional group of outburst coal was higher than that of the aromatic structural functional group. The aromatic ring polycondensation degree (*DOC*) of the outburst coal is higher than that of the primary coal and the hydrogen content is also higher than that of the primary coal, while A_{CH2}/A_{CH3} and maturity (*Csd*) are slightly less than native coal, indicating that native coal has more straight chains than side chains, while aliphatic hydrocarbons are mostly short chains and have high branched degree. The results are consistent with previous studies, i.e., coal degree is inversely proportional to A_{CH2}/A_{CH3}.

3.5. X-ray Diffraction (XRD) Analysis of Coal Samples

3.5.1. XRD Peak Fitting

In order to analyze the microstructure of outburst coal, the XRD patterns of coal samples were fitted by peak separation, as shown in Figure 11. The original spectrum was divided into 002 peaks, γ peaks and 100 peaks by Guassian [41] formula. The fitting formula is:

$$y = y_0 + \frac{A}{w \cdot \sqrt{\frac{\pi}{2}}} \cdot e^{-2\left(\frac{x-x_c}{w}\right)^2} \tag{10}$$

where y_0 is the baseline position, w is the diffraction peak width, A is the diffraction pattern area and x_c is the 2θ angle of the diffraction center.

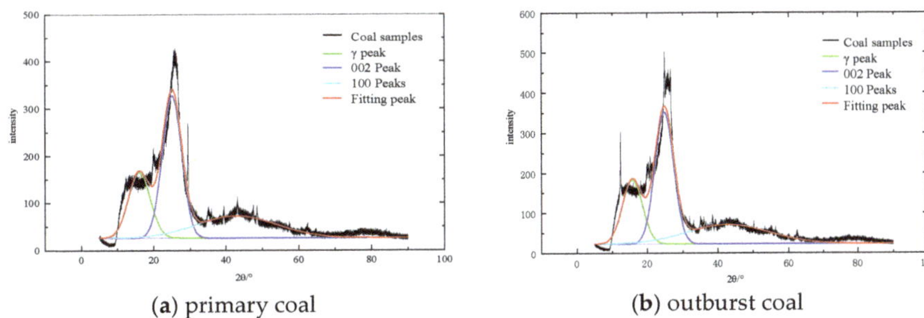

(**a**) primary coal   (**b**) outburst coal

Figure 11. Peak separation fitting results of coal samples.

It can be found from Figure 11 and Table 4 that the 2θ angle range of 002 diffraction peak is 24.92°–25.08°, with high intensity and good symmetry. Moreover, the 2θ angle range of 100 diffraction peaks is 42.73°–42.85° and the diffraction peaks are wide and low, indicating that the degree of aromatic ring condensation in coal is not high. For the γ band, it can be clearly seen from the graph that its area is large, reflecting the rich branched chain structure of aliphatic hydrocarbons and aliphatic hydrocarbons in coal.

Table 4. Diffraction peak fitting of the main 2θ parameters.

Coal Samples \ Diffraction Peak	002 Peak	100 Peak	γ Peak
outburst coal	24.92	42.73	15.87
primary coal	25.08	42.85	16.15

3.5.2. Structure Analysis of Aromatic Microcrystals

The layer spacing of d_{002} and d_{100}, the average stacking height of L_c and the ductility of L_a of the aromatic layer can be calculated from the diffraction angle and the half-peak width [49].

$$d_{002/100} = \frac{\lambda}{2 \sin \theta_{002/100}} \tag{11}$$

$$L_c = \frac{K_1 \lambda}{\beta_{002} \cos \theta_{002}} \tag{12}$$

$$L_a = \frac{K_2 \lambda}{\beta_{100} \cos \theta_{100}} \tag{13}$$

$$M_c = \frac{L_c}{d_{002}} \tag{14}$$

where λ is the wavelength of X-ray, and 0.15405 nm is used for the experiment with copper target irradiation; $\theta_{002/100}$ is the Bragg angle of 002 diffraction peak and 100 diffraction peak, respectively; β_{002} and β_{100} are the half-peak widths of 002 diffraction peak and 100 diffraction peak, respectively; K_1 and K_2 are Debye-Scherrer constants, with K_1 as 0.89, K_2 as 1.84 and M_c as the number of effective stacked aromatic slices.

In general, the interlayer spacing of the coal aromatic layer is between cellulose ($d_{002} = 3.975 \times 10^{-1}$ nm) and graphite ($d_{002} = 3.354 \times 10^{-1}$ nm). Therefore, the coalification degree P is used to determine the percentage of condensed aromatic ring, and the relative content of aromatic layer and aliphatic layer structure is obtained. The calculation formula is as follows:

$$P = \frac{3.975 - d_{002}}{3.975 - 3.354} \times 100\% \tag{15}$$

The calculation results are shown in Table 5. It can be found from the table that d_{002} of primary coal is 3.548, indicating that its coalification degree is high. The number of effective stacking aromatic slices of primary coal is about 4, and the coalification degree is 68.83%, indicating that the coal sample has a high degree of coalification, with fewer side chains and functional groups, and the internal arrangement of molecules is orderly and stable, and the condensation degree of aromatic nucleus is high. Combined with the data of d_{002} and d_{100}, the number of aromatic flakes in outburst coal is lower than primary coal, which suggests that the prominent heating effect increases the degree of aromatization, increases the d_{100} and d_{002} values and, finally, changes the dense ring structure in coal. Through the coal degree index P, it can be seen that the primary coal is slightly higher than the outburst coal. Combined with the analysis of d_{002} and d_{100} as basically unchanged, this shows that the content of side chain in the primary coal is lower than that of the outburst coal, resulting in the increase of the relative number of aromatic rings, which is finally reflected in the increase of P.

Table 5. XRD microstructure parameters of coal samples.

Structural Parameters Coal Samples	d_{002}	d_{100}	L_c	L_a	M_c	P
outburst coal	3.570	2.114	13.016	3.089	3.646	65.22%
primary coal	3.548	2.109	12.951	3.240	3.651	68.83%

4. Conclusions

(1) The proximate analysis results show that the volatile matter and moisture content of outburst coal are higher than those of primary coal, but that the ash content is lower than that of primary coal. The ultimate analysis shows that the proportions of N and S in outburst coal and primary coal are small.

(2) The aromatic structure absorption band analysis showed that the trisubstituted benzene ring and the disubstituted benzene ring in the primary coal were higher than that in outburst coal, and that the tetrasubstituted benzene ring and the pentasubstituted benzene ring in outburst coal were higher than that in primary coal. This may be due to the substitution reaction of fat chain cyclization and the orientation group of the aromatic ring. The absorption band of fat structure indicates that hydrogen volatilization or fat chain rupture is caused by thermal effect in the outburst process. Moreover, the hydroxyl absorption band shows that the internal structure of the molecule is tighter, which increases the probability of self-association hydroxyl synthesis.

(3) By analyzing and calculating the structural parameters of infrared spectroscopy, it is concluded that the aromatic hydrogen rate, aromatic carbon rate and I_1 and I_2 of outburst coal are higher than those of primary coal. The A_{CH2}/A_{CH3} raw coal is relatively low, and the maturity is slightly higher than that of the outburst coal, indicating that the raw coal has more straight chains than side chains and that the aliphatic hydrocarbons are mostly short chains and have high branched degree.

(4) Through the analysis and calculation of XRD microcrystalline structure parameters, the number of aromatic flakes in outburst coal is lower than in primary coal, which suggests that the prominent heating effect increases the degree of aromatization,

increases the d_{100} and d_{002} values and, finally, changes the dense ring structure in coal. Moreover, from the coal degree index P, it can be seen that the primary coal has a smaller increase than the outburst coal.

Author Contributions: Conceptualization: A.J., S.T.; Methodology: H.L.; Formal analysis and investigation: A.J., H.L.; Writing-original draft preparation: A.J.; Writing-review and editing: S.T.; Funding acquisition: S.T. All authors have read and agreed to the published version of the manuscript.

Funding: This work was supported by the National Natural Science Foundation of China under Grant number 52104079 and the Science & Technology Foundation of Guizhou Province under Grant number [2020]4Y050.

Informed Consent Statement: Informed consent was obtained from all subjects involved in the study.

Data Availability Statement: All data, models, and code generated or used during the study appear in the submitted article.

Acknowledgments: The authors appreciate the support from above funders.

Conflicts of Interest: The authors declare that they have no competing financial interests.

References

1. Li, X.-L.; Cao, Z.-Y.; Xu, Y.-L. Characteristics and trends of coal mine safety development. *Energy Sour. Part A* **2020**, *12*, 1–14. [CrossRef]
2. Tang, X.-G. Occurrence Regularity of Coal Resources in Guizhou Province. *Coal Geol. Explor.* **2012**, *40*, 1–5.
3. Zhao, Z.-H.; Kuang, S.-D. Thoughts on Development of Coal Resources in Guizhou. *China Min. Mag.* **2006**, *3*, 17–20, 28.
4. Jia, H.-F.; An, F.-H. Experimental Study on Influence of Gas Occurrence State on Strength of Granular Coal. *Saf. Coal Mines* **2020**, *51*, 29–33.
5. Wang, W.; Wei, Y.-Z.; Jia, T.-R.; Yan, J.-G. Deformation Characteristics and Gas Occurrence Regularity of Coal Measures in North. *Coal Geol. Explor.* **2021**, *49*, 121–130.
6. Kong, S.-S.; Yang, Y.; Jia, Y.; Song, Y.-Q.; Sun, T.-L.; Kang, R.-P. Study on Gas Occurrence Characteristics and Influence of Key Geological Factors in Coal Seam. *Coal Sci. Technol.* **2019**, *47*, 53–58.
7. Jia, T.-R.; Wang, W.; Yan, J.-G.; Tang, C.-A. Control law and zoning of gas occurrence structure in Guizhou coal mines. *Earth Sci. Front.* **2014**, *21*, 281–288.
8. Zhang, P.-X.; Long, Z.-G. Study on Gas Occurrence Law of Coal Mines in Guizhou Province. *J. Saf. Sci. Technol.* **2013**, *9*, 34–38.
9. Lin, H.-Y.; Tian, S.-X.; Jiao, A.-J.; Ma, R.-S.; Xu, S.-Q. Pore Characteristics and Fractal of Tectonic Coal and Primary Coal in Northern Guizhou. *Sci. Technol. Eng.* **2021**, *21*, 14451–14458.
10. Yan, M.; Yue, M.; Lin, H.-F.; Yan, D.-G.; Wei, J.-N.; Qin, X.-Y.; Zhang, J. Experimental Study on Effect of Functional Groups on Wettability of Medium and Low Rank Coal. *Coal Sci. Technol.* **2022**; *unpublished.*
11. He, X.; Wang, W.-F.; Zhang, X.-X.; Yang, Y.-T.; Sun, H. Distribution characteristics and differences of oxygen functional groups in macerals of low rank coal. *J. China Coal Soc.* **2021**, *46*, 2804–2812.
12. Chen, Y.-Y.; Mastalerz, M.; Schimmelmann, A. Characterization of chemical functional groups in macerals across different coal ranks via micro-FTIR spectroscopy. *Int. J. Coal Geol.* **2012**, *104*, 22–33. [CrossRef]
13. Guo, Y.; Bustin, R.-M. Micro-FTIR Spectroscopy of Liptinite Macerals in Coal. *Int. J. Coal Geol.* **1998**, *36*, 259–275. [CrossRef]
14. Karayigita, A.-I.; Bircan, C.; Mastalerz, M.; Oskay, R.-G.; Querol, X.; Lieberman, N.-R.; Turkmen, I. Coal characteristics, elemental composition and modes of occurrence of some elements in the İsaalan coal (Balıkesir, NW Turkey). *Int. J. Coal Geol.* **2017**, *172*, 43–59. [CrossRef]
15. Mastalerz, M.; Bustin, R.-M. Electron microprobe and micro-FTIR analyses applied to maceral chemistry. *Int. J. Coal Geol.* **1993**, *24*, 333–345. [CrossRef]
16. He, X.; Zhang, X.-X.; Jiao, Y.; Zhu, J.-S.; Chen, X.-W.; Li, C.Y.; Li, H.-S. Complementary analyses of infrared transmission and diffuse reflection spectra of macerals in low-rank coal and application in triboelectrostatic enrichment of active maceral. *Fuel* **2017**, *192*, 93–101. [CrossRef]
17. Arif, M.; Jones, F.; Barifcani, A.; Iglauer, S. Influence of surface chemistry on interfacial properties of low to high rank coal seams. *Fuel* **2017**, *194*, 211–221. [CrossRef]
18. Xin, H.-H.; Wang, D.-M.; Dou, G.-L.; Qi, X.-Y.; Xu, T.; Qi, G.-S. The infrared characterization and mechanism of oxygen adsorption in coal. *Spectrosc. Lett.* **2014**, *47*, 664–675. [CrossRef]
19. Jia, T.-G.; Li, C.; Qu, G.-N.; Li, W.; Yao, H.-F.; Liu, T.-F. FTIR characterization of chemical structure characteristics of coal samples with different metamorphic degrees. *Spectrosc. Spectr. Anal.* **2021**, *41*, 3363–3369.
20. Liang, C.-H.; Liang, W.-Q.; Li, W. Study on Functional Groups of Different Rank Coals Based on Fourier Transform Infrared Spectroscopy. *J. China Coal Soc.* **2020**, *48*, 182–186.

21. Wang, C.; Zhao, Y.-B.; Wang, Z.-W.; Lu, Z.-L. Improved method for characterizing coal metamorphic degree by XRD. *China Coal Geol. Explor.* **2019**, *47*, 39–44.
22. Li, X.; Liu, Z.; Qian, J. Analysis on the Changes of Functional Groups after Coal Dust Explosion at Different Concentrations Based on FTIR and XRD. *Combust. Sci. Technol.* **2020**, *193*, 2482–2504. [CrossRef]
23. Presswood, S.-M.; Rimmer, S.-M.; Anderson, K.-B.; Filiberto, J. Geochemical and petrographic alteration of rapidly heated coals from the Herrin (No. 6) Coal Seam, Illinois Basin. *Int. J. Coal Geol.* **2016**, *165*, 243–256. [CrossRef]
24. Yan, J.; Lei, Z.; Li, Z.; Wang, Z.; Ren, S.; Kang, S.; Wang, X.; Shui, H. Molecular structure characterization of low-medium rank coals via XRD, solid state 13CNMR and FTIR spectroscopy. *Fuel* **2020**, *268*, 117038. [CrossRef]
25. Kwiecinska, B.; Pusz, S.; Valentine, B.-J. Application of electron microscopy TEM and SEM for analysis of coals, organic-rich shales and carbonaceous matter. *Int. J. Coal Geol.* **2019**, *211*, 103203. [CrossRef]
26. Liu, M.; Bai, J.; Jianglong, Y.-U.; Kong, L.; Bai, Z.; Li, H.; He, C.; Ge, Z.; Cao, X.-I.; Li, W. Correlation between char gasification characteristics at different stages and microstructure of char by combining X-ray diffraction and Raman spectroscopy. *Energy Fuels* **2020**, *34*, 4162–4172. [CrossRef]
27. Li, X.; Yiwen, J.-U.; Hou, Q.; Li, Z. FTIR and Raman spectral research on metamorphism and deformation of coal. *J. Geol. Res.* **2012**, *2012*, 590857. [CrossRef]
28. Li, W.; Zhu, Y.-M.; Chen, S.-B.; Si, Q.-H. Coupling mechanism of hydrocarbon generation and structural evolution in low coal grade coal. *Spectrosc. Spectr. Anal.* **2013**, *33*, 1052–1056.
29. Li, W.; Zhu, Y.-M.; Wang, G.; Jiang, B. Characterization of coalification jumps during high rank coal chemical structure evolution. *Fuel* **2016**, *185*, 298–304. [CrossRef]
30. He, X.-Q.; Liu, X.-F.; Nie, B.-S.; Song, D.-Z. FTIR and Raman spectroscopy characterization of functional groups in various rank coals. *Fuel* **2017**, *206*, 555–563. [CrossRef]
31. Ibarra, J.; Muñoz, E.; Moliner, R. FTIR study of the evolution of coal structure during the coalification process. *Org. Geochem.* **1996**, *24*, 725–735. [CrossRef]
32. Sobkowiak, M.; Reisser, E.; Given, P. Determination of aromatic and aliphatic CH groups in coal by FT-IR:1. Studies of coal extracts. *Fuel* **1984**, *9*, 1245. [CrossRef]
33. Dela Rosa, L.; Pruski, M.; Lang, D. Characterization of the Argonne premium coals by using hydrogen-1 and carbon-13 NMR and FT-IR spectroscopies. *Energy Fuels* **1992**, *6*, 460. [CrossRef]
34. Alemany, L.-B.; Grant, D.-M.; Pugmire, R.-J. Solid state magnetic resonance spectra of Illinois No. 6 coal and some reductive alkylation products. *Fuel* **1984**, *63*, 513–521. [CrossRef]
35. Painter, P.-C.; Sobkowiak, M.; Youtcheff, J. FT-IR Study of hydrogen-bonding in coal. *Fuel* **1987**, *66*, 973–978. [CrossRef]
36. Orrego-Ruiz, J.-A.; Cabanzo, R.; Mejía-Ospino, E. Study of Colombian coals using photoacoustic Fourier transform infrared spectroscopy. *Int. J. Coal Geol.* **2011**, *85*, 307–310. [CrossRef]
37. Cao, X.; Chappell, M.-A.; Schimmelmann, A.; Mastalerz, M.; Li, Y.; Hu, W. Chemical structure changes in kerogen from bituminous coal in response to dike intrusions as investigated by advanced solid-state 13C NMR spectroscopy. *Int. J. Coal Geol.* **2013**, *108*, 53–64. [CrossRef]
38. Gupta, R. Advanced coal characterization: A review. *Energy Fuels* **2017**, *21*, 451–460. [CrossRef]
39. Machado, A.D.S.; Mexias, A.S.; Vilela, A.C.F.; Osorio, E. Study of coal, char and coke fines structures and their proportions in the off-gas blast furnace samples by X-ray diffraction. *Fuel* **2013**, *114*, 224–228. [CrossRef]
40. Song, D.; Yang, C.; Zhang, X.; Su, X.-B.; Zhang, X. Structure of the organic crystallite unit in coal as determined by X-ray diffraction. *Min. Sci. Technol.* **2011**, *21*, 667–671. [CrossRef]
41. Cartz, L.; Hirsch, P.-B. A contribution to the structure of coals from X-ray diffraction studies. *Philos. Trans. R. Soc. Lond.* **1960**, *252*, 557–602.
42. Li, M.; Zeng, F.; Chang, H.; Bingshe, X.; Wang, W. Aggregate structure evolution of low-rank coals during pyrolysis by in-situ X-ray diffraction. *Int. J. Coal Geol.* **2013**, *116–117*, 262–269. [CrossRef]
43. Saikia, B.-K.; Boruah, R.-K.; Gogoi, P.-K. XRD and FT-IR investigations of sub-bituminous Assam coals. *Bull. Mater. Sci.* **2007**, *30*, 421–426. [CrossRef]
44. Wu, D.; Liu, G.; Sun, R.; Xiang, F. Investigation of structural characteristics of thermally metamorphosed coal by FTIR spectroscopy and X-ray diffraction. *Energy Fuels* **2013**, *27*, 23–30.
45. Jiang, J.-Y.; Yang, W.-P.; Cheng, Y.-P.; Liu, Z.-D.; Zhang, Q.; Zhao, K. Molecular structure characterization of middle-high rank coal via XRD, Raman and FTIR spectroscopy: Implications for coalification. *Fuel* **2019**, *239*, 559–572. [CrossRef]
46. Boral, P.; Varma, A.-K.; Maity, S. X-ray diffraction studies of some structurally modified Indian coals and their correlation with petrographic parameters. *Curr. Sci.* **2015**, *108*, 384–394.
47. Yan, J.; Bai, Z.; Zhao, H.; Bai, J.; Li, W. Inappropriateness of the standard method in sulfur form analysis of char from coal pyrolysis. *Energy Fuels* **2012**, *26*, 37–42. [CrossRef]
48. Fu, Y.; Liu, X.-F.; Ge, B.-Q.; Liu, Z.-H. Role of chemical structures in coalbed methane adsorption for anthracites and bituminous coals. *Adsorption* **2017**, *23*, 711–721. [CrossRef]
49. Matthews, M.-J.; Pimenta, M.-A.; Dresselhaus, G.; Dresselhaus, M.-S. Origin of dispersive effects of the Raman D band in carbon materials. *Phys. Rev. B* **1999**, *59*, 6585–6588. [CrossRef]

Article

Evolution of Biomarker Maturity Parameters and Feedback to the Pyrolysis Process for In Situ Conversion of Nongan Oil Shale in Songliao Basin

Hao Zeng [1], Wentong He [2,*], Lihong Yang [1], Jianzheng Su [1], Xianglong Meng [1], Xueqi Cen [1] and Wei Guo [3]

[1] Production Project Division, Sinopec Exploration & Production Research Institute, Beijing 100083, China; zenghao.syky@sinopec.com (H.Z.); yanglh.syky@sinopec.com (L.Y.); sujz.syky@sinopec.com (J.S.); mengxl.syky@sinopec.com (X.M.); cenxq.syky@sinopec.com (X.C.)
[2] College of Earth Sciences, Jilin University, Changchun 130021, China
[3] College of Construction Engineering, Jilin University, Changchun 130021, China; guowei6981@jlu.edu.cn
* Correspondence: hewentong0510@163.com

Abstract: In the oil shale in situ conversion project, it is urgent to solve the problem that the reaction degree of organic matter cannot be determined. The yield and composition of organic products in each stage of the oil shale pyrolysis reaction change regularly, so it is very important to master the process of the pyrolysis reaction and reservoir change for oil shale in situ conversion project. In the in situ conversion project, it is difficult to directly obtain cores through drilling for kerogen maturity testing, and the research on judging the reaction process of subsurface pyrolysis based on the maturity of oil products has not been carried out in-depth. The simulation experiments and geochemical analysis carried out in this study are based on the oil shale of the Nenjiang Formation in the Songliao Basin and the pyrolysis oil samples produced by the in situ conversion project. Additionally, this study aims to clarify the evolution characteristics of maturity parameters such as effective biomarker compounds during the evolution of oil shale pyrolysis hydrocarbon products and fit it with the kerogen maturity in the Nenjiang formation. The response relationship with the pyrolysis process of oil shale is established, and it lays a theoretical foundation for the efficient, economical and stable operation of oil shale in situ conversion projects.

Keywords: oil shale; in situ conversion project; biomarker; pyrolysis process; organic geochemistry

Citation: Zeng, H.; He, W.; Yang, L.; Su, J.; Meng, X.; Cen, X.; Guo, W. Evolution of Biomarker Maturity Parameters and Feedback to the Pyrolysis Process for In Situ Conversion of Nongan Oil Shale in Songliao Basin. *Energies* **2022**, *15*, 3715. https://doi.org/10.3390/en15103715

Academic Editor: Reza Rezaee

Received: 1 April 2022
Accepted: 10 May 2022
Published: 19 May 2022

Publisher's Note: MDPI stays neutral with regard to jurisdictional claims in published maps and institutional affiliations.

1. Introduction

Oil shale is an important strategic resource, which has the characteristics of wide distribution and huge resources and has always been widely concerned [1]. The development methods of converting oil shale into usable petroleum resources are divided into ground retorting technology and in situ conversion technology [2–6]. The application of ground retorting technology is mainly for oil shale occurring at shallow burial depth, which limits the number of available resources, and the technology itself has problems such as environmental pollution [7–12]. Oil shale in situ conversion technology can exploit deep oil shale and has the advantages of green environmental protection, small footprint and low cost [12–19]. In Songliao Basin, Jilin University built two pilot test bases for oil shale in situ conversion and successfully produced oil in 2016 and 2020, respectively. The pilot experiments accelerated the industrialization of oil shale, but the difficulty in monitoring the process of pyrolysis of oil shale organic matter has become more and more prominent in the project [20–23].

Studies showed that the production and composition of organic products in each stage of oil shale pyrolysis change regularly [1–5,21,22]. Therefore, it is very important to master the process of pyrolysis in oil shale in situ conversion project for efficient cracking and maximizing economic benefits [24–28]. However, the process of pyrolysis reaction can be analyzed by the maturity of oil shale kerogen [29–32]. This method necessitates coring in the pyrolysis reaction formation with additional drilling, which can destroy the subsurface thermal reaction environment and cause serious adverse effects on the mining industry. In addition, the current research on the maturity of oil shale organic matter fails to fully consider the complex subsurface conditions and artificial thermal reaction environment of oil shale and is not suitable for in situ conversion of oil shale [33–36]. Based on the experience of in situ conversion pilot experiments, the study proposes that oil and gas products are to be obtained through production wells during the operation of the project, and the maturity information of oil and gas products can be fed back to the reaction process of oil shale pyrolysis, but related research has not yet been conducted in-depth development.

In this study, the research on the dynamic evolution of the maturity parameters of oil shale pyrolysis hydrocarbon products was carried out, and the applicable maturity parameters were selected to accurately feedback the maturity of kerogen in the Nenjiang formation. Then, the response to the oil shale pyrolysis process relationship was established, and the real-time situation of the yield and composition of organic products in the underground kerogen pyrolysis reaction was informed. This study is a necessary way to dynamically monitor the oil shale in situ cracking reaction process, and it is also the key to ensuring the high efficiency, economy and stability of the oil shale in situ conversion project.

2. Samples and Methods

2.1. Research Methodology

In this study, core samples from oil shale exploration drilling in the Nongan (NA) experimental area were selected for this research, which included core logging, stratigraphic age division, and high-precision geochemical logging (total organic carbon). The preliminary results revealed the target sampling horizon of subsequent experiments. Then, the experiment began with high-temperature and high-pressure reactions on oil shale samples from the Nenjiang Formation in the reactor. Vitrinite reflectance tests, rock-eval pyrolysis and elemental analyses were performed on the semicoke products. The pyrolysis oil extracted from oil shale semicoke was quantitatively separated and analyzed by gas chromatography–mass spectrometry (GC-MS). The related tests performed on pyrolysis oil samples collected from the Nongan oil shale in situ production product include group component separation and GC-MS analysis (Figure 1).

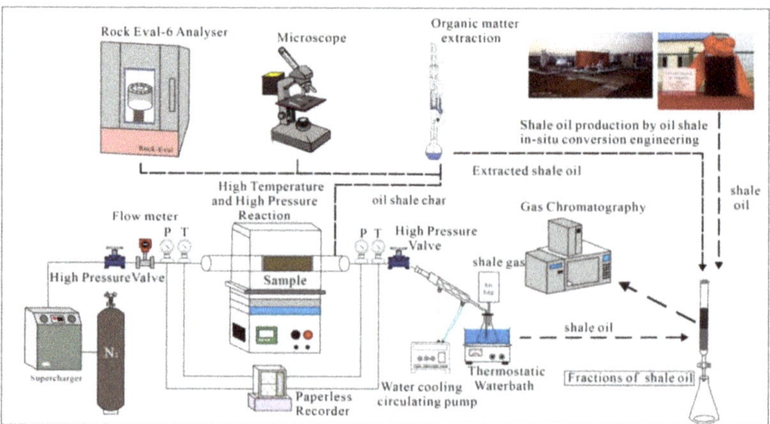

Figure 1. Experiment flow chart.

2.2. Selection of Oil Shale Samples

2.2.1. Stratigraphy of the Nenjiang Formation

The lower part of the Nenjiang Formation is a dark mudstone, and the upper part is composed of grey-green mudstone and glutenite deposits (Figure 2) [37,38]. These strata can be further divided into 5 members based on lithology. The first member of the Nenjiang Formation is composed of dark mudstone and thin oil shale. In the second member, the stable thick layer of oil shale in the whole area is at the bottom, and it transitions to grey mudstone and grey-green mudstone upward. The third to fifth members are eroded in the southeastern part of the basin, and the main parts are greyish green mudstone and greyish white sandstone, followed by a transition to purplish-red mudstone in the central part of the basin, and the Nenjiang group formation is the horizontal stratum of Shenzhen Lake phase in the research area [37–39] (Figure 3).

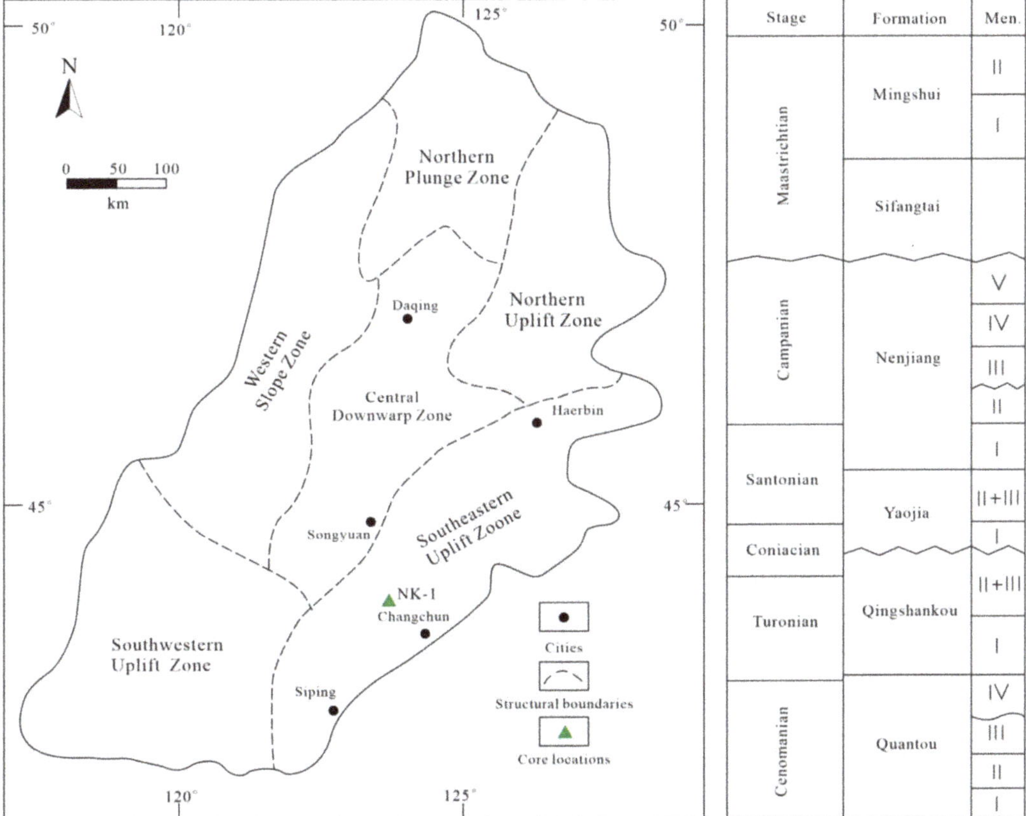

Figure 2. Geological map of the Songliao Basin.

The Nongan oil shale in situ conversion project is located 110 km north of Changchun. The total drilling depth is 135 m, and the drilling technique is full-core drilling. After a thin layer of Quaternary deposits, the strata that have been encountered are the Cretaceous upper Yaojia Formation (8 m) and the Nenjiang Formation (112 m). Based on core descriptions and lithofacies divisions of drilling cores in Nongan, we have a preliminary understanding of the overall features and sedimentary environment of the Nenjiang Formation and provide an important basis for selecting appropriate samples for testing and analysis. According to core records during drilling in Nongan, Quaternary sands are located 10 m from the surface to the ground, and no sampling has been carried out in this interval (Figure 3). Sampling was performed every 1 m for the cores until 133 m of depth. The main core samples from 66 to 74 m from the Nongan oil shale in situ conversion test area. The drilling fluid was washed off in distilled water and then dried at a constant temperature of 60 °C for 12 h [22,23].

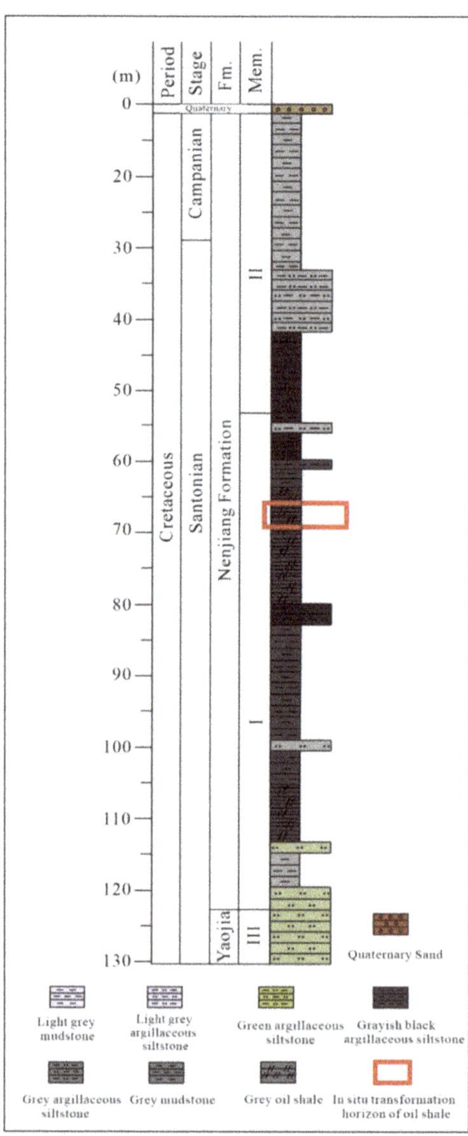

Figure 3. Stratigraphic log of the Nongan oil shale.

2.2.2. Sampling of Shale Oil Samples for In Situ Conversion Project

As shown in Figure 4, the process mainly involves drilling thermal injection wells and production wells in the ground facing the oil shale layer and performing conventional work, such as casing running and cementing after drilling. Then, the conventional hydraulic fracturing method is used for fracking the heat injection wells so that the heat injection wells and the production wells are connected through the fractures formed by fracturing. After the connection between the heat injection well and the production well is established, high-temperature nitrogen (450 °C) is introduced into the heat injection well on the ground surface, and nitrogen flows to the production well through the fractures formed by hydraulic fracturing. During the flow of high-temperature nitrogen, nitrogen contacts the oil shale layer to conduct convective heat conduction, thereby heating the oil shale layer [22,23], and this convection keeps it warm for the experiment. After heating for a period of time, the produced oil and gas and nitrogen flow to the production well and are recovered in the production well, and the generated oil, gas and mixed vapor and gas are sent to the surface. Two-stage cooling and separation are carried out to separate pyrolysis oil, pyrolysis gas and water. The heat carrier, nitrogen, used in the process of heating the oil shale layer, is obtained by separating the air from the pressure swing adsorption (PSA) air separation nitrogen generator and is heated to the designed temperature by ground heating equipment.

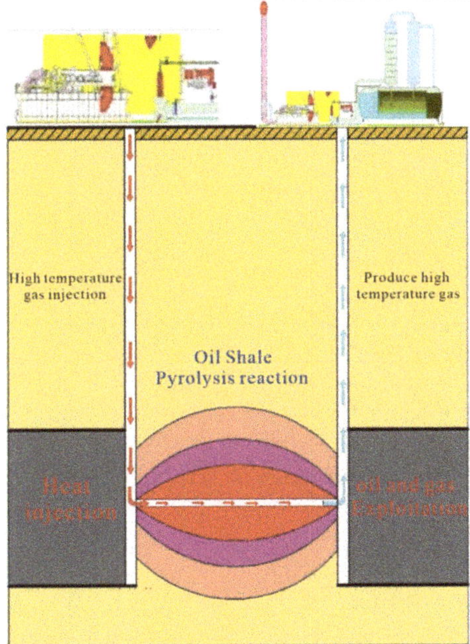

Figure 4. Scheme for the Nongan oil shale in situ conversion technology.

2.3. Experimental Methods

(1) High-pressure pyrolysis experiment

This experiment was carried out using a high-pressure nitrogen pyrolysis oil shale test device. The oil shale core samples were subjected to high-pressure heating treatment, and the samples were dried and put into the reactor, the reactor was filled with nitrogen and the pressure in the reactor was increased to the lithostatic pressure of the formation; the nitrogen was turned off, the reactor was heated up and the heating rate was 10 °C/min. The end-point temperatures of different samples were selected as 150 °C, 200 °C, 250 °C, 300 °C, 350 °C, 375 °C, 400 °C, 425 °C, 450 °C, 475 °C and 520 °C. After reaching the end temperature, keep the temperature for 6 h, then keep the pressure in the kettle unchanged for 2 h, open the inlet and outlet of the reaction kettle and pass nitrogen for 2 h, and the N_2 flow rate is 1 L/min. At the outlet, the cracked oil product was received through a water-cooled tube, and the cracked gas product was collected by an air bag. After reaching the termination time, the temperature of the reactor was lowered, the pressure was reduced, and the oil shale semicoke product in the reactor was taken out and stored in a sealed [16–18,27].

(2)   Rock pyrolysis evaluation experiment with oil shale semicoke

Different oil shale semicoke samples were crushed into size 200 mesh for pyrolysis evaluation. A *Rock Eval-6 analyzer* developed by the French Academy of Petroleum Sciences was used for the experimental analysis of rock pyrolysis. S1, S2, S3 and T_{max} were measured [40–42].

(3)   Total organic carbon

The total organic carbon (TOC) of all oil shale semicoke samples was determined by *vario PYRO cube* elemental analyzer produced by Elementar, UK, which conformed to the GBT-19145-2003 standard [43].

(4)   Separation of pyrolysis oil components

In the separation and analysis of group components in organic matter extracts of oil shale semicoke, the selected oil shale semicoke samples were extracted with a dichloromethane and methanol (93:7) mixture by an automatic Soxhlet apparatus. Then, sulfur was removed from the organic extracts of oil shale semicoke and other pyrolysis oil samples by active copper, and asphaltene was precipitated from n-hexane dichloromethane (DCM) solution (80:1). After centrifugation, the extracted organic matter was separated into saturated hydrocarbons and NSO by liquid chromatography. The experiment was carried out according to the SY/T5119-2008 standard [35,36,41–43].

(5)   Gas chromatography-mass spectrometry

After obtaining the saturated hydrocarbon components separated from the organic group components, the saturated components were dissolved in petroleum ether. The relative abundance of related biomarker compounds in saturated hydrocarbons and aromatic hydrocarbons was calculated from the manually integrated peak areas of the relevant ion chromatograms. Compared with the indicated range and the evolution law of natural hydrocarbon generation, the analysis process follows the GBT-30431-2013 standard [35,36,39].

(6)   Organic lithofacies observations and vitrinite reflectance (Ro) test

Petrological observations were performed using a high-power optical microscope (*Zeiss axioscope A1*) equipped with a photometric system with fluorescent lamps. According to ASTM standard d7708-14 (2014), the average random vitrinite reflectance (%), R_o was carried out [39,41].

3. Results and Discussion

3.1. Evolutionary Characteristics of Organic Matter at Different Pyrolysis Temperatures

3.1.1. Analysis of the High-Pressure Pyrolysis of Oil Shale

In the pyrolysis experiments of Nongan oil shale samples at different temperatures, all reaction temperatures can be divided into 20 °C to 300 °C, 300–475 °C and 475–520 °C. At 20–300 °C, the yield of pyrolysis oil is very low, reaching 0.76%, the total organic carbon content of semicoke decreases slightly, from 8.09% to 6.22%, and the amount of gas produced also increases slightly. At 300–475 °C, the main product of the emission was thermal solution oil, and the output increased significantly, up to 6.37%, and the output of water at the same time increased [44–47]. At 475–520 °C, the output of oil and water only increased slightly [46–50] (Figure 5, Tables 1 and 2).

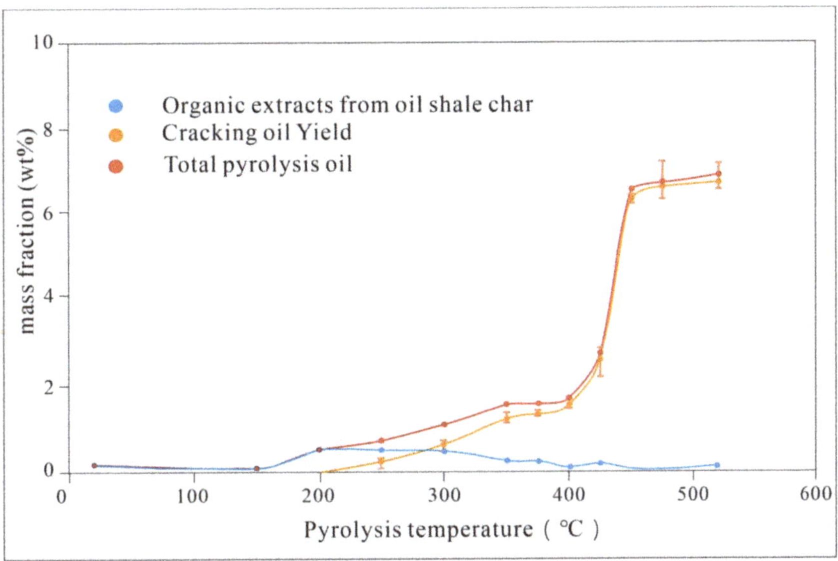

Figure 5. Nongan pyrolysis oil yield.

Table 1. Results of pyrolysis products at different temperatures.

Pyrolysis Temperature (°C)	Oil Yield (wt%)	Moisture Content (wt%)	Semicoke Yield (wt%)	Gas and Loss (wt%)
150	0	0.78	98.83	0.39
200	0	0.96	98.39	0.65
250	0.16	1.73	97.68	0.43
300	0.76	2.7	96.16	0.38
350	1.25	3.46	93.89	1.4
375	1.53	3.86	93.43	1.18
400	1.76	4.36	92.91	0.97
425	2.55	4.83	90.88	1.74
450	6.24	5.31	87.02	1.43
475	6.37	6.56	82.35	4.72
520	6.58	6.55	82.46	4.41

Table 2. TOC, R_o and T_{max} values of Nongan samples at different pyrolysis temperatures.

Pyrolysis Temperature, °C	TOC, wt%	R_o, %	Tmax, °C
original sample	8.09	0.266	430
150	7.46	0.284	433
200	6.84	0.455	437
250	6.59	0.584	439
300	6.22	0.643	440
350	4.55	0.846	440
375	3.78	0.864	441

Table 2. *Cont.*

Pyrolysis Temperature, °C	TOC, wt%	R_o, %	Tmax, °C
400	3.41	0.887	443
425	2.94	0.946	445
450	2.88	1.143	566
475	2.65	1.257	590
520	2.63	1.376	600

3.1.2. TOC Content

The average TOC content of the oil shale in this study is 8.09 wt% [35,36]. The TOC content of the 150 °C sample in each heating process is slightly lower than that of the original sample, which may be due to the fact that the water in the oil shale was dried in the previous test [33,34]. The TOC content of the pyrolyzed samples decreased slightly at 150–300 °C, and the S1 value gradually increased. It is because free hydrocarbon expulsion leads to a reduction in TOC [35,36,44,46]. In the 300–450 °C stage, TOC decreased significantly, corresponding to a large amount of oil generated from oil shale. In the 450–520 °C stage, TOC decreased slowly and tended to be stable (Table 2).

3.1.3. The Evolutionary Trend of R_o% and T_{max}

Vitrinite reflectance is used to judge the kerogen thermal maturity and is generally divided into five stages according to Ro: <0.5% immature, 0.5–0.7 early mature stage of oil generation, 0.7–1.0 mature middle stage of oil generation, 1.0–1.3 late mature stage of oil generation, >1.3 main gas generation stage [39–41,51–53]. In this experiment, the Ro value of agricultural dried samples was between 0.266% and 1.376%, indicating that the kerogen maturity from immature to mature (Table 2).

When T_{max} is lower than 435 °C, the organic matter is in the immature stage. When T_{max} is 435–440 °C, the kerogen is low mature. When T_{max} is 440–450 °C, kerogen is mature. When T_{max} is higher than 450 °C, it is in the high-maturity stage [54–56]. The T_{max} of Nongan oil shale is 430 °C, which indicates that the kerogen is immature [57]. The results show that the Tmax of the semicoke sample increases with the increase in the pyrolysis temperature. The T_{max} of samples below 450 °C varies between 430 and 445 °C, indicating the gradual maturity of kerogen, while the T_{max} from 450 °C to 520 °C is 566–600 °C. According to the spectrum, the peak value of S2 may be too low, while the instrument system calculates the maximum peak value, which may be caused by rapid pyrolysis. Therefore, when the pyrolysis temperature reaches 450 °C, it may be inaccurate to judge oil shale maturity by T_{max} [57].

3.1.4. Changes in Pyrolysis Oil Composition

Studies showed that the production and composition of organic products in each stage of oil shale pyrolysis change regularly, so it is very important to master the composition changes in pyrolysis oil in oil shale in situ conversion project for efficient pyrolysis [35,36,41–43]. In this study, it was found that the components of pyrolysis oil in the organic matter extract of oil shale semicoke showed regular changes. When the temperature is lower than 350 °C, the proportion of non-hydrocarbon components to asphalt is the highest, but when the pyrolysis temperature exceeds 375 °C, the value of the ratio is sharply reduced to 0.17. Additionally, when the pyrolysis temperature exceeds 375 °C, the oil yield starts to rise (Figure 6). Saturated hydrocarbons were the main component, followed by aromatic hydrocarbons, while NSO and asphaltene were the main components. A calculation of the composition of total pyrolysis oil from oil shale revealed that asphaltene accounted for the largest proportion in the pyrolysis oil at the initial stage of the pyrolysis reaction (room temperature to 250 °C, reaching 0.626%), while saturated hydrocarbons and aromatic hydrocarbons accounted for a small proportion (0.25% and 0.124%, respectively). Within

250–375 °C, the amount of pyrolysis oil increased gradually (0.16–1.35%). When the cracking temperature reaches 375 °C, the cracked oil is mainly saturated hydrocarbon, and the proportion of each component is relatively stable (Figure 6, Table 3).

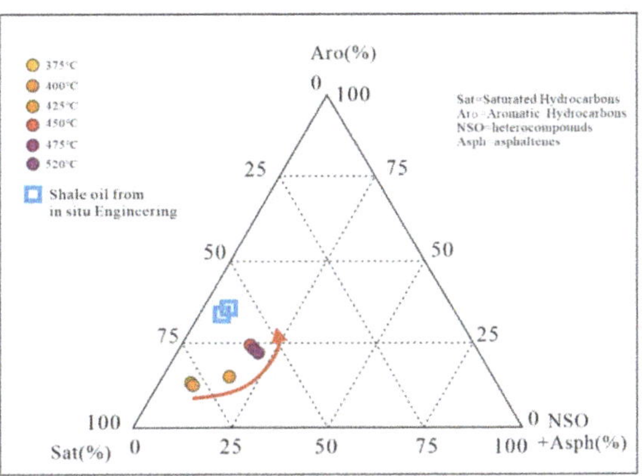

Figure 6. Variation in the pyrolysis oil composition.

3.2. Biomarker Maturity Parameter of Pyrolysis Oil

When the pyrolysis temperature is lower than 375 °C, a small amount of pyrolysis oil is sealed in the pores of oil shale because the pores in oil shale are nanopores [58–61]. In this study, oil shale semicoke at all temperature stages was pulverized and subjected to extraction of organics. The semicoke extracts and the pyrolysis oil collected from the simulation experiment were tested for group composition and analysed by GC-MS.

The carbon number distribution ranged from C_{11} to C_{33}, and the main carbon peak decreased with increasing temperature from C_{23} to C_{16}. Therefore, the main carbon peak also moved forward (Figure 7). The carbon preference index (CPI) (an odd and even dominance index) values of Nongan samples decreased with increasing pyrolysis temperature, gradually decreased to 1.025 at 400 °C, and reached relatively stable values, all at approximately 1.0 [60].

N-alkanes in organic matter rich in terrigenous clastic rock series have obvious odd carbon predominance. It is generally believed that these n-alkanes are derived from the wax of higher plants, are directly synthesized by plants or are from acid alcohol esters, even with carbon in early diagenesis [48,60,61]. The CPI represents the molecular ratio of odd carbon to even carbon of n-alkanes in the range of C_{25}–C_{33} [62].

$$\text{CPI} = 1/2 \left[\frac{\sum(C25 \sim C33)}{\sum(C24 \sim C32)} + \frac{\sum(C25 \sim C33)}{\sum(C26 \sim C34)} \right] \tag{1}$$

As organic matter matured, the carbon preference gradually disappeared and approached 1. The CPI can identify organic matter maturity; the index cannot differentiate between thermal evolution stages.

Table 3. Geochemical parameters of Nongan oil shale samples at different pyrolysis stages.

Area-Pyrolysis Temperature	Sample Properties	Oil Content	Pyrolysis Oil Group Components			N-Alkanes and Isoprenoids									Terpanes (m/z = 191)										Steranes and Rearranged Steranes (m/z = 217)			
			Saturated Hydrocarbon %	Aromatic Hydrocarbon %	Non Hydrocarbon + Asphaltene%	CPI				Ts/(Ts + Tm)				C_{32} 22S/(22S + 22R)				C_{29} 20S/(20S + 20R)				C_{29} ββ/(αα + ββ)			
						1	2	3	Average	1	2	3	Average	1	2	3	Average	1	2	3	Average	1	2	3	Average
NA-0	Oil shale semi coke	0.16	0.06	0.15	0.79	1.59	1.95	1.86	1.8	0.15	0.09	0.11	0.12	0.34	0.34	0.35	0.34	0.18	0.18	0.14	0.17	0.08	0.07	0.07	0.07
NA-150		0.11	0.08	0.02	0.9	1.54	1.73	1.78	1.68	0.15	0.15	0.17	0.15	0.35	0.34	0.36	0.35	0.17	0.18	0.16	0.17	0.06	0.11	0.08	0.08
NA-200		0.46	0.16	0.22	0.62	1.43	1.37	1.46	1.42	0.17	0.2	0.17	0.18	0.37	0.38	0.38	0.37	0.19	0.16	0.18	0.18	0.16	0.16	0.22	0.18
NA-250		0.47	0.25	0.124	0.626	1.31	1.29	1.38	1.32	0.26	0.24	0.27	0.26	0.37	0.38	0.39	0.38	0.22	0.22	0.21	0.21	0.19	0.21	0.21	0.2
NA-300		0.55	0.24	0.14	0.62	1.28	1.13	1.19	1.2	0.28	0.31	0.29	0.29	0.38	0.38	0.39	0.38	0.23	0.24	0.26	0.24	0.23	0.22	0.25	0.23
NA-350		0.39	0.26	0.23	0.51	1.16	1.06	1.03	1.08	0.3	0.29	0.31	0.3	0.41	0.41	0.43	0.42	0.27	0.27	0.23	0.26	0.25	0.24	0.27	0.25
NA-375	Pyrolysis oil		0.63	0.2	0.17	0.94	0.98	1.16	1.03	0.35	0.32	0.32	0.33	0.44	0.43	0.46	0.44	0.29	0.28	0.31	0.29	0.28	0.27	0.31	0.28
NA-400			0.63	0.28	0.09	0.85	0.77	1.05	0.89	0.44	0.43	0.43	0.43	0.45	0.44	0.46	0.45	0.32	0.34	0.31	0.32	0.29	0.34	0.34	0.32
NA-425			0.63	0.25	0.12	0.94	1.13	0.99	1.02	0.48	0.45	0.47	0.46	0.52	0.51	0.53	0.52	0.32	0.34	0.34	0.33	0.36	0.39	0.36	0.37
NA-450			0.66	0.21	0.13	1.08	1.05	0.91	1.02	0.57	0.55	0.54	0.55	0.53	0.55	0.56	0.55	0.42	0.41	0.44	0.42	0.4	0.4	0.37	0.39
NA-475			0.67	0.22	0.11	1.07	0.97	0.86	0.97	0.62	0.59	0.63	0.61	0.57	0.57	0.57	0.57	0.45	0.42	0.42	0.43	0.41	0.43	0.43	0.42
NA-520			0.67	0.14	0.19	0.95	1.08	1.16	1.06	0.63	0.63	0.61	0.62	0.58	0.58	0.59	0.58	0.49	0.44	0.43	0.45	0.5	0.46	0.53	0.5

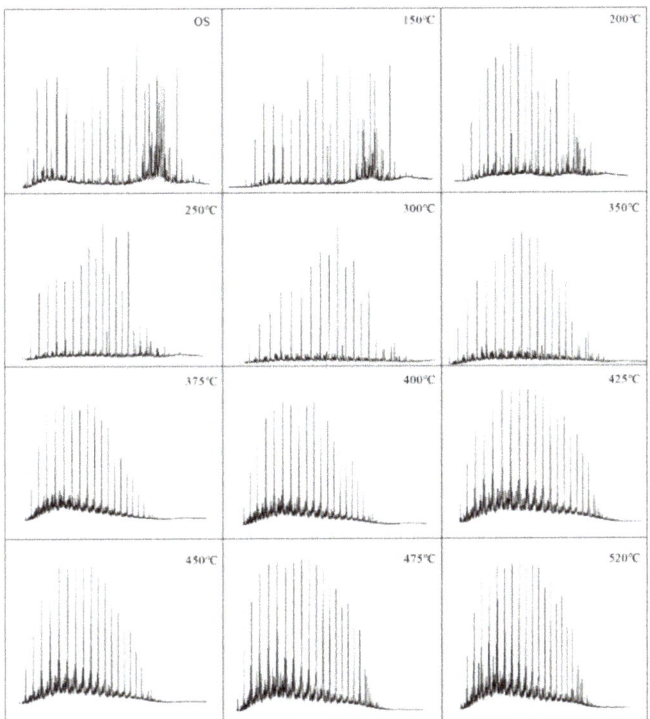

Figure 7. TIC diagrams in each stage.

Biomarkers are mainly used to study the source, maturity and paleosedimentary environment of organic matter in sediments [36,61,62]. In this study, the composition and biomarker characteristics of n-alkanes, isoprenoids, steranes, hopanes and other organic matter in the semicoke and pyrolysis oil of Nenjiang Formation oil shale in Nongan were investigated (Figures 8 and 9; Table 3) and selected to analyze the change in thermal.

A previous study found that the maturity parameter based on the relative stability of C_{27} hopanes ranged from immature to mature, but it was strongly dependent on the source [36]. The stability of C_{27} 17α-trinorhopane (Tm) is better than that of C_{27} 18α-trinorhopane II (Ts). As organic matter matures, the Ts/(Ts + Tm) value increases. Ts/(Ts + Tm) is a relatively reliable maturity index for evaluating samples from the same source rock location [62]. It was found that the Ts/(Ts + Tm) ratio increases with the increase in pyrolysis temperature. It seems that the value from the ratio increases gradually from 20 to 425 °C [61,62] (Figures 8 and 9, Table 3).

The isomerization index 22S/(22S + SSR) of hopane is feasible to evaluate the maturity from the immature to low maturity [36]. The isomerization of C_{31}–C_{35} 17α-hopane on C-22 can be used to evaluate the maturity of crude oil or asphalt. The biogenic hopane precursor has a 22R configuration, and it gradually transforms into a mixture of 22R and 22S isomers. After the oil generation stage reaches equilibrium, the 22S/(22S + 22R) value remains unchanged, so it is impossible to obtain further maturity information from it [36,61–64]. The change in hopane C_{32} 22S/(22S + 22R) in Nongan samples in the range of 20–520 °C can be divided into two stages, which increase from 0.341 to 0.447 within 0–400 °C. The hopane C_{32} 22S/(22S + 22R) in samples increases suddenly to 0.52 at 425°C and slowly from 0.52 to 0.584 within 425–520 °C (Figures 8 and 9).

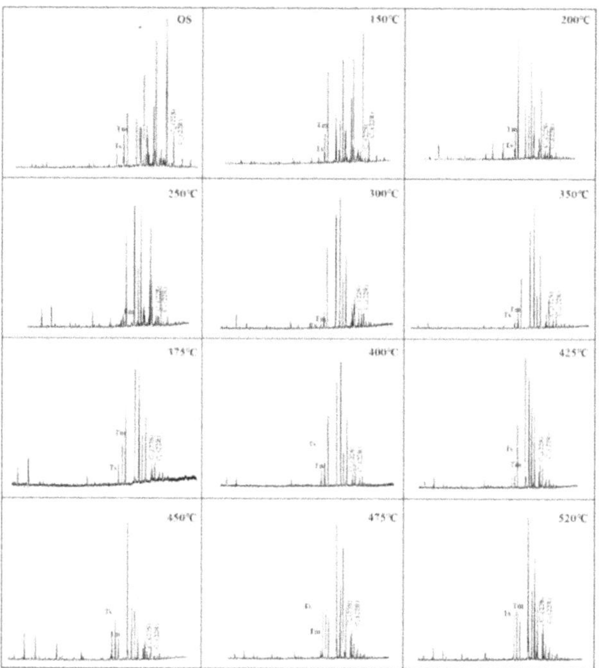

Figure 8. MS at different temperatures with mass chromatogram $m/z = 191$.

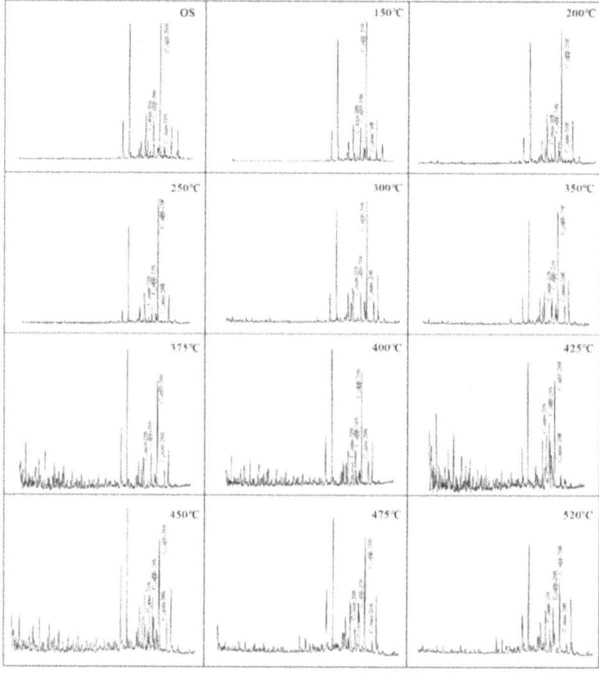

Figure 9. MS at different temperatures with mass chromatogram $m/z = 217$.

In this study, the CPI and hopane 22S/(22S + 22R) have mutations when the temperature reaches 350–400 °C, indicating the kerogen has reached the maturity stage. The value of Ts/(Ts + Tm) can indicate that the organic matter has never matured to the peak of the oil generation window.

The C_{29} $\alpha\alpha\alpha$20R of sterane is the biological configuration of the sterane precursor that exists only in living organisms, while C_{29} $\alpha\alpha\alpha$ 20S, C_{29} $\alpha\beta\beta$ 20R and C_{29} $\alpha\beta\beta$ 20S have stable chemical structures [36,65]. As the kerogen matures, the C-20 bond of the C_{29} $\alpha\alpha\alpha$20R configuration is isomerized. The bond converts the C_{29} $\alpha\alpha\alpha$20R configuration to a C_{29} $\alpha\alpha\alpha$20S configuration and forms a mixture of 20R and 20S, which increases the ratio of 20S/20 (R + S) of sterane from 0 to 0.5 [36,65,66]. The isomerization of 20 s and 20 R regular steranes at the C-14 and C-17 sites leads to the formation of isomers $\beta\beta/(\alpha\alpha + \beta\beta)$. The ratio increases from nearly 0 to approximately 0.7 with increasing thermal maturity [67–69]. This study found that with the increase in pyrolysis temperature, the evolution of sterane C_{29} 20S/20(R + S) and $\beta\beta/(\alpha\alpha + \beta\beta)$ can be subdivided into three stages. In the immature stage of the non-oil shale semicoke, the $\beta\beta/(\alpha\alpha + \beta\beta)$ ratio sample range from 0.165 to 0.214 and from 0.071 to 0.201, respectively, when the pyrolysis temperature is 0–250 °C. The temperature increases gradually and rapidly in the range of 250–450 °C leading to $\beta\beta/(\alpha\alpha + \beta\beta)$ values from 0.243 to 0.424 and from 0.233 to 0.390, respectively. As the pyrolysis temperature continues to increase to 520 °C, only C_{29} 20S/20(R + S) and $\beta\beta/(\alpha\alpha + \beta\beta)$ increase slowly, and this also shows that the kerogen is in the mature stage (Figures 8 and 9, Table 3).

3.3. Feedback of Maturity Parameters on the Progress of the Pyrolysis Reaction

The production and composition of organic products in each stage of oil shale pyrolysis change regularly. Therefore, it is very important to master the process of underground pyrolysis reaction in oil shale in situ conversion project for efficient cracking and maximization of economic benefits. Although the process of pyrolysis reaction can be analyzed by the maturity of oil shale kerogen, the existing theoretical research on oil shale organic matter maturity can only analyze the maturity of oil shale core samples Ro and T_{max} [29–34,36]. However, drilling and coring in the process of in situ conversion of oil shale not only greatly increases the cost of the project. Moreover, it destroys the underground thermal reaction environment and causes serious disadvantages to the exploitation. In order to realize the dynamic feedback of the in situ pyrolysis process of oil shale, this time, combined with the experience of in situ conversion pilot experiments, it is proposed that oil and gas products can be obtained through production wells during project operation. By testing the maturity information of biomarker compounds of oil and gas products, the reaction process of subsurface oil shale pyrolysis can be feedback.

As shown in Figure 10, as organic matter decreased, six different maturity indicators changed differently with thermal maturity. The T_{max} can define the reaction process of oil shale with pyrolysis temperatures at a temperature lower than 425 °C, but the small change in Tmax can be used only as an indicator to judge the maturity stage [29], and it is difficult to give feedback on the kerogen pyrolysis process. In the simulation experiment, the R_o and biomarker compounds at various temperature stages change regularly with the increase in the pyrolysis temperature, especially for the stage of rapid reduction in TOC; these thermal maturity parameters have obvious changes.

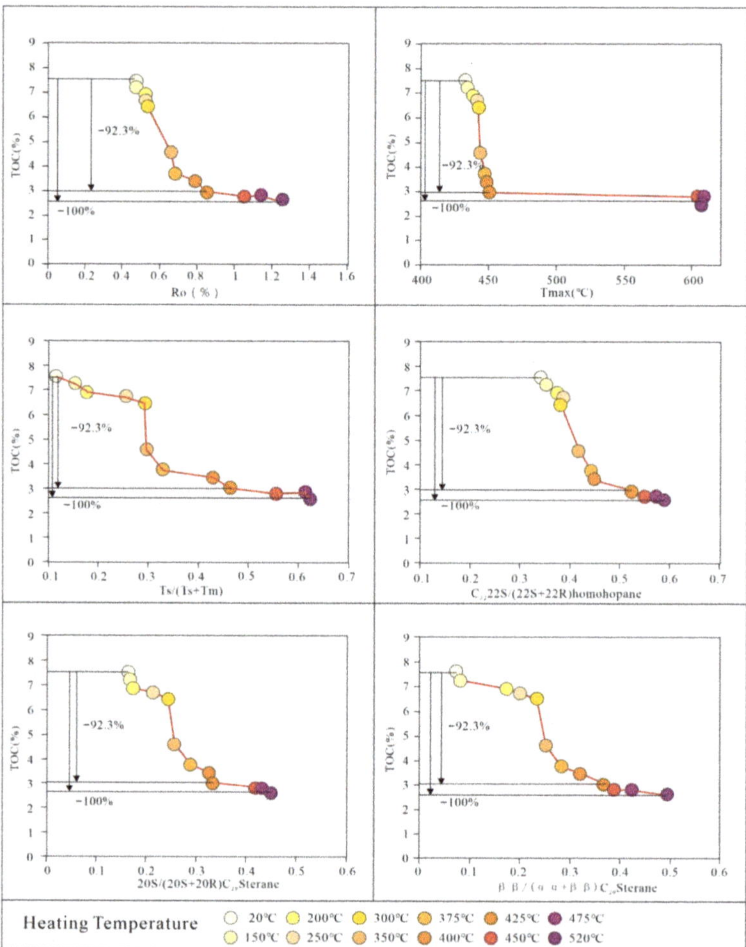

Figure 10. The changes in the six maturities and TOC contents of Nongan samples at different pyrolysis temperatures.

3.4. Analysis of Pyrolysis Process of Nong'an Oil Shale In Situ Conversion Project

The pyrolysis oil samples obtained on-site were geochemically tested, including group component separation and GC-MS analysis. The test results of the on-site cracking oil group components revealed that the sum of saturated hydrocarbons and aromatic hydrocarbons accounted for approximately 90%, non-hydrocarbon accounted for 10%, and asphaltene accounted for less than 1% (Table 4). The high-temperature and high-pressure simulation experiment of Nongan oil shale revealed that the proportion of non-hydrocarbon to asphaltene is more than 10% (Figure 6), but the sum of non-hydrocarbons and asphaltene in the group components of pyrolysis oil in the in situ conversion project is lower than 10%. When combined with field project analysis, it is necessary to pass through 66 m of strata from the thermal reaction strata at the bottom of the well to the wellhead of the mining well; this separation may have been due to the large molecular weight of asphaltene and non-hydrocarbon during the migration from the bottom of the well to the surface.

Table 4. Composition of pyrolysis oil in the Nongan in situ conversion project.

Sample	Asphaltene, %	Saturated Hydrocarbon, %	Aromatic Hydrocarbon, %	Non-Hydrocarbon, %
NA 1	0.51	60.26	30.42	8.80
NA 2	0.82	54.70	35.21	9.25

The two collected pyrolysis oils were subjected to group composition tests and GC-MS analysis together, with three parallel samples for each test and six samples in total. The results showed that the carbon number distribution range was between C_{11} and C_{33}, and the main peak carbon was between C_{19} and C_{22}. The CPI (the odd-even advantage index) was between 0.95 and 1.07, and the change was not significant. The average values of the two samples were 1.01 and 1.05, respectively, indicating that the organic matter in underground oil shale was mature (Figure 11).

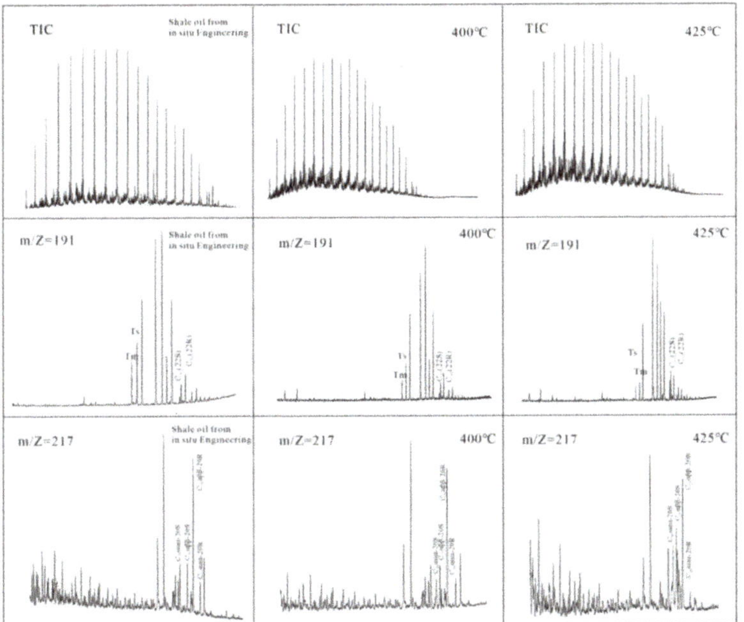

Figure 11. GC-MS analysis of field oil shale pyrolysis oil and 400–425 °C simulated pyrolysis oil samples.

The homohopane isomerization index 22S/(22S + SSR) has high feasibility for evaluating the maturity from the immature to early oil generation stages [65,66]. According to the identification integral of the mass spectrum at m/z = 191, the ratio of the homohopane isomerization index C_{32} 22S/(22S + SSR) of on-site pyrolysis oil samples remained between 0.41 and 0.49, and the average values were 0.45 and 0.47, respectively, reaching the hydrocarbon generation threshold [67,68]. The Ts/(Ts + Tm) ratios of the samples were maintained between 0.41 and 0.46, and the average values of the two samples were 0.43 and 0.44, respectively, which had reached the mature stage (Figures 11–13, Table 5). Additionally, this study found that sterane C_{29} 20S/20 (R + S) and $\beta\beta/(\alpha\alpha + \beta\beta)$ were 0.29–0.34 and 0.28–0.34, respectively, indicating that organic matter had entered the mature stage [62–66] (Table 5).

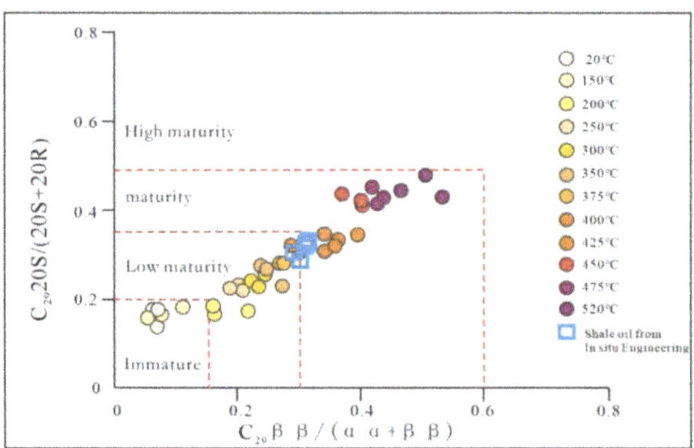

Figure 12. Analysis of C_{29} sterane $20S/(20S + 20R)$ and C_{29} sterane $\beta\beta/(\alpha\alpha + \beta\beta)$ on maturity.

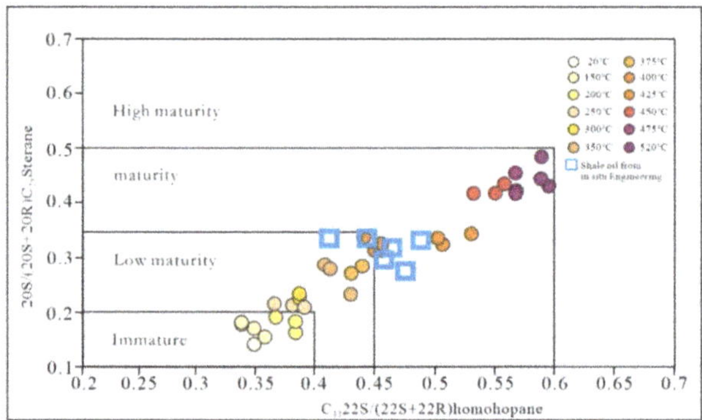

Figure 13. Analysis of C_{29} sterane $20S/(20S + 20R)$ and C_{32} hopane $22S/(22S + 22R)$ on maturity.

Table 5. Geochemical parameters of field pyrolysis oil.

Sample	N-Alkanes and Isoprenoids				Terpanes (*m/z* = 191)								Steranes and Rearranged Steranes (*m/z* = 217)							
	CPI				Ts/(Ts + Tm)				C32 22S/(22S + 22R)				C29 20S/(20S + 20R)				C29β β/(αα + β β)			
	1	2	3	Average	1	2	3	Average	1	2	3	Average	1	2	3	Average	1	2	3	Average
NA-1	1.07	1.04	1.05	1.05	0.46	0.41	0.46	0.44	0.46	0.47	0.41	0.45	0.31	0.33	0.33	0.32	0.28	0.31	0.33	0.31
NA-2	1.02	0.95	1.06	1.01	0.42	0.46	0.41	0.43	0.48	0.44	0.49	0.47	0.29	0.34	0.34	0.32	0.30	0.32	0.34	0.32

By comparing the maturity parameters of biomarkers in the simulation results of Nongan oil shale, the experimental field data are similar to the simulation data at 400 and 425 °C. The projection plots in Figures 12 and 13 also showed that the field cracking oil projection is uniform and concentrated for the maturity indicator, including the immature to overmature sterane C_{29} $20S/20(R + S)$ and $\beta\beta/(\alpha\alpha + \beta\beta)$ projection plots. In the isomerization diagram, the input point of hopanoids is relatively dispersed because the indicating ability is only to the hydrocarbon generation threshold, but the lateral concentration is also in the range of 400–425 °C. The target heating formation is in the maturity stage [62–66] (Table 5, Figures 11–13).

3.5. Calculation and Application Feasibility of the In Situ Conversion Degree of Nongan Oil Shale

Organic geochemistry of pyrolysis oil samples obtained in the in situ conversion project of the Nongan oil shale was carried out. In summary, the organic matter reaction process of subsurface oil shale in the in situ conversion project of Nongan oil shale is equivalent to the reaction of Nongan oil shale in the high-temperature and high-pressure simulation laboratory in the range of 400–425 °C. The TOC test of the Nongan subsurface oil shale reaction layer shows that the peak period of hydrocarbon generation is in the range of 400–425 °C, which also corresponds to the oil yield in the high-pressure thermal weightlessness experiment and high-pressure distillation experiment. Therefore, the pyrolysis oil is equivalent to the 400–425 °C range of the pyrolysis test samples and should be the best temperature for heating the oil shale formation to the oil production peak [35,36]. The previous paper on the area also points out that 425 °C is the key heating control point, and there is a turning point within 425–450 °C [39]. If heating continues, the input energy is wasted, and the economic benefits are reduced. Therefore, the research on the pyrolysis oil of the mining well in the in situ conversion project of Nongan oil shale reveals that the heating technology used in site construction corresponds to the original purpose of project experiments and has economic value.

In this paper, the analysis of biomarker compounds in pyrolysis oil is essential for the in situ mining of oil shale in other areas. Early resource evaluation should be undertaken, as well as tracking evaluation during the project and evaluation of surplus resource potential at the end of the project. It is necessary to conduct a detailed simulation study on the target horizon in early resource evaluation by comprehensive geochemical research. The geochemical characteristics of hydrocarbon gas and cracked oil samples discharged from mining wells were analyzed during project construction, and the reaction process of organic matter in subsurface oil shale was estimated. The total organic carbon reduction, as well as the oil and gas production rates, were calculated to aid in decision-making for the temperature increase and project process.

4. Conclusions

The results of high-pressure heating experiments show that when the pyrolysis temperature is 300–475 °C, the main emission products are pyrolysis oil, and the TOC of the semicoke sample decreases from 8.06% to 2.65%, and the yield is significantly improved. A transition point appeared between 425 °C and 450 °C, and the TOC of the oil shale above the temperature of the transition point decreased slowly. This temperature node is of great significance for the selection and control of the subsurface temperature during the original production of oil shale.

The research on the parameters of pyrolysis oil biomarker compounds found that the four parameters Ts/(Ts + Tm), C_{32} 22S/(22S + 22R), C_{29} 20S/(20S + 20R), and C_{29} $\beta\beta/(\alpha\alpha + \beta\beta)$ also have a good feedback effect on the progress of Nongan oil shale pyrolysis reaction.

A comprehensive analysis of the current in situ conversion project of Nongan oil shale shows that the pyrolysis stage of the Nongan oil shale in situ conversion project is equivalent to the 400–425 °C simulated experiment, which is at the peak of the oil window. The heating process that is currently used corresponds to the experimental purpose and has economic value.

If the evolutionary trend of maturity parameters of pyrolysis oil biomarker compounds during in situ conversion of oil shale is studied, and accurate feedback of the maturity of kerogen in the formation is obtained, it can reveal its response principle to the organic matter pyrolysis process. The study is of great significance to the efficient, economic and stable development of oil shale in situ conversion projects.

Author Contributions: Conceptualization, methodology, validation, formal analysis, investigation, resources, H.Z., L.Y., J.S., X.M. and X.C.; data curation, writing—original draft preparation, writing—review and editing, visualization, supervision, project administration, and funding acquisition, W.H. and W.G. All authors have read and agreed to the published version of the manuscript.

Funding: This work was supported by the Sinopec "Key Technologies for In-situ Conversion and Exploitation of Oil Shale" (Grant No. P20066), and the National Key R&D Program of China (Grant No. 2019YFA0705502, Grant No. 2019YFA0705503), and the National Natural Science Fund Project of China (Grant No. 41790453, 4210020395 and 42002153), and the China Postdoctoral Science Foundation (Grant No. 2021M700053).

Acknowledgments: Thanks to my sweetie Chenyang Wu for her contribution and assistance with this study.

Conflicts of Interest: The authors declare no conflict of interest.

References

1. Dyni, J.R. *Geology and Resources of Some World Oil Shale Deposits*; US Geological Survey Scientific Investigations: Reston, VA, USA, 2006.
2. Kok, M.V. Oil shale resources in Turkey. *Oil Shale* **2006**, *23*, 209–210.
3. Taciuk, W. Does oil shale have a significant future? *Oil Shale* **2013**, *30*, 1–5. [CrossRef]
4. Wang, S.; Jiang, X.M.; Han, X.X.; Tong, J.H. Investigation of Chinese oil shale resources comprehensive utilization performance. *Energy* **2012**, *42*, 224–232. [CrossRef]
5. Jiang, X.M.; Han, X.X.; Cui, Z.G. New technology for the comprehensive utilization of Chinese oil shale resources. *Energy* **2007**, *32*, 772–777. [CrossRef]
6. Han, X.X.; Niu, M.T.; Jiang, X.M. Combined fluidized bed retorting and circulating fluidized bed combustion system of oil shale: 2. Energy and economic analysis. *Energy* **2014**, *74*, 788–794. [CrossRef]
7. Altun, N.E.; Hicyilmaz, C.; Wang, J.Y.; Baggi, A.S.; Kok, M.V. Oil shales in the world and Turkey; reserves, current situation and future prospects: A review. *Oil Shale* **2006**, *23*, 211–227.
8. Pan, Y.; Zhang, X.M.; Liu, S.H.; Yang, S.C.; Ren, N. A review on technologies for oil shale surface retort. *J. Chem. Soc. Pak.* **2012**, *34*, 1331–1338.
9. Pan, S.; Wang, Q.; Bai, J.; Liu, H.; Chi, M.; Cui, D.; Xu, F. Investigation of Behavior of Sulfur in Oil Fractions During Oil Shale Pyrolysis. *Energy Fuels* **2019**, *33*, 10622–10637. [CrossRef]
10. Gavrilova, O.; Vilu, R.; Vallner, L. A life cycle environmental impact assessment of oil shale produced and consumed in Estonia. *Resour. Conserv. Recycl.* **2010**, *55*, 232–245. [CrossRef]
11. Velts, O.; Uibu, M.; Rudjak, I.; Kallas, J.; Kuusik, R. Utilization of oil shale ash to prepare PCC: Leachibility dynamics and equilibrium in the ash-water system. *Energy Procedia* **2009**, *1*, 4843–4850. [CrossRef]
12. Trikkela, A.; Kuusik, R.; Martins, A.; Pihu, T.; Stencel, J.M. Utilization of Estonian oil shale semicoke. *Fuel Process. Technol.* **2008**, *89*, 756–763. [CrossRef]
13. Bai, F.T.; Sun, Y.H.; Liu, Y.M.; Guo, M.Y. Evaluation of the porous structure of Huadian oil shale during pyrolysis using multiple approaches. *Fuel* **2016**, *187*, 1–8. [CrossRef]
14. Burnham, A.K. Porosity and permeability of Green River oil shale and their changes during retorting. *Fuel* **2017**, *203*, 208–213. [CrossRef]
15. Li, L.; Liu, Z.; Sun, P.; Li, Y.; George, S.C. Sedimentary basin evolution, gravity flows, volcanism, and their impacts on the formation of the Lower Cretaceous oil shales in the Chaoyang Basin, northeastern China. *Mar. Pet. Geol.* **2020**, *119*, 104472. [CrossRef]
16. He, W.; Sun, Y.; Guo, W.; Shan, X. Controlling the in-situ conversion process of oil shale via geochemical methods: A case study on the Fuyu oil shale, China. *Fuel Process. Technol.* **2021**, *219*, 106876. [CrossRef]
17. Cao, H.; Guo, W.; Shan, X.; Ma, L.; Sun, P. Paleolimnological environments and organic accumulation of the Nenjiang formation in the southeastern Songliao basin, China. *Oil Shale* **2015**, *32*, 5–24. [CrossRef]
18. Bansal, V.R.; Kumar, R.; Sastry, M.I.S.; Badhe, R.M.; Kapur, G.S.; Saxena, D. Direct estimation of shale oil potential by the structural insight of Indian origin kerogen. *Fuel* **2019**, *241*, 410–416. [CrossRef]
19. Burnham, A.K.; Mcconaghy, J.R. Semi-open pyrolysis of oil shale from the garden gulch member of the green river formation. *Energy Fuels* **2014**, *28*, 7426–7439. [CrossRef]
20. Jia, J.; Zhou, R.; Liu, Z.; Han, X.; Gao, Y. Organic Matter-Driven Electrical Resistivity of Immature Lacustrine Oil-Prone Shales. *Geophysics* **2021**, *86*, 165–178. [CrossRef]
21. Campbell, J.H.; Koskinas, G.H.; Stout, N.D.; Coburn, T.T. Oil shale retorting—Effects of particle size and heating rate on oil evolution and intraparticle oil degradation. *Situ* **1978**, *2*, 1–47.
22. Wang, Z.; Lü, X.; Li, Q.; Sun, Y.; Wang, Y.; Deng, S.; Guo, W. Downhole electric heater with high heating efficiency for oil shale exploitation based on a double-shell structure. *Energy* **2020**, *211*, 118539. [CrossRef]
23. Guo, W.; Wang, Z.; Sun, Z.; Sun, Y.; Lü, X.; Deng, S.; Qu, L.; Yuan, W.; Li, Q. Experimental investigation on performance of downhole electric heaters with continuous helical baffles used in oil shale in-situ pyrolysis. *Appl. Therm. Eng.* **2019**, *147*, 1024–1035. [CrossRef]
24. Burnham, A.K.; Ethan, B.H.; Mary, F.S. Pyrolysis kinetics for Green River oil shale from the saline zone. *Fuel* **1983**, *62*, 1199–1204. [CrossRef]
25. Jaber, J.O.; Probert, S.D. Non-isothermal thermogravimetry and decomposition kinetics of two Jordanian oil shales under different processing conditions. *Fuel Process. Technol.* **2000**, *63*, 57–70. [CrossRef]
26. Jiang, X.M.; Han, X.X.; Cui, Z.G. Progress and recent utilization trends in combustion of Chinese oil shale. *Prog. Energy Combust. Sci.* **2007**, *33*, 552–579. [CrossRef]

27. Zhao, S.; Sun, Y.; Wang, H.; Li, Q.; Guo, W. Modeling and field testing of fracturing fluid back-flow after acid fracturing in unconventional reservoirs. *J. Pet. Sci. Eng.* **2019**, *176*, 494–501. [CrossRef]

28. Kök, M.V. Heating rate effect on the DSC kinetics of oil shale. *J. Anal. Calorim.* **2007**, *90*, 817–821. [CrossRef]

29. He, W.T.; Sun, Y.H.; Shan, X.L. Geochemical characteristics of the Lower Cretaceous Hengtongshan Formation in the Tonghua Basin, Northeast China: Implications for Depositional Environment and Shale Oil Potential Evaluation. *Appl. Sci.* **2021**, *11*, 23. [CrossRef]

30. Cao, H.; He, W.; Chen, F.; Shan, X.; Kong, D.; Hou, Q.; Pu, X. Integrated chemostratigraphy (δ^{13}C-δ^{34}S-δ^{15}N) constrains cretaceous lacustrine anoxic events triggered by marine sulfate input. *Chem. Geol.* **2021**, *559*, 119912. [CrossRef]

31. Wang, L.; Yang, D.; Zhao, Y.; Wang, G. Evolution of pore characteristics in oil shale during pyrolysis under convection and conduction heating modes. *Oil Shale* **2020**, *37*, 224–241. [CrossRef]

32. Staplin, F.L. Interpretation of thermal history from color of particulate organic matter—A review. *Palynology* **1977**, *1*, 9–18. [CrossRef]

33. Wei, G.; Xie, Z.Y.; Bai, G.L.; Li, J.; Wang, Z.H.; Li, A.G.; Li, Z.S. Organic geochemical characteristics and origin of natural gas in Sinian-Lower Paleozoic reservoirs, Sichuan Basin. *Nat. Gas Ind.* **2014**, *34*, 44–49.

34. Dhaundiyal, A.; Singh, S.B. Study of Distributed Activation Energy Model using Various Probability Distribution Functions for the Isothermal Pyrolysis Problem. *Rud.-Geol.-Naft. Zb.* **2017**, *32*, 1–15. [CrossRef]

35. Peters, K.E.; Moldowan, J.M. The biomarker guide: Interpreting molecular fossils in petroleum and ancient sediments. *Choice Rev. Online* **1993**, *30*, 30–2690.

36. Peters, K.E.; Walters, C.C.; Moldowan, J.M. *The Biomarker Guide*; Cambridge University Press: Cambridge, UK, 2004.

37. Liu, Z.; Sun, P.; Jia, J.; Liu, R.; Meng, Q. Distinguishing features and their genetic interpretation of stratigraphic se-quences in continental deep water setting:A case from Qingshankou Formation in Songliao Basin. *Earth Sci. Front.* **2011**, *18*, 171–180. (In Chinese with English abstract)

38. Liu, Z.; Wang, D.; Liu, L.; Liu, W.; Wang, P.; Du, X.; Yang, G. Sedimentary Characteristics of the Cretaceous Songliao Basin. *Acta Geol. Sin.* **1992**, *66*, 327–338.

39. He, W.; Sun, Y.; Shan, X.; Cao, H.; Zheng, S.; Su, S. The oil shale formation mechanism of the Songliao Basin Nenjiang Formation triggered by marine transgression and OAE3. *Oil Shale* **2021**, *38*, 89–118.

40. He, W.; Sun, Y.; Shan, X. Organic matter evolution in pyrolysis experiments of oil shale under high pressure: Guidance for in-situ conversion of oil shale in the Songliao Basin. *J. Anal. Appl. Pyrolysis* **2021**, *155*, 105091. [CrossRef]

41. Caricchi, C.; Corrado, S.; Di Paolo, L.; Aldega, L.; Grigo, D. Thermal maturity of Silurian deposits in the Baltic Syneclise (on-shore Polish Baltic Basin): Contribution to unconventional resources assessment. *Ital. J. Geosci.* **2016**, *135*, 383–393. [CrossRef]

42. Li, C.; Huang, W.; Zhou, C.; Chen, Y. Advances on the transition-metal based catalysts for aqua thermolysis upgrading of heavy crude oil. *Fuel* **2019**, *257*, 115779. [CrossRef]

43. Li, L.; Liu, Z.; Jiang, L.; George, S.C. Organic petrology and geochemistry of Lower Cretaceous lacustrine sediments in the Chaoyang Basin (Liaoning Province, northeast China): Influence of volcanic ash on algal productivity and oil shale formation. *Int. J. Coal Geol.* **2021**, *233*, 103653. [CrossRef]

44. Puig-Gamero, M.; Fernandez-Lopez, M.; Sánchez, P.; Valverde, J.L.; Sanchez-Silva, L. Pyrolysis process using a bench scale high pressure thermobalance. *Fuel Process. Technol.* **2017**, *167*, dd345–dd354. [CrossRef]

45. Bai, F.; Sun, Y.; Liu, Y.; Li, Q.; Guo, M. Thermal and kinetic characteristics of pyrolysis and combustion of three oil shales. *Energy Convers. Manag.* **2015**, *97*, 374–381. [CrossRef]

46. Cao, H.; He, W.; Chen, F.; Kong, D. Superheavy pyrite in the upper cretaceous mudstone of the Songliao basin, NE China and its implication for paleolimnological environments. *J. Asian Earth Sci.* **2020**, *189*, 104156. [CrossRef]

47. Burnham, A.K. Thermomechanical properties of the Garden Gulch Member of the Green River Formation. *Fuel* **2018**, *219*, 477–491. [CrossRef]

48. Wang, S.; Jiang, X.; Han, X.; Tong, J.H. Effect of residence time on products yield and characteristics of shale oil and gases produced by low-temperature retorting of Dachengzi oil shale. *Oil Shale* **2013**, *30*, 501–516. [CrossRef]

49. Wang, S.; Jiang, X.; Han, X.; Tong, J.H. Effect of retorting temperature on product yield and characteristics of non-condensable gases and shale oil obtained by retorting Huadian oil shales. *Fuel Process. Technol.* **2014**, *121*, 9–15. [CrossRef]

50. Wang, L.; Zhao, Y.; Yang, D.; Kang, Z.; Zhao, J. Effect of pyrolysis on oil shale using superheated steam: A case study on the Fushun oil shale, China. *Fuel* **2019**, *253*, 1490–1498. [CrossRef]

51. He, W.; Sun, Y.; Shan, X.; Cao, H.; Li, J. Geochemical characteristics and oil generation potential evaluation of lower cretaceous Xiahuapidianzi formation shale in the southeastern sankeyushu depression, Tonghua basin: Evidence from shale pyrolysis experiments and biomarkers. *ACS Earth Space Chem.* **2021**, *5*, 409–423. [CrossRef]

52. Burnham, A.K.; Sweeney, J.J. A chemical kinetic model of vitrinite maturation and reflectance. *Geochim. Cosmochim. Acta* **1989**, *53*, 2649–2657. [CrossRef]

53. Sweeney, J.J.; Burnham, A.K. Evaluation of a simple model of vitrinite reflectance based on chemical kinetics. *Am. Assoc. Pet. Geol. Bull.* **1990**, *74*, 1559–1570.

54. Cao, H.; He, W. Correlation of carbon isotope stratigraphy and paleoenvironmental conditions in the cretaceous Jehol group, northeastern China. *Int. Geol. Rev.* **2020**, *62*, 113–128. [CrossRef]

55. Kalkreuth, W.; Sherwood, N.; Cioccari, G.; da Silva, Z.C.; Silva, M.; Zhong, N.; Zufa, L. The application of famm™ (fluorescence alteration of multiples macerals) analyses for evaluating rank of parana basin coals, brazil. *Int. J. Coal Geol.* **2004**, *57*, 167–185. [CrossRef]
56. Cheshire, S.; Craddock, P.R.; Xu, G.; Sauerer, B.; Pomerantz, A.E.; McCormick, D.; Abdallah, W. Assessing thermal maturity beyond the reaches of vitrinite reflectance and Rock-Eval pyrolysis: A case study from the Silurian Qusaiba Formation. *Int. J. Coal Geol.* **2017**, *180*, 29–45. [CrossRef]
57. Burnham, A.K.; Schmidt, B.J.; Braun, R.L. A test of the parallel reaction model using kinetic measurements on hydrous pyrolysis residues. *Org. Geochem.* **1995**, *23*, 931–939. [CrossRef]
58. Shuai, Z.; Xiaoshu, L.; Qiang, L.; Youhong, S. Thermal-fluid coupling analysis of oil shale pyrolysis and displacement by heat-carrying supercritical carbon dioxide. *Chem. Eng. J.* **2020**, *394*, 125037. [CrossRef]
59. Seifert, W.K.; Moldowan, J.M. Applications of steranes, terpanes and monoaromatics to the maturation, migration and source of crude oils. *Geochim. Cosmochim. Acta* **1978**, *42*, 77–95. [CrossRef]
60. Zhang, K.; Liu, R.; Ding, W.; Li, L.; Liu, Z. The influence of Early Cretaceous paleoclimate warming event on sedimentary environment evolution and organic matter sources in Yin'e Basin: Evidence from petrology and molecular geochemistry. *Int. J. Coal Geol.* **2022**, *254*, 103972. [CrossRef]
61. Jia, J.; Liu, Z. Particle-Size Fractionation and Thermal Variation of Oil Shales in the Songliao Basin, NE China: Implication for Hydrocarbon-Generated Process. *Energies* **2021**, *14*, 7191. [CrossRef]
62. Zhao, S.; Sun, Y.; Lü, X.; Li, Q. Energy consumption and product release characteristics evaluation of oil shale non-isothermal pyrolysis based on TG-DSC. *J. Pet. Sci. Eng.* **2020**, *187*, 494–501. [CrossRef]
63. Kolaczkowska, E.; Slougui, N.-E.; Watt, D.S.; Marcura, R.E.; Moldwan, J.M. Thermodynamic stability of various alkylated, dealkylated, and rearranged 17α and 17β hopane isomers using molecular mechanics calculations. *Org. Geochem.* **1990**, *16*, 1033–1038. [CrossRef]
64. Ensminger, A.; Albrecht, P.; Ourisson, G.; Tissot, B. Evolution of polycyclic alkanes under the effect of burial (Early Toarcian shales, Paris Basin). In *Advances in Organic Geochemistry 1975*; Enadimsa: Madrid, Spain, 1977; pp. 45–52.
65. Wang, L.; Yang, D.; Kang, Z. Evolution of permeability and mesostructure of oil shale exposed to high-temperature water vapor. *Fuel* **2021**, *290*, 119786. [CrossRef]
66. Mackenzie, A.S. *Application of Biological Markers in Petroleum Geochemistry*; Academic Press: London, UK, 1984.
67. Song, Y.; Li, S.; Hu, S. Warm-humid paleoclimate control of salinized lacustrine organic-rich shale deposition in the Oligocene Hetaoyuan Formation of the Biyang Depression, East China. *Int. J. Coal Geol.* **2019**, *202*, 69–84. [CrossRef]
68. Peters, K.E.; Moldowan, J.M. The biomarker guide: Interpreting molecular fossils. In *Petroleum and Ancient Sediments*; Prentice Hall: Hoboken, NJ, USA, 1993; Volume 363.
69. Peters, K.E. Guidelines for evaluating petroleum source rock using programmed pyrolysis. *Am. Assoc. Pet. Geol. Bull.* **1986**, *70*, 318–329.